The sort of praise that's making

PUTTING FOOD BY

the #1 bestseller in its field:

The daily press coast to coast:

"A down-to-earth guide for the do-it-yourself home-maker . . . Inspires a return to old-time skills and con-servation of the earth's bounty."
—*Christian Science Monitor*

"The style is chatty but authoritative, which makes for good reading and confidence in food preservation."
—*Cleveland Plain Dealer*

"Thoroughgoing 'how to.' An excellent book, written with clarity and precision."
—*Seattle Post Intelligencer*

Seed company:

"Best book we've seen on new ways to preserve all foods. Indispensable for the farmer, housewife, out-doorsman."
—*Gurney Seed Catalog*

Nutrition educator:

"One of the most complete and thoroughly practical books available on food preservation and safety. An excellent resource."
—*West Virginia University*

College bookstore:

"I thought I knew all I needed to know about preserv-ing food until reading this book. For beginners to experienced, *Putting Food By* is an excellent book."
—*Missouri Book Store*

Bantam Cookbooks
Ask your bookseller for the books you have missed

Putting Food By

By
Ruth Hertzberg, Beatrice Vaughan
and Janet Greene

Second Edition
Revised and Enlarged

BANTAM BOOKS
TORONTO · NEW YORK · LONDON · SYDNEY

PUTTING FOOD BY

*A Bantam Book / published by arrangement with
The Stephen Greene Press*

PRINTING HISTORY

Stephen Greene Press edition published January 1973

2nd printing . . . . June 1973	4th printing . . . . . . January 1974
3rd printing . . . . . . October 1973	5th printing . . . . May 1974
	6th printing . . . . July 1974	

2nd edition published July 1975

2nd printing . . . . July 1975   3rd printing . . . . . November 1975

*Woman's How-To Book Club edition
published February 1974*

*Better Homes & Gardens Family Book Service edition
published August 1975*
2nd printing . . . . . February 1976

Bantam edition / June 1976

2nd printing . . . . June 1976	4th printing . . . . March 1979
3rd printing . . . . . . August 1976	5th printing . . . . September 1979
6th printing . . . . . September 1980	

*All photographs and drawings not otherwise credited are from the U.S.
Department of Agriculture. The exceptions are: drawings on page 62—
John Devaney; photographs on pages 40, 42, 126, 295—Allan Seymour;
drawings on pages 7, 40, 66, 173, 176, 177, 273, 310, 333, 370, 371, 391,
418, 424, 432, 468, 471—Norman Rogers.*

This book, too, is for
    John D. Hutchens, who said to write it,
        and "P.R." who helped it so,
    and all the people everywhere who want
    the happiness of putting food by safely for
    their families and knowing what's in it.

WHAT "PUTTING BY" IS

To "put by" is an early nineteenth-century way of saying to "save something you don't have to use now, against the time when you'll need it." You still hear it today from old-time country people, and applied to food it is prudence and involvement and a return to the old simplicities. Putting food by is the antidote for running scared.

With this new second edition, I still feel that my main contributions to this book have been to agree with John Hutchens's suggestion to write it, and then to have the good sense to enlist Ruth D. Hertzberg for it.

Janet Greene
May 1975

Acknowledgments

For firsthand learned-by-successful-doing information, the authors are grateful to Guy Brunton, master of old and new building; John Caldwell, mathematician and sports coach; Robert Carpenter and Kenneth Carpenter, pharmacologists; Lona B. Chatterton, expert router of queries to the Extension Service; J. Y. Clowater of The Miramichi Salmon Association, Inc.; William H. Darrow, Jr., apple-grower and -storer; Val Dubal, D.V.M., of the Vermont Department of Agriculture; Kenneth G. Freund of the Freund Can Company; Hattie Hinkle, Home Economist of Kerr Glass Manufacturing Company; Peter Johnson of Dominion Glass Company, Limited; Jane Keely of the Good Housekeeping Institute; A. S. Kull, home-testing correlator; Merrill Lawrence of Lawrence's Smoke-house; Aidan Maloney of the Canadian Saltfish Corporation; Shirley Knight Morris of Burns & MacEachern Limited, Ontario, Canada; Petra Morrison and Gabrielle Pike, locators and senders of information on home-processing foods in Britain; Alexis Nason, geologist; Walter Needham, author of *A Book of Country Things;* Samuel R. Ogden, countryman and organic gardener; A. Rasmussen of Ives-way Products, Inc.; Norman Rogers, artist, world traveler and food researcher; Ronald Rood, naturalist; and Mary Lou Williamson of the Consumer Service of the Ball Corporation.

We thank Alan G. Barbour, Gerald Bee, Audrey C. Burkart, Aline Coffey, Fletcher Coolidge, Michele Daignault, Isabelle Downey, Marvin Eisenstadt, Frederick J. Francis, Donna Lee Funk, Yolan L. Harsanyi, Kirby Hayes, Evelyn Johnson, Ruth N. Klippstein, Jonathan Leff, Nicholas Pintacero, George Pollak, Thelma G. Russ, Evelyn B. Spindler, Marguerite Stetson, Winston A. Way, Isabel D. Wolf, and Robert Young.

And closer to home, we are beholden to Jon Anderson, Frances Bond, Albert and Mildred M. Dupell, Larry Feldman, Mary Lou Gould, Stephanie Greene, Thure Hertzberg, Hazel D. Howard, Florence Howe, Ann C. Johnson, Nancy Lent, Susan Mahnke, Ethel R. May, Esther Munroe, Annette Pestle, Neil Y. Priessman, Jr., Alice B. Robinson, Francis Rohr, and Mildred Wallace.

The authors formally acknowledge their debt to all the anonymous dedicated people in federal- and provincial- and state-run projects in the United States and Canada who are constantly researching better and safer ways to handle our food.

Contents

Things You'll Be Coming Back to Again and Again 1

Everything in this chapter applies to every safe method of putting food by at home. The material here is used specifically, even amplified if need be, in describing each process for each individual food in the rest of the book.

Meanwhile the following information has been corralled for ready reference because we believe that any newcomer to canning, freezing, making preserves, drying, root-cellaring, or curing can always keep the *How* of food safety in mind if the *Why* is handy and clear.

Why Put-by Foods Spoil

"Spoilage" is a very broad term, and in home-preserved foods it embraces conditions that range from ones that are merely disappointing or unappetizing right on through to those that endanger our health or even our lives.

Four kinds of things cause spoilage in preserved food: (1) *enzymes,* which are naturally occurring substances in living tissues; and three types of micro-organisms—(2) *molds,* (3) *yeasts* and (4) *bacteria*—all of which are present in the soil, water and air around us.

HOW ENZYMES ACT

Nature has designed each plant or animal with the ability to program the production of its own enzymes, which help to promote the organic changes necessary to the life cycle of all growing things. However, their action is reversible: they can turn around and cause decomposition, hence changing the flavor, texture and color of food, and making it unappetizing.

Their action slows down in cold conditions, increases most quickly between around 85 to 120 degrees Fahrenheit, and begins to be destroyed at about 140 F.

HOW MOLDS ACT

Molds are microscopic fungi whose dry spores (seeds) alight on food and start growing silken threads that can become slight streaks of discoloration in food or cover it with a mat of fuzz.

It used to be felt that "a little mold won't hurt you," but modern research has disclosed that only the mold introduced deliberately into the "blue" cheeses like Roquefort, Gorgonzola or Stilton is trustworthy. The others, as they grow in food, are capable of producing substances called mycotoxins, and some of them can be hurtful indeed.

In addition, molds eat natural acid present in food, thereby lowering the acidity that is protection against more actively dangerous poisons—but we'll have more to say about this in a minute.

Molds are alive but don't grow below 32 degrees Fahrenheit; they start to grow above freezing, have their maximum acceleration between 50 and 100 F., and then taper off to inactivity beginning around 120 F.; they start to die with increasing speed at temperatures from 140 to 190 F.

HOW YEASTS ACT

The micro-organisms we call yeasts are also fungi grown from spores, and they cause fermentation—which is delightful in beer, necessary in sauerkraut, and horrid in applesauce. As with molds, severe cold holds them inactive, 50 to 100 F. hurries their growth, and 140 to 190 F. destroys them.

HOW BACTERIA ACT

Bacteria also are present in soil and water, and their spores, too, can be carried by the air. But bacteria are often far tougher than molds and yeasts are: certain ones actually thrive in heat that kills these fungi, and in some foods there can exist bacteria spores which make hidden toxins that will be destroyed in a reasonable time only if the food is heated at from 240 to 250 degrees Fahrenheit—*at least 28 degrees higher than the boiling temperature for water, and obtainable only under pressure.*

Of the disease-causing bacteria we're concerned with mainly, the most fragile are members of the genus Salmonella, which are transmitted by pets, rodents, insects and human beings, in addition to existing in our soil, water and air. Salmonellae live in frozen food, are relatively inactive up to 45 F., and are killed when held at 140 F., with destruction much quicker at higher temperatures.

More heat-resistant are the transmittable bacteria that cause "staph" poisoning, the *Staphylococcus aureus* whose growth makes a toxin that is destroyed only by many hours of boiling or 30 minutes at 240 F., although the growth of the bacteria themselves is checked if the food is kept above 140 F. or below 40 F. Thus "staph" germs are halted in their tracks fairly easily at temperatures that kill Salmonellae, but the poison they make merely by growing takes a very long time at boiling, or a much higher temperature (240 F.), more briefly, to destroy.

Most dangerous of the tough bacteria is *Clostridium botulinum*, which deserves a section of its own.

Botulism

The scientific books describe *Cl. botulinum* as a "soil-borne, mesophilic, spore-forming, anaerobic bacterium." Which, translated into everyday language, means that it is present in soil that is carried into our kitchens on raw materials, on implements, on clothing, on our hands—you name it. Next, it thrives best in the middle range of heat—beginning at about room temperature, 70 degrees Fahrenheit, on up to 110 F. Next, it produces spores (and the *spores* are extremely durable: whereas the bacterium is destroyed in a relatively short time at 212 F., the temperature of briskly boiling water, the *spores* are not destroyed unless they are subjected to at least 240 degrees F. for a sustained length of time). And finally, the bacterium lives in the *absence* of air (and also in a very moist environment: which are combined in the condition existing in a container of canned food).

This description does not mention the poison thrown off by the spores as they grow: the toxin is so powerful that one teaspoon of the pure substance could kill hundreds of thousands of people.

The grave illness caused by eating the toxin in canned food is comparatively rare—rare compared with the cases of "staph" or salmonellosis—but, unlike them, it is often fatal. The symptoms are blurred vision, blurred speech, inability to hold up the head, and eventual respiratory arrest unless the victim is given help to breathe until medication can reverse the progress of the disease.

The botulinum toxin can be destroyed by brisk boiling for 15 minutes. This is why, throughout the Canning chapter, we give the following warning, with 20 minutes required for the foods that are more dense:

> **Before tasting canned food, boil it hard for 15 minutes to destroy any hidden toxins (corn, greens, meat, poultry and seafood require 20 minutes). If it looks spoiled or foams or has a bad smell during boiling, destroy the food completely so it can't be eaten by people or animals.**

Botulinum is held inactive at freezing, starts to grow even in the refrigerator, and comes into its own at room temperatures, as mentioned earlier.

As with all micro-organisms, a moisture content of below 35 percent directly inhibits its growth.

Dealing with the Spoilers

ENTER *pH*—THE ACIDITY FACTOR

The strength of the acid in any food determines to a great extent which of the spoilage micro-organisms each food can harbor: therefore acidity is a built-in directive that tells us what temperatures are necessary to destroy these spoilers within it, and make it safe to eat. (Heat alone, not natural acidity, controls the action of enzymes.)

Acid strength is measured on the *pH Scale*, which starts with strongest acid at 1 and declines to strongest alkali at 14, with the Neutral point at 7, where the food is considered neither acid nor alkaline. The *pH* ratings appear to run backwards, since the larger the number, the less the acid, but it may help to think of the ratings as like sewing thread: size 100 cotton thread is smaller (finer) than size 60. Or like a shotgun: the barrel of a 16-gauge is smaller in diameter than that of a 12-gauge shotgun.

pH RATINGS

This listing indicates in a general way the natural acid strength of common foods on the *pH* scale. Different varieties of the same food will have different ratings, of course, as will identical varieties grown under different conditions. Foods with a *pH* of 4.5 or higher are considered to be *low-acid,* and are handled accordingly; those below 4.5 are regarded as *strong-acid.*

Lemons	2.2-2.8	Squash	5.0-5.4
Gooseberries	2.8-3.0	Beans, string,	
Plums	2.8-4.0	green, wax	5.0-6.0
Apples	2.9-3.7	Spinach	5.1-5.9
Grapefruit	3.0-3.7	Cabbage	5.2-5.4
Strawberries	3.0-3.9	Turnips	5.2-5.6
Oranges	3.0-4.0	Peppers, green,	
Rhubarb	3.1-3.2	bell	5.3
Blackberries	3.2-4.0+	Sweet potatoes	5.3-5.6
Cherries	3.2-4.2	Asparagus	5.4-5.8
Raspberries	2.8-3.6	Potatoes, white	5.4-6.0
Blueberries	3.3-3.5	Mushrooms	5.8-5.9
Sauerkraut	3.4-3.7	Peas	5.8-6.5
Peaches	3.4-4.0+	Tuna fish	5.9-6.1
Apricots	3.4-4.4+	Beans, lima	6.0-6.3
Pears	3.6-4.4+	Corn	6.0-6.8
Pineapple	3.7	Meats	6.0-6.9
Tomatoes	4.0-4.6+	Salmon	6.1-6.3
Pimientos	4.6-5.2	Oysters	6.1-6.6
Pumpkin	4.8-5.2	Milk, cow's	6.3-6.6
Carrots	4.9-5.4	Shrimp	6.8-7.0
Beets	4.9-5.8	Hominy	6.8-8.0

Testing for *pH*

The abbreviation "*pH*" stands for "potential of hydrogen"—which means a great deal to a chemist, who can determine the *pH* of any raw or cooked food by using a sophisticated testing device. For the interested homemaker we suggest *pH* test papers that are rather like the litmus paper we used in high-school science courses, but are a good deal more refined (such litmus papers have too coarse a reaction for what's needed here). The *pH* papers come from laboratory supply houses—look them up in the Yellow Pages, or ask your druggist or the high-school science teacher to let you use a supplier's catalog.

The papers come in a range of 3.0 to 5.5, and allow you to read with an accuracy of 0.5 *pH* unit. They come ½ and ¼ inch wide—the former is easier to read—in rolls like scotch tape, with or without a dispenser. On the side of the dispenser is a guide to *pH* colors for comparison.

THE IMPORTANCE OF SANITATION

Even though the spoilage micro-organisms in a food are rendered inactive by cold, they start functioning with a vengeance as soon as they warm up.

Or, adequately preheated food can be contaminated by airborne spores while it, or its unfilled container, stands around unprotected; and, unless it is cooled quickly or refrigerated, the new batch of spoilers can start growing—and growing fast. In this connection, it's interesting to trace back the notion that one must not put hot food in a refrigerator in order to cool it quickly. This idea is a holdover from the days of the wooden ice chest, which was kept cool by a big block of ice: of course the hot food warmed the inside of the ice chest, and there was a long lag before the ice (or what was left of it) could reduce the internal temperature of the cabinet to a safe-holding coolness. With modern refrigerators, though, a container of warm food merely causes the thermostat to kick on, and the cooling machinery goes to work immediately.

Another good way to deal with the spoilers is simply to wash them off the food as carefully as possible, and to keep work surfaces and equipment sanitary at every stage in the procedure. Food scientists refer to the "bacterial load" in

TEMPERATURE vs. THE SPOILERS

F. = Fahrenheit, C = Celsius
(under 1000 ft. alt.)

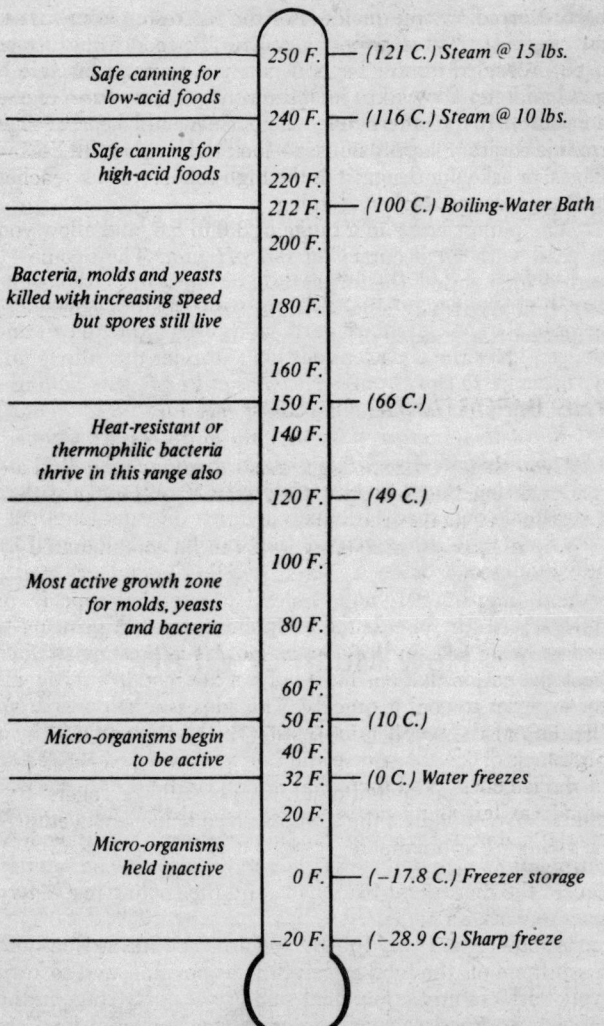

	250 F. — *(121 C.) Steam @ 15 lbs.*
Safe canning for low-acid foods	
	240 F. — *(116 C.) Steam @ 10 lbs.*
Safe canning for high-acid foods	220 F.
	212 F. — *(100 C.) Boiling-Water Bath*
	200 F.
Bacteria, molds and yeasts killed with increasing speed but spores still live	180 F.
	160 F.
	150 F. — *(66 C.)*
Heat-resistant or thermophilic bacteria thrive in this range also	140 F.
	120 F. — *(49 C.)*
	100 F.
Most active growth zone for molds, yeasts and bacteria	80 F.
	60 F.
	50 F. — *(10 C.)*
Micro-organisms begin to be active	40 F.
	32 F. — *(0 C.) Water freezes*
	20 F.
Micro-organisms held inactive	0 F. — *(−17.8 C.) Freezer storage*
	−20 F. — *(−28.9 C.) Sharp freeze*

describing the almost staggering rate of increase in bacteria if the food is handled in an unsanitary manner or allowed to remain at the optimum growing temperature for the spoilers. The procedures described in this book have been established after years of research by food technologists, and naturally there is an allowance made for extra micro-organisms that must be destroyed. But in many cases the food would have to be processed almost beyond palatability if the bacterial load had been allowed to increase geometrically to an enormous extent; the alternative of course would be that the treatment was not intensified, so the food spoiled after all.

HEAT + ACIDITY = CANNING SAFETY

Combining (1) the temperatures that control life and growth of spoilage micro-organisms with (2) the *pH* acidity factor of the foods that are particularly hospitable to certain spoilers, gives the conscientious home-canner this rule to go by: *It is safe to can strong-acid foods at 212 F. in a Boiling-Water Bath, but low-acid/nonacid foods must be canned at 240 F.—a temperature possible only in a Pressure Canner at 10 pounds' pressure—if they are to be safe.*

Processing times vary according to acidity and density of the food concerned. Adequate temperature and length of processing time are given in the specific instructions for individual foods.

How Much of What

YIELDS IN CANNED OR FROZEN PRODUCE

Since the legal weight of a bushel of fruits or vegetables differs between States, the weights given below are average; the yields are approximate.

Fruits	Fresh	Quarts Canned	Pints Frozen
Apples	1 bu. (48 lbs.)	16–20	32–40
	2½–3 lbs.	1	2
Apple juice	1 bu. (48 lbs.)	10	
Applesauce	1 bu. (48 lbs.)	15–18	30–36
	2½–3½ lbs.	1	2

Fruits	Fresh	Quarts Canned	Pints Frozen
Apricots	1 bu. (50 lbs.)	20–24	60–72
	2–2½ lbs.	1	2–3
Berries (excluding strawberries)	24-qt. crate	12–18	32–36
	5–8 cups	1	2
Cherries, as picked	1 bu. (56 lbs.)	22–32	36–44
	2–2½ lbs.	1	2
Cranberries	1 peck (8 lbs.)		16
Figs	2–2½ lbs.	1	2
Grapes	28-lb. lug	7–8	14–16
	4 lbs.	1	2
Grapefruit	4–6 fruit	1	2
Nectarines	18-lb. flat	6–9	12–18
	2–3 lbs.	1	2
Peaches	1 bu. (48 lbs.)	18–24	32–48
	2–2½ lbs.	1	2
Pears	1 bu. (50 lbs.)	20–25	40–50
	2–2½ lbs.	1	2
Pineapple	2 average	1	2
	5 lbs.	2	4
Plums and Prunes	1 bu. (56 lbs.)	24–30	38–56
	2–2½ lbs.	1	2
Rhubarb	15 lbs.	7–11	15–22
	2 lbs.	1	2
Strawberries	24-qt. crate	12–16	38
	6–8 cups	1	2
Tomatoes, cut up	1 bu. (53 lbs.)	15–20	
	2½–3 lbs.	1	

Vegetables			
Asparagus	1 bu. (45 lbs.)	11	8–11
	3–4 lbs.	1	3–4
Beans, lima, in pods	1 bu. (32 lbs.)	6–8	12–16
	4–5 lbs.	1	2
Beans, snap/green/wax	1 bu. (30 lbs.)	15–20	30–45
	1½–2 lbs.	1	2
Beets, without tops	1 bu. (52 lbs.)	17–20	35–42
	2½–3 lbs.	1	2
Broccoli	25-lb. crate	10–12	24
	2–3 lbs.	1	2
Brussels sprouts	4 qts.	1–1½	6
	1 lb.	½	1
Carrots, without tops	1 bu. (50 lbs.)	16–20	32–40
	2½–3 lbs.	1	2
Cauliflower	2 medium heads	1½	3
	1 bu. (12 lbs.)	4–6	8–12
Corn, in husks	1 bu. (35 lbs.)	8–9	14–17
	4–5 lbs.	1	2
Eggplant	2 average	1	2
Kale	1 bu. (18 lbs.)	6–9	12–18
	2–3 lbs.	1	2
Okra	1 bu. (26 lbs.)	17	34–40
	1½ lbs.	1	2

Vegetables	Fresh	Quarts Canned	Pints Frozen
Peas, green, in pods	1 bu. (30 lbs.)	6–8	12–15
	2–2½ lbs.	1 pt.	1
Peppers, sweet	⅔ lb. (3 peppers)		1
Potatoes, white	1 bu. (50 lbs.)	20	
	2½–3 lbs.	1	
Potatoes, sweet	1 bu. (50 lbs.)	20	
(and yams)	2½–3 lbs.	1	
Pumpkin	50 lbs.	15	30
	3 lbs.	1	2
Spinach	1 bu. (18 lbs.)	6–9	12–18
	2–3 lbs.	1	2
Squash, summer	1 bu. (40 lbs.)	16–20	32–40
	2–2½ lbs.	1	2
Squash, winter	3 lbs.		2

Conversions and Adjustments

METRIC CONVERSIONS FOR TEMPERATURE

In 1714 Gabriel Daniel Fahrenheit proposed a thermometer that fixed Zero at the lowest temperature reached by a freezing mixture of ice and salt. He divided the interval between Zero and the temperature of the human body into 96 parts. Using this scale, pure water at sea level freezes at 32 degrees Fahrenheit and boils at 212 F.

Nearly three decades later, Anders Celsius invented the Centigrade thermometer, using the freezing and boiling points of water at sea level as its fixed points and putting 100 degrees between them. Thus in the Fahrenheit scale there are 180 degrees between freezing and boiling temperatures of water, compared with the 100-degree interval in the Celsius scale, where water freezes at Zero C. and boils at 100 C. The two scales intersect only at −40 degrees F. and C.

THE ARITHMETIC

The list of equivalents includes the temperatures cited frequently in this book. To convert other temperatures, you can use the following formulas.

If you know *Fahrenheit*, you can find *Celsius* (Centigrade) by subtracting 32, then multiplying by 5/9 (i.e., multiply by 5 and divide by 9). For example, 200 F. = (200 − 32) × 5 ÷ 9 = 93 C.

If you know *Celsius* (Centigrade), you can find *Fahrenheit* by multiplying by 9/5, then adding 32. For example, 93 C. = 93 × 9 ÷ 5 + 32 = 200 F.

Equivalent Temperatures
Fahrenheit and Celsius (Centigrade)

F.	C.	F.	C.	F.	C.
−40	−40.0	40	4.4	205	96
−35	−37.2	50	10	210	99
−30	−34.4	60	15.5	212	100
−25	−31.7	70	21	220	104
−20	−28.9	80	26.6	225	107
−15	−26.1	90	32.2	228	109
−10	−23.3	100	37.8	230	110
− 5	−20.6	110	43	235	113
0	−17.8	120	49	240	116
5	−15.0	130	54	245	118.5
10	−12.2	140	60	250	121
15	− 9.4	150	66	255	124
20	− 6.7	160	71	260	127
25	− 3.9	165	74	265	129.5
30	− 1.1	170	77	270	132
32	0	175	79.5	275	135
35	1.7	180	82	280	138
36	2.2	185	85	285	140.5
37	2.8	190	88	290	143
38	3.3	195	90.5	295	146
39	3.9	200	93	300	149

METRIC CONVERSIONS FOR VOLUME

The terms *teaspoon, tablespoon, cup, pint, quart* and *gallon* used throughout this book and in the list that follows are U.S. measurements. Although their volumes differ in some cases from those of similar British or European terms, their metric equivalents provide the means for translating the differences.

For everyday use and for the purposes of this book, it is precise enough to say that 1 ounce = 30 milliliters. However, because 1 ounce is actually just short of 30 milliliters (being 29.573 milliliters), a slight adjustment is made at the pint level, where 1 pint is closer to 470 than to 480 milliliters.

Another thing: Householders in the United States are used to seeing the term *cubic centimeter* (*cc.*) for volume in pharmaceutical prescriptions and solutions, etc. For their benefit we note that *1 cubic centimeter = 1 milliliter.*

THE ARITHMETIC

If you know *ounces* (*oz.*), you can find *milliliters* (*ml.*) by multiplying by 30.

If you know *milliliters*, you can find *ounces* by multiplying by 0.034.

If you know *cups* (*c.*), you can find *liters* (*l.*) by multiplying by 0.24.

If you know *liters*, you can find *cups* by multiplying by 4.2.

If you know *pints* (*pt.*), you can find *liters* by multiplying by 0.47.

If you know *liters*, you can find *pints* by multiplying by 2.1.

If you know *quarts* (*qt.*), you can find *liters* by multiplying by 0.95.

If you know *liters*, you can find *quarts* by multiplying by 1.06.

If you know *gallons* (*gal.*), you can find *liters* by multiplying by 3.8.

If you know *liters*, you can find *gallons* by multiplying by 0.26.

1 teaspoon = 5 ml.	2 qt. = 1.90 *l*.
1 tablespoon = 15 ml.	3 qt. = 2.85 *l*.
1 oz. = 30 ml.	4 qt. (1 gal.) = 3.80 *l*.
4 oz. = 120 ml.	100 ml. = 3.4 oz.
1 c. (8 oz.) = 240 ml./0.24 *l*.	500 ml. = 17 oz.
1½ c. (12 oz.) = 360 ml./0.36 *l*.	1 *l*. = 1.06 qt.
2 c. (1 pt.) = 470 ml./0.47 *l*.	1.5 *l*. = 1.59 qt.
3 c. (1½ pt.) = 720 ml./0.72 *l*.	2 *l*. = 2.12 qt.
4 c. (1 qt.) = 950 ml./0.95 *l*.	5 *l*. = 1.30 gal.

METRIC CONVERSIONS FOR WEIGHT

Because 1 ounce is actually slightly more than 28 grams (being 28.350 grams), slight adjustments are made in the following list to approach the nearest whole number of grams.

THE ARITHMETIC

If you know *ounces* (*oz.*), you can find *grams* (*gr.*) by multiplying by 28.

If you know *grams*, you can find *ounces* by multiplying by 0.035.

If you know *pounds* (*lb.*), you can find *kilograms* (*kg.*) by multiplying by 0.45.

If you know *kilograms*, you can find *pounds* by multiplying by 2.2.

1 oz. = 28 gr.	10 lb. = 4.535 kg.
4 oz. = 113 gr.	15 lb. = 6.803 kg.
8 oz. = 226 gr.	100 gr. = 3.5 oz.
12 oz. = 340 gr.	1,000 gr. = 1 kg. = 2.2 lb.
16 oz. (1 lb.) = 454 gr./ 0.454 kg.	2 kg. = 4.4 lb.
2 lb. = 0.907 kg.	5 kg. = 11.02 lb.
5 lb. = 2.27 kg.	10 kg. = 22.05 lb.

METRIC CONVERSIONS FOR LENGTH

If you know *inches* (*in.*), you can find *centimeters* (*cm.*) by multiplying by 2.5.

If you know *centimeters*, you can find *inches* by multiplying by 0.4.

If you know *feet* (*ft.*), you can find *centimeters* by multiplying by 30.

If you know *centimeters*, you can find *feet* by multiplying by 0.033.

If you know *feet*, you can find *meters* (*m.*) by multiplying by 0.305.

If you know *meters*, you can find *feet* by multiplying by 3.28.

¼ in. = .63 cm.	3 in. = 7.5 cm.
½ in. = 1.25 cm.	4 in. = 10.0 cm.
1 in. = 2.5 cm.	6 in. = 15.0 cm.
2 in. = 5.0 cm.	12 in. = 30.0 cm.

WHERE IS BOILING?

At sea level, which for all practical purposes includes up to 1,000 feet/305 meters in altitude, water boils at 212 F./ 100 C.—and by *boils* we mean big bubbles charging to the surface in a boisterous romp that you can hear as well as see. But above 1,000 feet the atmospheric pressure decreases so much that we must do something about it, because it lets water boil furiously at temperatures critically *lower* than the 212 F. we count on—208 F. at 2,000 feet, 203 F. at 5,000 feet, a mere 198 F. at 7,500 feet, and so on. Therefore you compensate for less actual heat in the Boiling-Water Bath by *adding time*—for the same reason that a cook in Denver, the "Mile-high City," must boil potatoes longer than a cook in San Francisco does. But you *add pressure* (which means actual heat) when using a Pressure Canner.

Processing adjustments for altitude

The Pressure adjustments below are for *dial*-gauge canners only. Read the manufacturer's directions for adjusting with a *weighted* gauge.

For the B-W Bath you *add minutes* to processing time— how many, depends on whether the sea-level time required is *20 minutes or less,* or *more than 20 minutes.* For Pressure-processing you *add pounds*—but not varying the progression even though the sea-level pressure required is 5 or 10 (used most often) or 15 pounds. Thus:

IN B-W BATH ADD:		AT THESE ALTITUDES:		IN PRESSURE CANNER RAISE POUNDS TO:		
If 20 mins. or less	If over 20 mins.	Feet	Meters	If 5 lbs.	If 10 lbs.	If 15 lbs.
1 min.	2 mins.	1,000	305	5½ lbs.	10½ lbs.	15½ lbs.
2 mins.	4 mins.	2,000	610	6 lbs.	11 lbs.	16 lbs.
3 mins.	6 mins.	3,000	914	6½ lbs.	11½ lbs.	16½ lbs.
4 mins.	8 mins.	4,000	1,219	7 lbs.	12 lbs.	17 lbs.
5 mins.	10 mins.	5,000	1,524	7½ lbs.	12½ lbs.	17½ lbs.
6 mins.	12 mins.	6,000	1,829	8 lbs.	13 lbs.	18 lbs.
7 mins.	14 mins.	7,000	2,134	8½ lbs.	13½ lbs.	18½ lbs.
8 mins.	16 mins.	8,000	2,348	9 lbs.	14 lbs.	19 lbs.
9 mins.	18 mins.	9,000	2,743	9½ lbs.	14½ lbs.	19½ lbs.
10 mins.	20 mins.	10,000	3,048	10 lbs.	15 lbs.	20 lbs.

As for "simmer"—

There are many descriptions of *simmer*, ranging from the French cook's "making the pot smile" on up to the one we mean when we say to simmer: which is 185–200 F./85–93 C. at sea level. In this range small bubbles rise gently from the bottom of the pot, and the surface merely quivers, instead of dancing as in the full boil at 212 F./100 C.

We mention this now because we use a *Hot*-Water Bath in processing a number of canned fruit juices. The individual instructions specify the simmer temperature needed—most often it's 190 F./88 C.—but again this is a sea-level reading, so increase the H-W Bath processing time for such foods according to the adjustments for altitude given above.

Pasteurizing

Pasteurization is the *partial* sterilization of food, and it was devised in the nineteenth century by Louis Pasteur as a means of controlling the fermentation of wine. As applied to milk, one method raises the temperature to 142–145 F./ 61–63 C., holds it there for 30 minutes, then quickly reduces the temperature to well below 50 F./10 C., where it is held for storage. The other way—called the "flash" method and used commercially by dairies—raises the temperature to 160–165 F./71–74 C. for a mere 15 seconds, followed by rapid cooling to well below 50 F.

The longer time at the lower temperature range is the method given in the instruction for pasteurizing milk in Chapter 9 because we think it is less hair-trigger, and therefore less chancy, for the average homemaker to use for achieving the result desired.

Here, too, it is important to follow the "Processing Adjustments for Altitude," above.

In addition we use the term *pasteurizing* in connection with the *Hot*-Water Bath for processing canned fruit juices (Chapter 2); and also in Chapter 5 in recommending a "finishing" treatment by dry heat at 175 F./79.5 C. for 10 or 15 minutes for foods dried in the sun.

Common Ingredients

WATER

There's hardly any method of putting food by that does not involve water somewhere along the line, beginning with the first washing of raw materials, continuing to the various water-based solutions used in canning and freezing and curing—which includes brining of all sorts, to ferment or to preserve—and even extending to some steps in drying.

The most important single thing about water used in any process for preserving food is this: THE WATER MUST BE FRESH AND *AT LEAST* OF DRINKING QUALITY. A staggering number of spoilage micro-organisms are *added to food by impure water*. Therefore:

—Don't assume that "it's-going-to-be-cooked-anyway" will counteract *all* the extra contamination: remember that an excessive bacterial load can tax your preserving method beyond the point where it is effective. (Nor is there any magic sterilizing substance you can tuck into a container or package of food that has not been handled properly.)

—Don't wash food in any water that you would not care to drink; change wash water often; high-protein, low-acid foods like meats and poultry, and especially seafoods, should be washed under running water.

DEALING WITH MINERALS IN WATER

"Hard" water has above-average mineral content (calcium is often an offender here). Hard water can shrivel pickles or toughen vegetables.

You can check for hardness by shaking a small amount of soap—*not* detergent—in a jar of water: if it makes a good head of suds in your water, hardness is not a problem.

Or ask your municipal water department to tell you the composition of the water that comes from your tap. In rural areas, your health officer can tell you how to have your private water supply tested for mineral content as well as for bacterial count.

If your water is hard, and you don't want the expense of buying distilled water, collect rainwater *in the open*—not as run-off from dusty roofs—and strain it through several layers of cheesecloth.

Or if you know that hardness is caused by calcium or magnesium carbonates, boil the water for 20 to 30 minutes to settle out the mineral salts; then pour off and save the relatively soft water, taking care not to disturb the sediment.

Iron compounds in your water will darken the foods you put by.

Don't use water that contains sulfur if you can avoid doing so. The sulfates in water do settle out with boiling, but become more concentrated as the water evaporates. Sulfur will darken foods.

SALT

Several forms of common salt (sodium chloride, chemical formula NaCl) are given in this book as ingredients: ordinary table salt of the supermarket variety, and pure pickling salt.

Table salt is finely ground and contains small amounts of anti-caking agents that keep it free-flowing; if it also contains iodine the label will say so. It is used simply for flavor in canning, and of course is optional. It should not be used in pickling because the additives that prevent caking could cloud the liquid; the little bit of salt used in canning vegetables and meats won't affect the liquid, but the iodine in iodized (or sea) salt could cause sulfur-y vegetables like cabbage and cauliflower, and shellfish, to discolor.

Pure pickling salt is coarser than the table seasoning—and a good deal cheaper; it also can be got at any supermarket. Our older folks remember this as "Turks Island Salt," named for the Caribbean islands where it was produced years ago in great quantity. It is recommended for canning because of its low cost and lack of iodine or fillers; it is *required* for pickling and curing.

Double warning: (1) Never use halite salt, the sort used to clear ice from sidewalks, because it is not food-pure. And (2) never use a salt substitute as a seasoning in preparing food that will be heat-processed, because cooking could give it an unwanted aftertaste; add it just before serving the food.

Kosher salt, also gettable at any supermarket, makes a good pickling salt; sometimes it contains polysorbates. And so-called "dairy" salt is good: any farm-supply store should have it.

BRINING

Brine is salt dissolved in liquid, and for the purposes of this book there are two kinds of brine. One is the result of adding pure pickling salt to water by volume or proportion. The other is the solution that results when dry salt is added to plant or animal material to draw out the juices from the tissues, and the salt combines with the juices.

When we describe brine as being of a certain *percent*, we are referring to the proportion of weight of salt to the volume of liquid—*NOT* to the sophisticated salinometer/salometer reading used in laboratories or in industry (and occasionally confused in some bulletins with the simple weight/volume rule-of-thumb that is adequate for use in putting food by at home). Which is: 1 part of salt to 19 parts of water (totaling 20 parts) = 5 percent, allows some benign fermentation; 1 part of salt to 9 parts of water (totaling 10) = 10 percent, the strongest solution generally used in pickling at home, prevents the growth of most bacteria.

Brine of 2½ percent actually encourages benign fermentation; brine of 20 percent controls the growth of even salt-tolerant bacteria. (See Chapter 7.)

These percentages are offered for your ready reference—but are not actually needed in this book: at every stage in every procedure involving brine, we tell you how much salt and how much liquid you need to achieve the result desired. (See also Chapters 2, 4 and 5.)

SWEETENERS

The relatively small amount of sugar (or alternative natural sweetener) used in canning or freezing fruit helps keep the color, texture and flavor of the food, but it is optional (Chapters 2 and 3).

The sugar in jams and jellies (Chapter 4) helps the gel to form, points up the flavor, and, in the large amount called for, is a preservative—and as a preservative and for the gel it is *not* optional. (The so-called sugarless or "diet" confections rely on gelatin, not pectin; are made with artificial non-nutritive sweeteners; and must be refrigerated or frozen.)

Sugar is also an ingredient in some vinegar solutions used for vegetable pickles, where it enhances the flavor (also Chapter 4).

In curing meats like ham, bacon, etc., a relatively small amount of sugar (perhaps brown) is combined with the salt; but the sugar is added more to feed flavor-producing bacteria than to provide flavor on its own (Chapter 7).

In canning, freezing and drying fruits, the sugar is in the form of a sirup of varying concentrations of sugar; in addition, dry sugar is used for some types of packs of fruit for freezing.

THE ALTERNATIVES

Sugar. White sugar—refined, granulated cane or beet sugar —is the sweetening implied in almost all the instructions, but other sweeteners may be substituted for it, at least in part.

    Brown sugar—semi-refined and more moist than white—is called for in some directions. It also may be used in place of white sugar where its color and flavor will not affect the looks and taste of put-by food in a way you dislike. In substituting, use it measure-for-measure instead of white sugar, but pack the brown sugar firmly in the measuring cup.

Corn sirup. Corn sirup comes in light and dark forms; use light only when substituting for sugar. Generally, ¼ of the sugar in a recipe can be replaced by light corn sirup without changing the flavor of the food.

Honey. Honey has nearly twice the sweetening power of white sugar. Because of this property, plus a distinctive taste, use mild-flavored honey; you will get best results by replacing only part (no more than ½) of called-for sugar with honey.

Maple sirup—which is usually too hard to come by for routine use in the kitchen—should replace only about ¼ of the required sugar, because of its pronounced flavor and color.

Sorghum and molasses. Sorghum and molasses are not recommended for most food-preservation because of their strong flavors.

ARTIFICIAL SWEETENERS

Also called "non-nutritive" sweeteners, these are available in liquid or tablets in the United States. The sweeteners vary slightly in their chemical composition, but American trade-name products gettable at all drugstores and virtually all food markets have a saccharin compound as their base. Saccharin is 300 to 500 times sweeter than sugar.

Saccharin is prohibited as a sweetener in commercial food products in Canada, and home economists in the Dominion recommend strongly against its use in food prepared and served at home.

At this writing (Spring 1975) the safety of saccharin is still under investigation by government agencies in the United States, who advise limiting the consumption of saccharin by the average, healthy adult to 1 gram daily. This amount would include saccharin in all commercially prepared foods and the amount used by the individual to sweeten foods prepared at home.

Also at this writing, cyclamates—banned in 1969—are still under investigation by the Food and Drug Administration; the FDA also is investigating aspartame, a substance about half as sweet as saccharin.

All of which boils down to the importance of the injunction repeated so often: *Never use any artificial sweetener unless it is approved by your doctor*.

This warning is given in the directions for making sugarless, or "diet," jellies and jams in Chapter 4. The instructions also say to read the labels carefully, because the sweetening power varies from product to product—and what you want is the effect, not volume, in the sweetener you use.

Artificial sweeteners do not help retain color and texture of home-preserved fruits, and they sometimes produce an unwanted aftertaste when cooked or processed in canning.

ANTI-DISCOLORATION TREATMENTS

Special treatments are given the cut surface of certain fruits to prevent oxidizing in the air and turning brown when they are canned, frozen or dried. Vegetables are treated to prevent discoloration from enzymatic action when they are frozen or dried. Some cured meats may be treated to retard the inevitable loss of their appetizing pink color during storage. Canned shellfish are given a pre-canning treatment to prevent discoloration from the natural sulfur in their flesh. Fatty fish are given a special treatment to forestall some of the oxidation that causes them to turn rancid in the freezer after awhile.

C-enamel and R-enamel tin cans (see Chapter 2) have special linings that prevent naturally sulfur-y or bright-colored foods from changing color in contact with the metal. If glass jars of red or bright fruits are not stored in the dark or wrapped in paper, the light will bleach the contents. Fruit canned with too much headroom or too little liquid is likely to darken at the top of the jar; for the same reason, keep fruit submerged in its juice, sirup, etc., during freezer storage.

Specifically for fruits

Ascorbic acid. This is Vitamin C, and volume for volume it is the most effective of the anti-oxidants. You should be able to get pure crystalline ascorbic acid at any drugstore and perhaps at any natural-food store; certainly either is likely to have Vitamin C in tablet form. There are about 3,000 milligrams of ascorbic acid in 1 teaspoon of the fine, pure crystals, so buy 400-mg. or 500-mg. tablets to get the maximum amount of Vitamin C with the minimum of filler, then crush them between the nested bowls of two spoons. Ascorbic acid dissolves readily in water or juice (both of which should be boiled and cooled before the solution is made).

It is used most often with apples, apricots, nectarines, peaches and pears; in a strong solution, it is a coating for cut fruit waiting to be processed or packed.

Generally speaking, the ascorbic-acid solution is strongest for drying fruits (Chapter 5), less strong for freezing them (Chapter 3), and least strong when they're canned (Chapter 2).

The crystals also may be added to the canning sirup or to the Wet packs of frozen fruit.

Citric acid. Get the pure crystals (so fine they're almost a powder) at your drugstore, special-ordering it in advance if you have to. Or get large crystals bottled like spices as "sour salt" from a Kosher food store or the special-foods section of large supermarkets; these crystals are easily pulverized between the nested bowls of two spoons or in a mortar and pestle. (See also "Acids to Add" later on.)

Volume for volume, it's about ⅓ as effective as ascorbic acid for controlling oxidation (darkening), and therefore enough to achieve the same result could mask delicate flavors of some fruits.

Lemon juice. Contains both ascorbic and citric acids. Average acid-strength of fresh lemons is about 5 percent (also usually the strength of reconstituted bottled lemon juice; some strains of California lemons are more strongly acid, however).

Being in solution naturally, it's about 1/6 as effective volume for volume as ascorbic acid for preventing darkening. Even more of a flavor-masker than citric acid, it also adds a distinctive lemony taste to the food.

Commercial color-preservers. Gettable at supermarkets alongside paraffin wax and commercial pectins; most often comes in 5-ounce tins. The best-known brand has a sugar base, with ascorbic acid and an anti-caking agent. Expensive to use because of the relatively small proportion of ascorbic acid in the mixture. The label tells how much to use for canning or freezing.

Another brand contains sugar and citric acid, plus several other ingredients, but no ascorbic acid. *Moral:* Read labels to learn what you're paying for.

Both preparations, and similar ones, also have directions for using with fresh fruit cocktail, etc., that is made well in advance and chilled.

Mild acid-brine holding bath. In Chapter 2 this is one of the choices for treating apples particularly, but also apricots, nectarines, peaches and pears while they wait to be packed in containers and processed. It's a solution in the proportions of 2 tablespoons salt and 2 tablespoons white (distilled) vinegar to each 1 gallon of water. Cut fruit is held for no longer than 20 minutes in the acid-brine (which does leach nutrients), then is well rinsed (to remove salt taste), drained and packed.

A similar treatment before freezing these fruits, particularly apples, omits the vinegar (Chapter 3). We recommend against using any such holding bath before drying (Chapter 5) because it adds some liquid to fruit you're trying to take moisture out of.

A similar treatment before freezing these fruits, given in Chapter 3, omits the vinegar.

Never use the regular cider vinegar (especially if it's unpasteurized) or wine vinegar: either could add its own color and, perhaps, sediment to the fruit.

Steam-blanching. Used sometimes before freezing fruits (especially apples) likely to darken when cut, and always before drying (even though you'll also sulfur fruit that is to be sun-dried). Blanch in a single layer held in strong steam over briskly boiling water, 3 to 5 minutes depending on size of pieces. (See Chapters 3 and 5.)

Sirup-blanching. Again a pre-drying treatment to hold color (Chapter 5), but the sirup is so heavy—1 cup sugar or natural substitute to each 1 cup water—and the fruit is in it so long (about 30 minutes all told) that you end up with a confection.

Sirup holding bath. This is simply dropping each piece of cut fruit into the sirup it'll be canned or frozen in, and to which you may have added ascorbic acid (Chapters 2 and 3). Fruit held this way before canning is usually packed Hot (precooked); if Raw pack, the fruit must be fished out, put in the containers, and covered with the sirup after it has been brought to boiling.

Sulfuring before drying. In addition to steam-blanching most fruits in order to slow enzymatic action, all fruits and berries (except grapes) that are to be air- or sun-dried are the better for exposure to sulfur dioxide, which is the fumes from burning pure sulfur. "Better" means protection of Vitamins A and C (both hurt by air), protection against insects, etc., protection against molds and yeasts; sulfuring also protects color. Sulfur is a natural substance, long used in many "spring tonics" and mineral waters.

Get "sublimed" sulfur (also called "sulfur blossoms") from the drugstore; it's pure, and looks like soft-yellow powder. *Don't use fumigating compounds*, even though they contain sulfur. A 2-ounce box will sulfur 16–18 pounds of prepared fruit (for how to burn it, see Chapter 5).

Note: We don't like the sulfide soaks because they leach nutrients, distribute the sulfur compound unevenly in the tissues, waterlog the fruit.

Specifically for vegetables

Blanching. All vegetables to be frozen or dried are blanched
beforehand in order to slow down enzymatic action and
produce the side-effect of helping to protect natural
color. Before freezing, most vegetables are blanched in
boiling water (Chapter 3); before drying, all are
blanched in steam (Chapter 5).

    Before canning, white potatoes are held in a mild
salt solution; salsify (oyster plant) is held in the mild
acid-brine bath we mentioned earlier for certain cut
fruits.

Specifically for meats

Saltpeter. In the United States in its pure form this is usually
sodium nitrate, occasionally potassium nitrate, both gettable at drugstores. For generations it has been added to
the salt mixture in order to hold the appetizing pink color
of cured meats that may or may not be smoked after-
wards (ham, bacon, etc., corned beef, pastrami) and in
frankfurters, bologna, and so on. It's merely cosmetic: it
has no preservative action. And its side-effects on human
health are being scrutinized by the Food and Drug Administration and the U.S. Department of Agriculture
(USDA), spurred on by consumer activists (Chapter 7).

    In individual instructions we include it as an optional
ingredient after saying our say about it.

Ascorbic acid. Good Vitamin C again. It will hold the color
of cured meat—not so brightly and certainly more briefly
than saltpeter does. But both these detriments add up to
a virtue: Ascorbic acid will lose its hold on the color well
before the meat has taken on a bacterial load that makes
it unfit to eat (whereas saltpeter will keep nice pink color
in meat that would give health departments fits).

    Use ¼ teaspoon pure crystalline ascorbic acid for
every 5 pounds of dressed meat to be cured; add it to the
salt mixture or to the brine.

Specifically for seafood

Citric acid (and lemon juice). For canning, the picked meat
of crabs, lobsters, shrimp and clams is given a brief dunk

in a fairly tart solution of citric acid or lemon juice as a way to offset the darkening action of minerals naturally present in such foods (otherwise the meats would be likely to discolor during processing and storage). The dip lasts about 1 minute, and the meat is pressed gently to remove excess solution. (See Chapter 2.)

White (distilled) vinegar may be used too, but it might contribute a slight flavor of its own.

Ascorbic acid. For freezing fatty fish, a 20-second dip in a cold solution of 2 teaspoons crystalline ascorbic acid dissolved in 1 quart of water will lessen the chance of rancidity during freezer storage. The fish to be treated include mackerel, pink and chum salmon, lake trout, tuna and eel, plus all fish roe.

ACID

Three acid substances—citric acid, lemon juice and vinegar—are added to foods in this book for reasons that have little to do with the cosmetic purpose of helping to control oxidation or color changes in put-by food (see "Anti-discoloration Treatments," above).

ADDING ACIDS TO PRESERVES AND PICKLES

We add acid to enhance flavor, making it brighter or more tangy, in condiments like ketchup or chili sauce or chutney. We add it to help create the balance that makes a gel in combination with pectin and sugar, in cooked jellies and jams. And we add it much more lavishly to aid preservation of a number of pickles served as garnishes. How much of which particular acid is added to all these foods is given in the specific instructions in Chapter 4, "The Preserving Kettle."

ADDING ACID TO DECREASE pH

The relationship between a food's natural acidity (*pH* rating) and its ability to provide hospitable growing conditions for spoilage micro-organisms is discussed at some length in the first part of this Chapter 1. In Chapter 2, small amounts of acid are added for canning specific foods: figs, berry juices, the nectars and purées of apricots, peaches and pears; tomatoes, sweet green peppers and pimientos.

Acid is added *ONLY to these foods before canning them by the method specified and for the specified processing time*.

Added acid *is NOT a crutch:* it *does not let you short-cut* any step in safe canning procedure, it *does not let you fiddle* with canning methods.

The three acids

Citric acid. Pure crystalline citric acid, U.S.P. (meeaning "United States Pharmacopoeia" and therefore of uniform stability and quality) is the acid which the USDA Extension Service nutritionists recommended in April 1975 be added in canning tomatoes. It is sold by weight, is gettable at most drugstores, is not expensive—especially when you consider that 4 ounces (the consistency is like finely granulated sugar) will do about 45 quarts or slightly more than 90 pints of tomatoes. (As an anti-oxidant it is cheaper than ascorbic acid, though less effective, see "Anti-discoloration Treatments," above.)

Food-pure citric acid is also sold in Kosher food stores or the Kosher section of large supermarkets as "sour salt." It comes often in bottles the size of those in which spices are sold; it is often in crystals the size of small peas, but it is easily pulverized—in a blender, by pressing between the nested bowls of two spoons, by rolling with any glass or metal cylinder.

Citric acid is preferred for increasing acid-strength of foods because it does not contribute flavor of its own to food (unlike lemon juice and vinegars, which can alter flavor if used in large enough amounts).

Fine citric acid may be substituted for a 5-percent acid solution (the average for store-bought vinegar or for the juice of most lemons) *whenever the called-for measurements of the solutions are by the spoonful*, in this general proportion: ¼ teaspoon citric acid powder = a generous 1 tablespoon of 5-percent lemon juice/vinegar; ½ teaspoon citric acid power = a generous 2 tablespoons of the vinegar or lemon juice. (The equivalents actually are ¼ : 4 teaspoons and ½ : 8 teaspoons, but 1 and 2 tablespoons are easier measurements to make in the usual household's kitchen.)

To reverse the coin and make a 5-percent solution of citric acid, use the rule-of-thumb for making salt brines: dissolve 1 part fine citric acid in 19 parts of boiled (and cooled) water. Translated into measurements used in the average kitchen, this means dissolving 2 tablespoons fine citric acid in 1 pint (2 cups) of boiled water, or, if you want to be metric, dissolving 28 grams of fine citric acid in ½ liter (470 ml/cc) of boiled water. Either translation will produce a solution around 6 percent instead of 5— but the result will serve the purpose we're after.

Lemon juice. This is recommended over the other acids for use in canning fruit juice in Chapter 2, and for increasing the acidity of fruit juices to ensure a good gel for jellies and jams in Chapter 4.

The people who live in citrus-growing regions are likely to know the virtues and relative acidities of the different strains of lemons; the rest of us buy anonymous lemons at the market and assume that their average acidity equals a 5-percent solution.

If you squeeze lemons ahead of time to have the juice handy, don't hold the strained juice in its sterilized, tightly capped jar in the refrigerator for more than a couple of days: its flavor tends to change as it sits around.

Even just-squeezed juice can alter the flavor of foods, especially when used in fairly large amounts. Therefore rank it below citric acid on this score—but above vinegar.

Vinegar. Vinegar is acetic acid, and all vinegars corrode metal, so when you use them in larger-than-spoonful amounts—in making pickles and relishes (Chapter 4)— make sure your kettle or holding vessel is enameled, stainless steel, ceramic or glass. Acetic acid reacts badly with iron, copper and brass, and with galvanized metal (which we don't like to use with *any* food because of the possibility of contamination from cadmium in connection with the zinc used in galvanizing).

The cider vinegar bought in supermarkets usually runs about 5 to 6 percent acid. Its pronounced flavor can be an asset with spicy condiments and pickles, but a drawback elsewhere. Its color is unimportant for dark relishes, but hurts the looks of light-colored pickles. Because of its flavor and color, it should not be used with fruits or bland foods. The minerals present in any vinegar that has not been distilled can react with compounds in the water or the foods' tissues to produce undesirable color changes. All of which boils down to: *do not use* cider (or malt) vinegar to reduce the *pH* rating of the specific fruits and vegetables cited above—unless it's a last resort, and unless you're prepared for changes in color and flavor.

Use *white (distilled) vinegar* for decreasing the *pH* of the foods mentioned above if you don't have citric acid or lemon juice—though it is close to interchangeable with lemon juice. White vinegar sometimes is slightly less acid than cider vinegar, but not enough to oblige you to alter measurements. Certain pickle recipes call for white vinegar specifically. See "Citric Acid" above for translating.

Avoid using "raw" or "country" vinegar for the purposes cited here—it's likely to have sediment, and its flavor is pronounced. And save wine vinegar for making salad dressings, because it also will result in disappointing products.

FIRMING AGENTS

Having raw materials in prime condition and perfectly fresh, plus handling them promptly and carefully add up to the best *natural* means of ensuring that your put-by foods are firm and appetizing. (See also "Water," page 16, and in the Postscript that starts on page 529.)

Salt. In canning, freezing or curing fish for smoking, a short stay in a mild brine (proportions vary, and are given in Chapters 2, 3 and 7, respectively) that is kept ice-cold will not only draw diffused blood from the tissues but also will firm the flesh and result in a better product.

In drying fish (Chapter 5), a larger concentration of salt is generally used to draw moisture from the tissues and make the flesh firm.

In pickles, the original "short-brine" (q.v. Chapter 4) firms the vegetables. Also helpful is shaving off the vestigial remains of the blossoms of pickling cucumbers, because this is where enzymes concentrate; and unless enzymatic action is halted, such foods will soften as they start to decompose.

Cold. Refrigerating meats, poultry and produce, and holding seafoods on ice until they are prepared for processing will do much to ensure food with good, firm texture.

Alum. This is any of several allied compounds, and was called for in old cookery books to make pickles crisp—usually watermelon or cucumber chips. If you see it in a responsible modern cookbook, follow the directions for its use carefully: when a fairly large amount is eaten, it often produces nausea and even severe gastro-intestinal trouble. This book does not include alum in any recipe for pickles.

Slaked lime. Another old-style additive for firming pickles like the ones above, this is calcium hydroxide. If you do use it (it's not an ingredient in any pickle recipe in this book), make sure that it is food-pure; and *never* use the so-called "rock" lime, which is the highly caustic quick-lime.

Calcium chloride. Some people find this more acceptable than alum or slaked lime but we do not include it any pickle recipe or canning instruction in this book. It is an ingredient often used by commercial canners, especially in tomatoes.

   If you feel impelled to use it, get it from a drugstore in a food-pure form—*not* as sold at farm and garden supply centers for settling dust on roads or for dehumidifying closets, etc., or for fireproofing. And, because too much of it could leave a bitter aftertaste, never substitute it measure-for-measure for regular salt (sodium chloride, $NaCl$). Instead, figure how much salt you'll need for a batch of, say, tomatoes, and in advance mix *not more than 1 part calcium chloride* with 2 parts regular salt. Then add the mixture in the amount of optional salt seasoning that the canning instructions call for.

". . . ask your County Agent"

   We say this at almost every turn in *PFB*, and for many people in the United States it is direction enough for finding a source of practical help cheerfully given. For other people, though, we offer some general background and some suggestions for finding a County Agent to ask.

Big, rich counties have many agents in their Extension Service offices, listing at least one agent each for Agriculture, Forestry, Youth (4-H) and Home Economics: and this last department will have separate specialists in foods and nutrition, family economics, textiles and consumer education. Less well-endowed counties have one agent for each department. Really sparsely settled counties have only the Agriculture and Home Ec agents, but they will field all questions. Ably.

A huge metropolis may not have agents listed for every county it embraces geographically, but it is likely to have an Expanded Food and Nutrition Education Program (EFNEP) center, with branches in boroughs or settlement houses. Try them.

Telephone directories don't have an across-the-board system for indexing their entries; this is especially true for their Yellow Pages. Nevertheless look in your local telephone book under plain "Extension Service," or under "[Your county] Extension Service" or in the listings under your county government. The Yellow Pages may list County Agents under "Vocational and Educational Guidance."

Write to the information director of the Co-operative Extension Service at your land-grant college, which is usually the State University and often its branches as well. (Among the notable exceptions are Rutgers University in New Jersey and Cornell University in New York, both of which are land-grant colleges with thriving Extension Services; Massachusetts Institute of Technology is too, but it doesn't have an Extension Service.)

Ask at your public library (a good reference librarian makes Sherlock Holmes look like a bunny when it comes to chasing down a source of information).

Ask at your Chamber of Commerce.

Ask the Practical Arts department at your nearest vocational high school.

Ask local farm-supply dealers.

Ask any up-to-date honest-to-goodness farmer.

Ask your local newspaper; if it's a big metropolitan daily, ask the Woman's-page editor or the Consumer Affairs editor or the Garden editor.

Call radio stations that air farm-and-home programs, especially the ones that have shows very early in the morning (when they're likely to run taped interviews with various County Agents).

There's an annual *County Agents Directory* that a friend of yours in the country could ask his/her County Agent to look *your* County Agent up in—if you follow us.

WHERE TO GET GOVERNMENT BULLETINS

Having found your County Extension Service office with the help of the foregoing suggestions, visit your Home Ec agent to learn what government publications are available, whether they are free for the asking or require a small fee, and where to send off for them.

Some of the bulletins mentioned below or cited in the rest of the book may be had free for the asking from your County Agent—who also is likely to have extensive catalog-lists of U.S. Government Printing Office publications (which are sold by the Superintendent of Documents, U.S. Government Printing Office, Washington, D.C. 20402). *PFB* yields to none in admiration for the work done, and the material published, by the various divisions of the Agriculture department, but sometimes their manuals take forever and a day to come through channels when you send for them; not always, but often enough so you should allow a month for delivery. Back-and-forthing makes it worse: be sure to send the right amount of money, using the most recent GPO list to order from.

Run through the listings from divisions other than the U.S. Department of Agriculture—such as those from the Food and Drug Administration; and write to Consumer Product Information, Public Documents Distribution Center, Pueblo, Colorado 81009, for their Index.

All in all, material published by university Extension Services is more uneven in quality than Federal publications (as why wouldn't it be, with fifty States?), but it's a lot faster in arriving. And of course your state university can be more specialized about regional conditions or problems than the USDA in Washington is.

In Canada visit or write your Provincial Agriculture Department, or write to Canada Department of Agriculture Information Division, Sir John Carling Building, 930 Carling Avenue, Ottawa, Ontario K1A OC7.

Some good food-handling publications

As we go along through the various ways to put food by we cite Federal or State bulletins, and occasionally general trade books, that we found helpful. The following are listed here because they are good all-round "how-to's" or references. Extension Services have current prices on all U.S. government publications. The Ball and Kerr manuals can usually be purchased where their equipment is for sale (or write to the addresses we've furnished).

USDA *Home and Garden Bulletin No. 8, Home Canning of Fruits and Vegetables*

USDA *H&G Bulletin No. 106, Home Canning of Meat and Poultry*

USDA *Farmers Bulletin No. 2209, Slaughtering, Cutting and Processing Beef on the Farm*

USDA *Farmers Bulletin No. 2138, Slaughtering, Cutting and Processing Pork on the Farm*

USDA *H&G Bulletin No. 10, Home Freezing of Fruits and Vegetables*

USDA *H&G Bulletin No. 70, Home Freezing of Poultry*

USDA *H&G Bulletin No. 56, How to Make Jellies, Jams and Preserves at Home*

USDA *H&G Bulletin No. 92, Making Pickles and Relishes at Home*

USDA *H&G Bulletin No. 119, Storing Vegetables and Fruits in Basements, Cellars, Outbuildings, and Pits*

USDA Research Service *Agricultural Handbook No. 8, The Composition of Foods—Raw, Processed, Prepared*

USDA *H&G Bulletin No. 162, Keeping Food Safe to Eat*

USDA *H&G Bulletin No. 30, Pressure Canners, Use and Care*

Alabama Co-operative Extension Service HE-01, *Food Preservation in Alabama*, by Isabelle Downey, Auburn University 36830

The Technology of Food Preservation (Third Edition), by Norman W. Desrosier, Ph.D., Avi Publishing Co., Inc., Westport, Connecticut 06880

Ball Blue Book and *Ball Freezer Book*, both from Ball Brothers Company, Inc., Muncie, Indiana 47302

Kerr Home Canning Book, Kerr Glass Manufacturing Corporation, Sand Springs, Oklahoma 74063

Canning   2

Canning is a comparatively recent way to put food by—
only freezing, with its offshoot of freeze-drying, is newer—
and canning is the means most often used in North America
for preserving perishable foodstuffs safely at home.

For its effectiveness, canning relies on sterilization and
the exclusion of air. Both these functions are accomplished
by heat, which destroys the things in the food that cause
spoilage or poisoning, and drive out air from the contents,
thus creating a vacuum that seals the containers completely
against outside contamination during storage.

How *much* heat depends on the acidity of each particular
food you intend to can, as was discussed at length in the first
part of Chapter 1. At home in our kitchens this means that
every carefully prepared food with a *pH* rating below 4.5
may be canned safely in a Boiling-Water Bath at 212 degrees
Fahrenheit or 100 degrees Celsius, and that every carefully
prepared food with a rating of 4.5 or higher on the *pH* scale
must be processed in a Pressure Canner—which produces
temperatures ranging from 1 degree hotter than the boiling
point of water on up to 250 F./121 C., and beyond.

How *long* to apply the necessary heat depends on the
size of the food containers and the density of the food itself:
obviously it takes longer for heat to reach the center of a
quart jar than to reach the center of a smaller container;
and it stands to reason that firm-textured foods take longer
to heat than softer ones do.

There's no need to guess about the right heat and the
length of time to apply it, because these specifics are given
in the individual directions for canning each food. First,
though, we'll examine the methods and how they work. Then
we'll turn to the containers to use, and how to fill them.

Shopping note: If you're looking for a new cookstove
and like the idea of a double-oven range whose upper
oven partially overhangs the cooking-top, check the
clearance above the burners before you buy. Some

33

models don't have room for a really big double-boiler
—much less a canner tall enough to process quart jars.

Community Canning Kitchens

Scarcities or brutally high food prices—and they often
go together—have always stimulated interest in the community canning kitchen, just as they inspire the householder
to grow his own produce or to join his neighbors in a community garden project, as witness the co-operative ventures
during the Great Depression and the Victory gardens and
kitchens of World War II. In the late 1960's the Extension
Service of the U.S. Department of Agriculture started its
Expanded Food and Nutrition Education Program primarily
to serve low-income populations, and it is from this source
that explicit information on starting such a kitchen is available (ask your state university Extension Service for the address of the EFNEP unit for your region).

FEASIBILITY TEST

A workable, equable neighborhood garden requires
thoughtful co-operation to be a success, and a safe and truly
thrifty community canning kitchen demands even more in
the way of clear-headed dedication from the people involved. Therefore several questions must be answered with
a resounding Yes before any group undertakes to start one
up.

Food supply. Is there a large enough amount of fresh, in-
season food coming into your town or county to make the
kitchen worth its cost in time, effort and money?

Beneficiaries. Are there enough people benefiting to
make it worth-while?

Location and space. Because the type of project we're
talking about is a lot more sophisticated than the occasional
work bee in one's home, is there available a convenient insti-
tutional-size kitchen where the equipment can be set up and
used regularly? (A school that prepares hot lunches, say, or
a well-equipped grange hall or church.) And is there plenty
of good running water, and enough space for counters and
work tables?

Stove. Will the stove-top accommodate two 2-to-2½ feet-in-diameter processing vats (two at least, to avoid waiting around between batches)? Can the stove's heat be regulated easily to ensure correct processing at every stage?

Note on equipment: At this writing, the Ball Corporation of Muncie, Indiana 47302, has in operation a Food Preservation Program that sells units of community-canning equipment designed for groups of various sizes. Drop them a line.

THE PEOPLE TO WORK IT

A well-trained leader is essential if the kitchen is to turn out *safe* canned products in a consistently smooth, economical manner. This person must have a number of competencies: (1) thorough grounding in the bacteriology of food spoilage—either from contaminants in food itself, or in the course of handling or processing it; (2) experience in preparing and processing large batches of food, and in managing large-scale equipment; and (3) the ability to organize the work-flow, ranging from the preparation of raw materials to packing to processing to cooling to storage (and including cleaning all equipment and work surfaces at the end of each session).

Note: It is the leader's responsibility to ensure safe processing for each batch of food, and to see that *pH* tests are run on all raw foods having borderline natural acidity, in order to set the exact method to use. (See "Testing for *pH*" in the first chapter.)

Next on the list of key personnel is *one qualified, regular operator per shift* to handle the actual processing—and in addition to the duties of timing and temperature control, this "brewmaster" should be strong enough to heft the vats and remove their contents to cooling tubs of water if cans are used, or to counters for jars.

A team to examine and prepare containers for filling.

A team whose duties are broken down to *sorting, paring/cutting* and *packing the food* in containers.

A clean-up team.

And of course a clerk-of-the-works to *check in supplies, keep records, and distribute allotments of canned food.*

BASIC EQUIPMENT FOR HOME CANNING

Some specialized canning equipment is essential for turning out *safe* and attractive products in return for your efforts, and it includes large canners for processing filled containers, and the containers (with their fittings), which are made to withstand the required heat treatments, and to seal well. These few special items—and their *Why's*—are described fully in the following pages; for all the rest, your regular kitchen utensils will be adequate.

Essential:

> A deep Water-Bath Canner for processing strong-acid foods
>
> A steam Pressure Canner for processing low-acid foods
>
> 6- to 8-quart enameled or stainless-steel kettle for pre-cooking or blanching foods to be canned
>
> Jars or "tin" cans in prime condition, with lids/sealers/gaskets ditto
>
> Sealing machine (hand-operated), if you're using cans
>
> Alarm clock, to time processing longer than 1 hour—
>
> *Plus* a minute-timer with warning bell (more accurate than an alarm clock and better for processing less than 1 hour)
>
> Pencil-shaped glass food thermometer (you'll need it for meats, poultry and seafoods in jars, and for all foods in cans)
>
> Shallow pans (dishpans will do)
>
> Wire basket or cheesecloth to hold foods for blanching
>
> Ladle or dipper
>
> Perforated ladle or long-handled slotted spoon, for removing food from its precooking kettle
>
> Wide-mouth funnel for filling jars
>
> Jar-lifter
>
> Sieve or strainer, for puréeing
>
> Colander, for draining
>
> Large measuring cups, and measuring spoons
>
> Muslin bag, for straining juices
>
> Plenty of clean, dry potholders, dish cloths and towels

Nice, but not absolutely necessary:

> Long-handled fork
>
> Household scales
>
> Food mill, for puréeing (a blender's the ultimate in luxury)
>
> Large trays

The Canning Methods

WHAT THE BOILING-WATER BATH DOES—

The Boiling-Water Bath has limitations: *it is suitable ONLY for canning strong-acid foods*—fruits and vinegared things, which include most tomato products—and for *"finishing"* almost all cooked specialties from the Preserving Kettle (q.v.). With such foods, the B-W Bath does these things *if it is used correctly:*

—In raw or blanched (merely partly precooked) strong-acid foods, it kills yeasts and molds and the bacteria that cannot live at 212 F./100 C., the temperature reached in the middle of the containers of food. Thus it sterilizes the food and, in passing, sterilizes all inner parts of the containers that could have got contaminated by airborne spoilers.

—Drives out the air naturally present in the tissues of the foods and in the canning liquid, as well as any air trapped in the container as it is filled. Otherwise the air can prevent a perfect seal and permit spoilage.

—Creates a vacuum that enables the *jars to seal themselves.* (As we'll see, *cans* are sealed by hand after the air in them is exhausted; then they go in the B-W Bath, which sterilizes the contents.)

The Boiling-Water Bath canner

People with long memories recall the covered wash-boiler filled with jars of orchard goodness and boiling merrily on the stove in the summer kitchen. These lovely vessels are collectors' items now: today we buy our B-W Bath canner from the housewares sections of hardware, farm supply or department stores.

It is round, usually made of heavy enameled ware (also called "granite" ware)—and this finish is a blessing because it allows us to use the canner for cooking or treating loose foods that are acid or heavily salted. Stainless steel is good too, but it's more expensive; aluminum, while fine for processing filled containers, will react with acid or salt in loose food.

It has a cover.

It has a rack to hold containers off the bottom of the kettle, thus letting boiling water circulate under them as they
process. The large-size canners often have a second rack
that's supposed to hold a second tier of containers. Sometimes the racks are like shallow baskets with folding handles
that let you lift out bodily the entire batch of containers you
have processed.

There are many B-W Bath canners offered for sale, and
it should be easy to get one that's right—but sometimes it
isn't.

Note: To use a Pressure Canner for the B-W Bath, see
page 46.

AND "BOILING" MEANS <u>BOILING</u>

In order to do its job of processing strong-acid
foods safely, the B-W Bath (1) must be kept constantly *at a full rolling boil* for the entire time stipulated, and (2) must be filled so deep that the fiercely
boiling water *constantly covers the tops* of the food
containers by at least 1 to 2 inches.

So set your timer just after the covering water
reaches a full boil around and above the containers of
food. *Don't fudge:* the *processing* times have been ar-
rived at through careful research by independent food
scientists, and the times *do not include* the minutes it
takes to bring the canner to boiling. See "What Is Boiling?" in Chapter 1; and if you live more than 1,000
feet/305 meters above sea level, add minutes to your
processing time according to the table "Processing
Adjustments for Altitude."

And don't ever let anyone try to persuade you that
boiling water need not cover your jars adequately
"because the steam in the canning kettle will sterilize
the stuff in the jars." IT WON'T: the steam rising from
the water in your B-W Bath *is not under pressure,* and
therefore it cannot get the food hot enough to sterilize
at 212 F./100 C.

IT'S GOT TO BE <u>DEEP</u> ENOUGH

The most popular sizes are billed as 21-quart and 33-quart, and we happen to have both of them. These capacities actually mean loose contents; which is fair enough, since the labels also say, of the respective sizes, "7 jar rack" and "9 jar rack."

But this 7-jar/9-jar bit is *not* fair, because the reasonable inference is that they hold 7 pints or quarts, or 9 pints or quarts, and *even the 33-quart one is not deep enough to process quart jars correctly*—meaning, for our money, *safely*. When you're in a store looking at canners, and if the salesperson has enough experience with canning to point out that a particular kettle is too short for certain jars, pay attention—and ignore what the manufacturer's label implies.

Here's the arithmetic. Starting at the bottom of the 33-quart canner, you should have ¾ to 1 inch between the holding rack and the bottom of the kettle—call it 1 inch for simplicity. Then you must have between 1 and 2 inches of boiling water covering the tops of your containers: 2 is better, so call it 2. Now you have 3 inches accounted for.

Then you must have a *minimum* of 1 more inch of "boiling room" between the bubbling surface of the water and the rim of the kettle. If you don't, the briskly boiling water will keep slopping out onto your stove-top and drenching the well of your heating element (making a mess even if it doesn't extinguish a gas flame, and offering a temptation to reduce your boil to a polite, and inadequate, simmer). Be cagey and remember that the close-fitting cover on your canner will increase its tendency to slop over: allow 1½ inches for headroom to boil in. Now you have 4½ inches—and your containers haven't even gone in yet.

The height of a quart jar (we'll use jars here, as the manufacturer and the great majority of home-canners do) is 7 inches if it's a modern mason with a 2-piece screwband lid, or 7½ inches if it's the old bailed type with a domed glass lid.

Add 4½ inches and 7–7½ inches and you get 11½–12 inches.

But the inside depth of our 33-quart Boiling-Water Bath canner is only a scant 10 inches. Therefore—though it's O.K. for pints, which are 5–5½ inches tall, and of course for the even shorter ½-pint jars—*we cannot use it for processing*

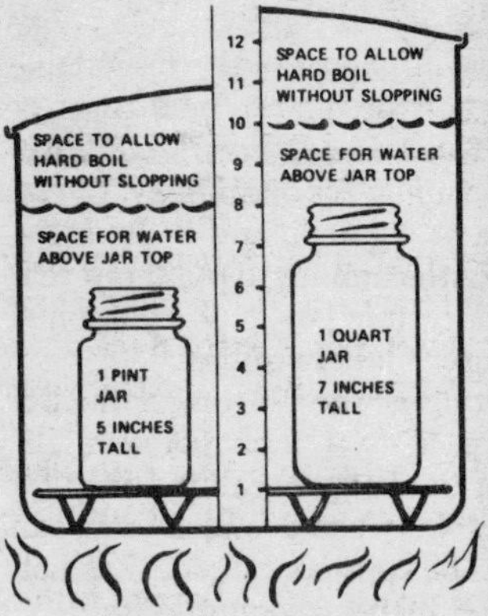

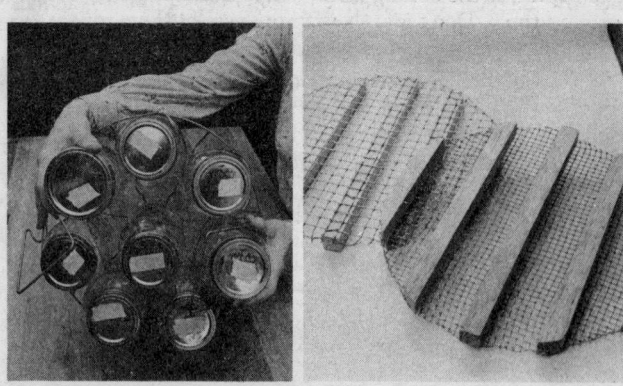

Above left, the cutaway drawing shows the room needed above jar tops for correct processing in a Boiling-Water Bath; the canner at right is too shallow for the quart jar (even though it was sold as O.K. for quarts): there's only 1 inch between jar top and kettle rim. At left below, jars are wedged tight together and boiling water can't circulate around them. Guy Brunton's sturdy canner-racks are made of hardware cloth stapled to ¾- x ¾-inch softwood strips.

quart jars; and it's barely tall enough to deal safely with pints.

There are plenty of canners around, so don't despair. Insist first on depth when you're looking for a B-W Bath kettle. Know the height of your jars, add $4\frac{1}{2}$–5 inches, and measure the height of every canner you see. Stick to your guns, and you'll find the one that will do the job right.

How to use the B-W Bath

Put the canner on the stove and fill it halfway with water; put a rack in the bottom; turn on the heat. If the food you're canning takes relatively little time to prepare for packing in the containers, start heating a large teakettle of extra water now too.

Prepare the food for canning according to the individual instructions, packing it in clean, scalded containers and putting on lids as directed.

Jars of *raw* (cold) food *must not be put in boiling water*, lest they crack; you may need to add cool water to the canner. However, if the food in the jars is very hot, the filled jars may go into boiling water without fear of breakage. Cans will always be hot, because they've just been exhausted.

Process in each load only *one type of food,* in *one size of container*.

Use your jar-lifter to lower jars/cans into the hot water. Place them away from the sides of the canner and about 1 inch apart, so the boiling water will be able to circulate freely around them. Don't jam them in or even let them touch: jars might break as they expand ever so slightly while processing. If your batch is too small to make a full canner-load, submerge open mayonnaise jars, or whatever, in the empty spaces to keep your capped jars of food from shifting around as they boil.

If your canner is tall enough to let you process two tiers of small jars/cans at a time, use a wire cake-cooling rack— or one of Guy Brunton's (described fully below)—to hold the upper layer of containers. Ensure adequate circulation by staggering them as you do when baking on several racks in your oven.

Pour enough more hot water around the jars/cans to bring the level 2 inches above the tops of all containers; be

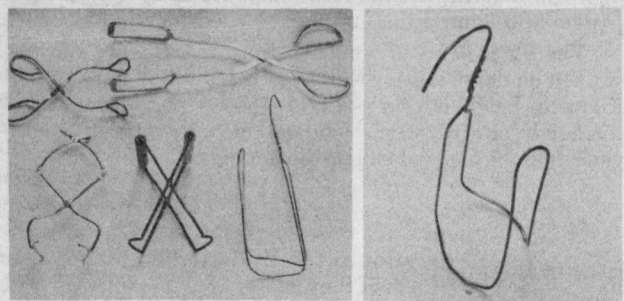

Use these to lift hot jars from canners. Left to right at top: usual small kitchen tongs, big barbecue tongs (their extreme tips bent slightly inward to ensure that they catch under the flange of the jar's neck); below, a patented gadget that pulls snug with the weight of the jar, then a fine old-style lifter that's worth hunting hard for, and finally a desperation lifter made from a very stout coathanger. At right, another view shows the coathanger bent so it will grab under the jar's flange; just hold it upright over the jar, squeezing a wire loop "handle" tight in each hand to get enough purchase—and lift.

careful not to dump boiling water smack on top of jars, particularly if their contents are cool.

WHAT TO DO ABOUT RACKS

The racks that come in store-bought B-W Bath kettles can be infuriating. Made of heavy-gauge wire, they may resemble a flimsy wagon wheel or they look like the skeleton of a basket; at any rate the "spokes" or dividers are so wide apart that even pint jars sometimes fall between them, instead of resting on them. Some lack support strong enough to hold containers high enough off the bottom of the canner. Many actually cause jars to teeter and topple—and you have broken glass to contend with.

So buy several large round cake-cooling racks. Their wire is more fragile, but it's closely crisscrossed; and for support toward the middle, cluster on the bottom of the kettle five or six screwbands that have become slightly rusted or bent, and therefore aren't fit to use on jar lids. Measure the bottom diameter of your B-W Bath canner (not the top, which usually is about an inch larger), and choose your racks accordingly: ½ inch of leeway all around is about right.

Our favorites are the racks made for us by Guy Brunton (who also figured the materials needed for the homemade indoor dryer in Chapter 5). He cut ¾-×-¾-inch softwood in graduated lengths, the longest being 1 inch shorter than the bottom diameter of the canner. He laid them parallel a couple of inches apart, the longest in the center and the outer ones diminishing in size, and on them he staple-gunned a larger circle of ½-inch or ¼-inch hardware cloth, which he bent down to make a smooth finish around the edge. Guy says the racks are likely to warp after a while, being made with softwood scraps instead of the harder-to-get poplar or butternut. The weight of the filled jars will hold them flat for some time, though; after that, we'll whomp the replacements together ourselves.

HOT-WATER BATH (PASTEURIZATION)

We have bent over backwards in referring to the *Boiling*-Water Bath because it's necessary to distinguish clearly between the B-W—which maintains a real boil at 212 F.—and the *simmering* Hot-Water version, which is a form of pasteurizing.

The Hot-Water Bath is recommended *only for certain sweet, acid fruit juices*. Despite the casual swapping of these terms in some older manuals, the processes themselves are not interchangeable.

Specific temperatures for the simmering *Hot*-Water Bath and processing times are given under "Fruit Juices" in the blow-by-blow procedures for canning particular fruits.

THE *WHY* OF PRESSURE CANNING

Every vegetable except tomatoes (q.v.), every meat, every seafood that is canned at home MUST BE CANNED IN A PRESSURE CANNER.

Or put it this way:

Pressure Canning is the ONLY process that is able to destroy the tough spores of bacteria like *Clostridium botu-*

linum which can grow and produce deadly poison in jars or cans of *any* low-acid food.

—Or this way:

Dangerous spoilers can live through even a long Boiling-Water Bath at 212 F./100 C. in containers of food with a natural acidity of 4.5 on the *pH* scale, but *they are killed by higher temperatures that can be reached only in a Pressure Canner*.

Please look at the *pH* ratings of selected foods and the section called "Dealing with the Spoilers" in Chapter 1.

And before we talk about how to use a Pressure Canner, we'll make four points:

(1) The Pressure Canner we're talking about is not to be confused with a *steamer*—either the old-style arrangement that swirled steam at Zero pounds/212 F. around containers of food, or today's compartmented kettles for cooking clams or lobsters, or whatever.

(2) The Pressure Canner we're talking about is an honest-to-goodness, regular, conventional, *big* Pressure *Canner* —not the little pressure saucepan that you may already use for cooking loose food. Don't can food in a pressure saucepan: its heating-up and cooling-down periods are different from those of a real Pressure Canner, so the results are not the same. Nor do we know of any safe rule-of-thumb for tinkering with processing times in order to compensate for using a saucepan.

(3) Independent food scientists, who certainly have no axes to grind, consider that 10 pounds (240 F./116 C.) is the safest pressure to use: 5 pounds (228 F./109 C.) is generally likely to be ineffective for the results desired, and 15 pounds (250 F./221 C.) is very near the caution zone for operating canners sold for use at home. If we've counted correctly, in this book there are only four exceptions to the 10-pound stipulation.

(4) And the pressures given in the individual instructions are based on atmospheric conditions roughly at sea level: do follow "Processing Adjustments for Altitude" in Chapter 1 if you live at an altitude greater than 1,000 feet/305 meters.

Buying a Pressure Canner

Pressure Canners are usually made of cast aluminum. Like B-W Bath kettles, they come in sizes measured according to loose-contents capacity; unlike B-W Bath kettles, though, their diameters are all roughly the same: it's their height that varies. The most popular sizes are 16-quart (actually taking 7 quart jars or 9 pint jars), and 22-quart (actually taking 7 quarts, or 18 pints in two layers, or 34 ½-pints stacked in three tiers).

They're expensive, but they do their job for years and years if you take good care of them and keep their sealing rims/gaskets and pressure gauges, vents and safety valves clean and in good working order.

It may make economic sense to buy a Pressure Canner if you plant a good-sized garden or can get lots of fresh produce cheaply in season, and have surplus meat to put by. But if your family is small and you have just a few surplus foodstuffs to store, maybe your answer is freezing (Chapter 3).

Borrowing?—usually not feasible. But something that is: Go shares with a thoroughly compatible friend in buying a large one (estimating the size carefully beforehand) and pool your resources and energies for canning bees. (See also "Community Canning Kitchens," pages 34–35.

Note: You can use your Pressure Canner for a Boiling-Water Bath if its inside height is 4 to 5 inches greater than the height of your tallest jars/cans (see pages 39–41). Then *lay* the cover on the rim—*don't lock* it, or pressure will start to build after 10 minutes or so, even with the vent open—and *don't close the vent.* Start timing when you hear a hard boil inside, and steam vents strongly.

The anatomy of a Pressure Canner

The base, or kettle, is covered with a tight-fitting lid that contains the controls. The lid is fastened down with clamps or a system of interlocking ridges and grooves. It may or may not have a rubber gasket.

Controls are: (1) a pressure gauge—a dial, or a weight like those usually seen with pressure saucepans that has 5/10/15-pound holes that seat on the vent; (2) an open vent to let air and steam exhaust before processing time begins, and is closed with a petcock or separate weight (if gauge is a dial) or with the weighted gauge, to start raising pressure; and (3) a safety valve/plug that blows if pressure gets unsafe.

The canner also will have a shallow removable rack to keep jars/cans from touching the kettle bottom, or a strong wire basket that serves as a rack and lets you lift all the hot containers out in one fell swoop.

Using a Pressure Canner

First, read the operating manual that comes with it.

Next, clean the canner according to directions, to remove any factory dust or gunk. Use hot sudsy water for the kettle, avoiding strong cleansers or scouring powders. Do *not* immerse dial in water: instead, wipe lid clean with a soapy cloth, and follow with a clean damp one to remove the soap.

Check the openings of the vent, safety valve and pressure gauge to make sure they're unclogged and clean. Take a small sharp-pointed tool (like a large darning needle or a bodkin) to the openings if they need it; clean the vent by drawing a narrow strip of cloth through it.

Be sure the rims of kettle and lid, and rubber gasket if there is one, are smoothly clean, so they'll make a perfectly tight seal.

CHECKING THE DIAL GAUGE

A weighted gauge remains accurate (within its limitation of being hard to "read"), provided it is kept clean and undamaged. But well before each canning season—or more often if you put by large quantities—a dial gauge on your Pressure Canner should be checked for accuracy against a master gauge. Just because the dial may rest at, say, 2 pounds when the canner is not in use does not mean that the gauge is simply 2 pounds high, or that it is uniformly 2 pounds off throughout its range. *So check it.*

The problem is to find the nearest master gauge and the people able to make the test. But first unscrew the gauge at home—carefully, so the threads do not strip—and have it ready to mail off or to take by in person. Likely sources of help or information:

The manufacturer's instruction booklet. Or a store that carries your brand of canner (it should have catalog material and a listing of the nearest regional sales or servicing centers).

Call the Vocational or the Practical Arts departments of a district high school: someone in either section might be able to steer you. The Home Economics department at your state technical college is another idea.

Telephone your county Agricultural Extension Service offices. The County Agent in Home Ec there just may have the equipment to do it for you—or at least she can tell you where to go. The director of the EFNEP (Expanded Food and Nutrition Education Program) center for your area might be able to offer similar information.

ADJUSTING FOR AN INACCURATE DIAL GAUGE

You need a new dial gauge if yours is 5 pounds or more off an accurate reading in either direction.

For lesser variations, make the following adjustments— but *make them ONLY for processing at 10 pounds,* because the variations might not be consistent at 5 pounds or 15 pounds:

If the gauge reads high by 1 pound—process at 11 pounds; by 2 pounds—process at 12; by 3 pounds—process at 13; by 4 pounds—process at 14.

If it reads low by 1 pound—process at 9; by 2 pounds— process at 8; by 3 pounds—process at 7; by 4 pounds— process at 6.

THE PRESSURE CANNER AT WORK

Here are the mechanics of using a Pressure Canner (for filling jars/cans see pages 69–81, and directions for individual foods). And a common-sense rule: don't leave your canner unattended while it's doing its job.

WATER IN—

Put about 1½ inches of warm water in the thoroughly clean canner. If you know that your canner leaks steam slightly when the vent is closed, add an extra inch of water to ensure that the canner will not boil dry if the processing time is more than 30 minutes.

START IT HEATING—

Place the uncovered kettle (bottom section) over heat high enough to raise the pressure quickly *after* the lid is clamped tight. (See how to estimate this heat in "Leary of Pressure Canners?" on page 51.)

COVER ON TIGHT—

When the batch is loaded, put the lid in position, matching arrows or other indicators. Turn the cover to the Closed position or tighten knobs, clamps, etc., to fasten the closure so no steam can escape at the rim.

LET IT VENT—

It is important to consider the *Why* of this next step, because sometimes it isn't stressed hard enough in the instruction booklet that comes with your canner—and your processing could be thrown off whack.

Two "musts" first and last—time carefully, and lift the lid this way.

Air that is trapped in your Pressure Canner will expand and exert extra pressure—that is, pressure in addition to that of the steam—and your gauge will give you a false indication of the actual temperature inside your canner. Therefore you must make sure that *all air is displaced by steam before you close your petcock or vent.*

For canners with either type of gauge, leave the vent open on the locked-in-place lid until steam has been issuing from it in a *strong, steady* stream—*7 minutes* for a canner that holds 7 quart or 9 pint jars, or at least *10 minutes for the larger ones* that hold 18 pint jars. Use your minute-timer here. And don't get stuck on the telephone or something: after 10 minutes of strong venting, internal pressure can reach 1–3 pounds (enough to affect processing time). When your canner has vented, close the vent by the means specified for dial or weighted gauge.

MANAGE THE CONTROLS—

Follow carefully the manufacturer's instructions. For processing adjustments at altitudes higher than 1,000 feet/ 305 meters, see page 14.

For a canner with a *dial-faced gauge,* close the vent and let pressure rise quickly inside the kettle until the dial registers about ½ pound *under* the processing pressure that is called for. At this point, reduce the heat moderately under your canner to slow down the rate at which the pressure is climbing: in a minute the gauge will have reached the exact poundage you want, so you adjust the heat again to keep pressure steady at the correct poundage.

For a canner with a *weighted gauge* (which is harder to keep track of than a dial is), set the gauge on the vent and then carry on as for the canner with a dial gauge—but of course keeping track of the frequency of the jiggles as the maker's instructions say to do.

WATCH THE GAUGE—

Pressure, once the right level is reached, *must be constant.*

You're using a Pressure Canner so your food will be safe, so if the pressure sags you should bring it back up to the mark and start timing again from scratch. Don't guess—turn your timer right back to the full processing period. Your food will be overcooked, of course—but next time you'll keep your eye on the pressure, right?

Fluctuating pressure also can cause liquid to be drawn from jars. And this in turn can prevent a perfect seal.

AND WATCH THE CLOCK—

Count the processing time from the moment pressure reaches the correct level for the food being canned. *Be accurate:* at this high temperature even a few minutes too much can severely overcook the food.

REDUCE TO ZERO

At the end of the required processing time turn off the heat or remove the canner from the stove. It's heavy and hot, so take care.

If you're using *glass jars,* let the canner cool until the pressure drops back to Zero. Then open the vent very slowly.

CAUTION: If the vent is opened suddenly or before the pressure inside the canner has dropped to Zero, liquid will be pulled from jars, or the sudden change in pressure may break them.

If you're using *tin cans,* the pressure does not need to fall naturally to Zero by cooling: open the vent gradually when processing is over and the canner is off heat, to let steam out slowly until pressure is Zero.

LIFT THE LID—

Never remove the lid until after steam has stopped coming from the vent.

Open lid clamps or fastenings.

Lift the back rim of the cover first, tilting it to direct remaining heat and steam *away* from you.

TAKE OUT JARS OR CANS—

If the canner doesn't have a basket rack, use a jar-lifter and *dry* pot-holders (wet ones get unbearably hot in a wink) to remove jars/cans.

Cool cans quickly in cold water (pages 68–69).

Ignore the "afterboil" still bubbling in jars and complete seals if necessary. Stand jars on a wood surface, or one padded with cloth or paper, to cool away from drafts.

Test seals when containers have cooled overnight (page 76).

LEERY OF PRESSURE CANNERS?

If you're a beginner, and fearful, prepare the canner and take it through a practice run on your stove before processing any food in it. Thus:

Put a couple of inches of water in it, secure the cover, and put it on heat; exhaust the air by venting according to directions on pages 48–49. Close the vent and watch the gauge until it indicates 10 pounds pressure (the pressure used most often in canning), and hold it there—fiddling with the heat under the canner to determine how much is needed to hold a steady 10 pounds—for at least five minutes, to get the feel of it. Then slide the canner off heat, let pressure fall back to Zero, open the vent to release any last gasp of steam, and remove the cover, tilting it away from you.

You've done it! Now you'll know what to expect, and you know you can boss the beast.

Summarizing: Pressure Canners are nothing to be afraid of if you treat them with respect and follow the manufacturer's directions exactly. Much research by government and private agencies has determined the best techniques for Pressure-processing: *don't try shortcuts.*

MAVERICK CANNING METHODS

There's nothing like the *Morbidity and Mortality Weekly Report* for discouraging nostalgic ways of canning food at home, because in almost every account of an outbreak of food-borne botulism, the Editorial Note suggests that "inadequate processing" or "inadequate heating" allowed the toxin to form.

Spelling it out, this means that low-acid foods that should have been Pressure-processed were merely given a Boiling-Water Bath; and that strong-acid foods—which should have been given a B-W Bath to sterilize container and contents adequately—were canned instead by "open-kettle," or were treated with dry heat in an oven, or were put in a tank affair and surrounded by swirls of steam *not* under pressure.

Goodbye to old "open-kettle"

The method called "open-kettle" was the first to be used in canning, and it's the one whereby hot jars were filled to brimming with boiling-hot, fully cooked food, the lids were clapped on, and a vacuum was expected to form a seal as the jars were allowed to cool.

The open-kettle procedure was developed in 1810 by a Frenchman named Nicolas Appert, who was given a fat prize by Napoleon for his innovation. Since then, microbiologists and food scientists have chipped away at it so successfully that today it is considered universally safe only for filling and sealing containers of cooked jelly.

From the beginning, a lot of spoilage occurred—why not, since Louis Pasteur didn't establish the reason for bacterial contamination until the 1860's?—but people who were canning by open-kettle were grateful for any containers of appetizing food; half a loaf is always better than none.

By World War II, home-canning in the United States was most of a century old and had undergone numerous refinements. It had long been established that not only was the maximum temperature reached in the open-kettle method *not hot enough for low-acid foods,* but also that these foods must be Pressure-processed. As for the strong-acid favorites canned by open-kettle, research into the causes of spoilage had persuaded many a housewife to can even these in a Boiling-Water Bath. Still, many people had a very high batting-average with open-kettle for things like peaches and tomatoes and relishes, and they continued to use their old method: you may lose a few but you win a lot, and anyway "a little mold won't hurt you."

By 1970 open-kettle was in such disrepute among the experts that all up-to-date manuals dismissed it flatly with the injunction that it should not be used. Nevertheless, *PFB* realized that the method *was* being used by a new generation of beginning homemakers who were anxious to turn back the clock in their search for a simpler way of life. Therefore in our first edition we thought it was necessary to describe how to can by open-kettle the sweet, acid preserves and cooked relishes and pickles with a high vinegar content, and recommended a finishing B-W Bath to ensure sterilization and a seal for any such strong-acid product where storage was less than ideal.

However, now—in 1975—we say: *open-kettle is safe ONLY FOR COOKED JELLIES*. In the previous two years there were published widely the results of investigations which proved that *"a little mold" CAN hurt you*, by turning a strong-acid medium into one that is low-acid enough to harbor poison-producing spores of *Cl. botulinum*, and by throwing out mycotoxins of their own as well. Do see the discussion of botulism in Chapter 1, and the introductory comments to Canning Tomatoes later in this chapter, and to Pickles and Relishes in Chapter 4.

"No" to oven-canning

The folklore of homemaking has another bad method that continues to surface: *oven*-canning—which is attempting to process strong-acid foods by putting the filled and capped containers in a moderate oven for perhaps more time than is required to do it right in a proper Boiling-Water Bath.

Dry heat *does not penetrate* adequately, even if the oven could heat the contents to more than 212 F./100 C. to compensate (and it can't).

In addition to the danger of inadequate processing, there is also the likelihood of having a jar of food burst in your face as you open the oven door.

Oven-canning just plain won't do the job, even with strong-acid foods. And think how much fuel it costs you to try it, compared with the one-burner worth to do the batch safely, on all counts, in a Pressure Canner or B-W Bath!

Note about microwave ovens: The energy shortage is being used to strengthen the sales pitches for using microwave ovens at home, and the matter of oven-canning has cropped up here too. We say only that there has not been enough research reported out to convince us that a microwave oven will prevent the faults of inadequate heating which we object to about oven-canning in general—or that it will prevent other faults in the product that could result from the unique character of the microwave process. We understand that irresponsible claims for "microwave canning" have toned down considerably anyway.

And another "No" to steamer-canning

As far as we can find out, steamer-canning (*NOT to be confused with Pressure Canning—PLEASE*) was conducted, during its brief and bygone day, in a tank arrangement that allowed plain steam to waft around the jars placed on racks inside it.

Even though the steam was expected to peak at 212 F./ 100 C. doesn't mean that it stayed that hot, or maintained a strong, steady flow. Letting jars of food putter along in a gentle vapor simply cannot result in a sealed and safe product; nor did it, 'way back then, because the things apparently weren't long on the market.

("But steam is *steam,* isn't it?" Nope. It isn't. As any youngster studying practical science in grade-school can tell us.)

As for rigging up a clam-steamer or even an honest B-W Bath kettle and trying it—*Don't!*

And maybe a "Maybe" to cold-water canning

However, there is an old back-country way of canning raw cranberries and raw green gooseberries—both almost as low on the *pH* scale as natural foods come—and the method works if its limitations are observed. Otherwise be prepared to heave out some jars of nasty berries.

Use jars that have rubber seals and screw caps or lids held with bails. Fill sterilized jars with picked-over, washed, perfect cranberries or green gooseberries. Next, run cold drinking water into a scrupulously clean and scalded pail to a level some 6 inches above the height of the jars. Then fill the jars of fruit to overflowing with cold drinking water and submerge them—still open—in the pail and, *under water,* cap them tightly. Store immediately in a cool place.

THE LIMITATIONS

Cranberries and green gooseberries must be strongly acid, and be tough-skinned and firm, so they don't break down readily during storage. Old-time country housewives used to can fresh lemons this way when they could get a supply fairly cheaply in February or March.

Because there's no heat processing anywhere along the line, you must wash the fruit well and sterilize utensils and jars. If you're leary of your water you can always boil it in advance (cooling it well before canning with it)—but then so much boiled water would be needed for canning by this method that you're off easier, and better, using a Boiling-Water Bath in the first place.

These berries that are cold-water canned require really cool storage; having no added sugar to retard freezing, though, means they should not get below 32 F.

Remember that the seal is not a perfect one, and that some air remains in the tissues of the food: therefore it can't store as long as properly heat-processed food does.

USING THE BERRIES

If you're going to cook the berries with sugar for a preserve or conserve, drain off the canning water and boil it down to at least 1/3 its volume before adding the fruit and sugaring to taste.

If you're making a pie or tart, etc., as with fresh berries, discard the water.

About Jars and Cans

Either glass jars *specially made for home-canning* or metal cans may be used for putting food by.

Over-all, canning jars are more versatile than cans are: (1) anything that is canned in cans may also be canned in glass; (2) Boiling-Water Bath and Pressure-processing are used for either jars or cans of food; but (3) there are certain foods that should be home-canned only in glass—the individual instructions later on will tell you which ones, but among them are jellies, preserves, pickles, relishes, a couple of fruits and vegetables, and all seafoods.

Another advantage of jars is that they usually require one less step in the packing procedure than cans do; the step is called "exhausting," and will be described in detail in a minute.

Still another plus-mark for jars is that they're generally

a lot easier to get, being sold at hardware and farm-supply stores and in supermarkets.

And of course jars show off their contents, and hence are a requisite for exhibiting your food at the fair. Jars of food must be stored in a cool, dry place that's also dark, otherwise the contents will fade in the light.

And last, jars are re-usable as long as they have no nicks or scratches, even minute ones, and as long as you can get the proper fittings for them.

Aside from having to be special-ordered from the distributor, cans require an expensive hand-operated machine for crimping their lids onto the sides to form a perfect seal.

Cans are not re-usable.

But cans of food require only dry, cool storage—not darkness, since the contents are not exposed to light. And cans stack easily. And also, when they're piping hot from the processing kettle they don't need to be handled so gingerly as glass does. (It's a mistake to be rough with them, though: a dented rim can mean a damaged seal.)

THE TYPES OF JARS FOR HOME-CANNING

The jars recommended for use to North American householders are all alike in three highly important respects:

(1) *The jars are manufactured specifically for canning food AT HOME.* Generally speaking, this means that they are designed to withstand more cavalier treatment during filling, processing and storage than their commercial counterparts undergo at the hands of industrial processors and shippers. Compared to the jars that we buy mayonnaise and peanut butter, etc., in, jars made specially for home-canning have slightly heavier glass, and sometimes have been subjected to higher heat-tempering.

North America's largest manufacturers of home-canning jars—two in the United States and one in Canada—have told *PFB* that, though they all three also make commercial glass containers, their commercial jars are "one-trip" containers, and therefore are not reliable for re-use in canning food at home.

(2) *The jars come in three standard capacities—½-pint, 1-pint and 1-quart—and therefore can be processed*

safely according to the specific instructions given in this chapter.

The manufacturers also make 1-pint and 1-quart jars with dimensions slightly different from those of their standard sizes above, but these variations would make no appreciable difference in processing—except as more height could mean that these jars would not fit in your Boiling-Water Bath Canner with enough extra room to ensure a safe product.

The manufacturers also offer ½-gallon jars for canning foods at home, *but we give no instructions for ½-gallons, because dense low-acid foods don't process well in them.* In addition, ½-gallons could be too tall for your canners anyway.

(3) *The jars have types of closures (described below) for which fresh sealers are available before each canning season,*[*] and our filling/processing/sealing instructions take these closures into account.

Modern mason with 2-piece screwband metal lid

"Mason" goes back to the name of the originator of the screwtop canning jar, and denotes any jar with a neck threaded to take a closure that is screwed down.

Today's modern mason is the highly refined home-canning jar being produced in the millions each year by such makers as Ball Corporation and by Kerr Glass Manufacturing Corporation in the United States (with Anchor Hocking in Ohio introducing a jar for home use in late winter of 1975), and by Dominion Glass Company Limited in Canada.

The jar comes with standard and wide-mouth openings, and, depending on the maker, has a variety of shapes and sizes, including: straight-sided ½-pint; standard and straight-sided pint (the latter taking a wide-mouth closure); straight-sided 1½-pint, which uses a wide-mouth closure; standard and wide-mouth 1-quart. The straight-sided and wide-mouth jars are especially handy for large pieces of food that you'd like to remove more or less intact.

The cartons will say which of these jars is O.K. for freezing.

---

[*]See Postscript, page 529 ff.

The jars come packed with their closures. The lids also are sold separately, with or without screwbands.

HOW THE 2-PIECE METAL CLOSURE WORKS

The lid—called "dome" or "self-sealing" or "snap" by the individual makers, but of one basic design—is a flat metal disc with its edge flanged to seat accurately on the *rim* of the jar's mouth; the underside of the flange has a rubber-like sealing compound; the center surface next to the food is enameled, often white.

The lid is sterilized (more about this in a minute), placed on the clean-wiped rim of the jar of food, and then is held in place by the screwband, which is screwed down on the neck of the jar *firmly tight—AND IS NEVER TIGHTENED FURTHER.* "Firmly tight" means screwed down completely but without using full force, or without being yanked around as with a wrench.

The capped jar is processed, during which the "give" in the metal lid allows air in the contents to be forced out to form the seal. As the jar cools, the pliant metal will be sucked down by the vacuum until the lid is slightly concave. You often hear the small *plink* as the lid snaps down, thus indicating that the jar is sealed.

Note: For several months in 1974 one manufacturer publicized a variation in capping jars with its 2-piece screwband lid. However, in the spring of 1975 this maker assured *PFB* that henceforth its procedure would be the same as that of all other manufacturers: screw down the band firmly tight before processing, and never tighten it further.

To be on the safe side, though, pay attention to each maker's instructions as given in the package of closures.

STERILIZING JARS AND 2-PIECE LIDS

Wash jars, screwbands and lids in hot soapy water, rinse well in scalding water. Containers and closures that will be processed in a Boiling-Water Bath or Pressure Canner need not be sterilized further: but do let them stand filled/covered with the hot water until used, to protect them from dust and airborne spoilers.

Containers of food that will not be processed at 212 F./ 100 C. (this means jelly, etc., glasses and all things done in the pasteurizing *Hot*-Water Bath, and canned by the maverick "cold-water" method sometimes used for green gooseberries and cranberries) must be sterilized. Wash jars and closures as above. Stand open jars upright in a big kettle, fill with hot water until the jars are submerged; bring the whole thing to boiling, and boil for 15 minutes. Remove from heat but let jars stay in the hot water.

Sterilize screwbands the way you sterilize jars.

But sterilize *the metal lids according to the manufacturer's directions.* This isn't a cop-out: the makers give different instructions here, and all of us consumers must assume that they know what's best for their own product. The result in every case is to sterilize, though; and all the lids are left in the hot water as a protection against dust, etc.

WHAT IS RE-USABLE

You must *NEVER use the lid itself again* for canning: the sealing compound on the lid will never seal right a second time around. And besides, it is ever so slightly warped now because you pried it off (and it may be punctured to boot). Even when a lid looks unbent after it comes off, we scratch a big fat "X" on its enameled underside, then it's washed and tossed in a catch-all drawer to use sometime on a refrigerator-storage jar.

You of course can use the jars over and over again as long as they have perfectly smooth sealing rims, and have no cracks or scratches anywhere.

You can use the screwbands again for canning—*unless* they're rusted, or bent, or the threads are marred. (Screwbands too tired to be safe in canning can hold foil or plastic wrap tight on jars of food in the refrigerator, and the like. We often use "retired" screwbands for extra support under a springy wire rack in the bottom of our B-W Bath kettle, or to hold jars well apart in the canner.)

Old-style masons with a one-piece cap

We don't know of anyone making these jars any longer, but there apparently are enough of them still in use to warrant at least one leading producer of home-canning materials to list the caps in its 1974 instruction booklet.

The cap is zinc, porcelain-lined to protect the food from the metal, and it is screwed down on a cushioning rubber ring which you put around the neck of the jar at the base of the threads, *near the shoulder*. The cap fits standard-mouth jars only.

Never use a lid whose porcelain is damaged in even the slightest way: exposure to zinc could mean that the food is exposed to cadmium. Henry A. Schroeder, M.D., founder and director of the Trace Element Laboratory, wrote in his *Pollution, Profits & Progress:* "Cadmium is our most insidious pollutant—mammals have not learned to handle it; an industrial contaminant, it is always found with zinc."

HOW THE ONE-PIECE CAP WORKS*

Gently, and only slightly, stretch the rubber ring—which is still wet from washing/sterilizing—and fit it over the neck of the jar and seat it carefully on the ledge of glass at the base of the threads. Screw the washed/sterilized cap down *tight* on the rubber ring—and *then give it a ¼-inch counter-turn:* this slight loosening will allow air in the contents to vent during processing.

The moment the jars are removed from the processing kettle, *retighten the cap slowly* (a quick jerk is likely to shift the rubber, breaking the seal). And never tighten the cap again, especially after the jar has cooled. This gentle retightening immediately after the jar is removed from the processing kettle is the action referred to in the individual instructions as *"complete the seals if necessary."*

STERILIZING & RE-USING

Wash lids and fresh new rubbers in hot-soapy water, rinse well. Unless the food is to be processed at 212 F./100 C. or higher, sterilize the clean lids by boiling for 15 minutes and letting them sit covered by the hot water until used; *do not boil rubbers*—pour boiling water over them in a dish, and let them sit in the water until used.

---

* See Postscript, page 533.

Put rubbers on the jars wet, and be sure to wipe them clean of dribbles with a fresh, damp cloth.

What's re-usable. Only perfect jars and perfect lids are re-usable. Rubbers lose their sealing ability in just one use: *discard rubbers.*

—BUT DON'T USE THESE

Supplies of new, good, home-canning jars and fittings are beginning to catch up with the demand, even though local stocks may be sold out when August rolls around. So be forehanded. Start looking in *March.* Ask your friendliest storekeeper to let you know when he expects a shipment. And protect your family by using only the jars and fittings that are GRAS, as the Food and Drug Administration might put it (meaning "generally recognized as safe"). Don't be stampeded into using unreliable or makeshift closures (see Postscript, page 529). As for containers, there are enough to go around without taking a chance with these:

(1) Any jars imported from Europe or Asia *for which you don't have full and explicit directions* that live up to USDA or Agriculture Canada standards of safety. We have a number of such jars, from several countries— they're charming as cannisters—but on the whole they're too large for dense, low-acid foods or for many strong-acid raw foods; their bodies are often too fragile for our handling; and the sealing arrangements (mostly a super-strong latch like the ones on a foot-locker) can either be next to impossible to vent during processing, or are so strenuous that the seal comes not from a proper vacuum, but rather from external pressure applied by the clamp. We were only 50 percent successful with the ½-liter size. And there's one brand that claims its rubber sealer is re-usable!

(2) Commercial glass containers—the one-trip jars holding baby food, salad dressing, peanut butter, pickles, fancy fruit, whatever. Not even if good, safe, modern home-canning closures will fit them (and especially not if we'd have to re-use the commercial closure—which cannot seal safely a second time). Save these for jelly and the like, which needs *protection* instead of processing.

(3) Very old jars for which new fittings/closures are not gettable. There are a number of books about antique canning jars—like the ones for collectors of old bottles— and you may have some real prizes (to keep or to sell, but not to can in). Look in the *Readers' Guide to Periodical Literature,* at every library, for articles in antiques magazines.

Old-style bailed jars with glass lids

These haven't been manufactured since the early 1960's as far as we know, but there are still countless thousands of these sturdy performers around. Sometimes called "lightning" or "ideal" type, they have a domed glass lid cushioned on a separate rubber ring that seats on a glass ledge a scant ¼-inch down on the neck of the jar. The lid is held in place during processing by the longer hoop of the two-part wire clamp.

As you remove the jars from the canner after processing, snap down the shorter spring-section of the clamp so it rests on the shoulder of the jar. Therefore this is another case where *"complete the seals if necessary"* applies in the individual instructions.

FITTINGS FOR BAILED JARS

The glass lids and the two-part wire bail are no longer made, but occasionally you can find a trove of unused glass lids or separate bails in the storeroom of an old-fashioned neighborhood store.

Discard any lids that have rough or chipped rims, or whose top-notch (which holds the longer hoop in place) is worn away.

Sometimes the wire bails are so rusted or old or tired that they have lost their gimp and can't hold the lid down tightly. *Please* don't go in for makeshift tightening by bending the wires or padding the lids. Retire the jars.

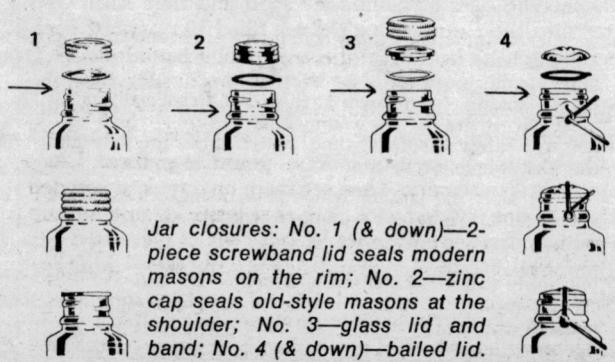

Jar closures: No. 1 (& down)—2-piece screwband lid seals modern masons on the rim; No. 2—zinc cap seals old-style masons at the shoulder; No. 3—glass lid and band; No. 4 (& down)—bailed lid.

Rubbers are still sold in boxes of twelve, and come in standard and wide-mouth sizes. *Never re-use rubbers*, or use old, stale rubbers that have been hanging around for years. Stretch rubbers gently and only enough so they'll go over the neck of the jar.

To sterilize. Boil jars and lids for 15 minutes, and let them wait, covered by the hot water, until used. But *don't boil rubbers:* Pour boiling water over washed rubbers in a dish, and let them stand in the water until used.

Masons with 3-piece glass lids

These are like the modern mason with the 2-piece screw-band *metal* lid, and are still made in Canada. The center of the glass lid protrudes down inside the jar; the seal occurs on the *rim* of the jar, where the edge of the lid is cushioned by a separate rubber ring applied, wet, to the lid.

The screwband is turned down *tight, then given a* $\frac{1}{4}$-*inch counterturn* (as with the one-piece zinc cap above) to allow air in the contents to vent during processing. *Retighten the band carefully as you remove each jar* from the processing kettle, and never retighten it further. (Another example of what you do to "complete seals if necessary.")

In and out of the canner

Don't put cold jars of food into very hot water; don't fill a canner with jars of boiling-hot food and then slosh cold water into the canner. Don't clunk filled jars against each other—especially if they're filled with boiling-hot food.

After processing time is up, remove jars at once from the B-W Bath, but let the Pressure Canner return to Zero naturally before you lift out the jars. Take care that you don't knock the jars against each other as you unload them. Complete seals if necessary. Then set them on a rack, a wooden surface, or one padded with cloth or newspaper, and be sure it's not in a draft of cold air. And *never* invert processed jars in the mistaken idea that you're helping the seal—quite the contrary!

Cooling. Jars must cool naturally: *don't* drape a towel over them with the idea of protecting them from air currents, because keeping them warm will invite flat sour. Let them sit undisturbed for 12 hours before you check the seals and perform the other chores described in "Post-processing Jobs."

THE TYPES OF CANS FOR HOME-CANNING

The containers called "tin cans" are made of steel and are merely coated with tin inside and outside. This tin coating is satisfactory for use with most foods. These are called *plain* cans.

But certain deeply colored acid foods will fade when they come in long contact with plain tin coating, so for them there is a can with an inner coating of special acid-resistant enamel that prevents such bleaching of the food. This is called an *R-enamel* can.

Then we have still other foods that discolor the inside tin coating—perhaps because some of them are high in sulfur, and act on the tin the way eggs tarnish silver; or because some are so low in acid as to nudge or straddle the Neutral line. Although there's no record of any damage to these foods from tin coating, there's a can for them that is lined with *C-enamel*.

A leading manufacturer of cans for home-canning says that we may use *plain* cans and *C-enamel* cans interchangeably—except for meats, which always take *plain* cans—with the sole detriment that sometimes the insides of plain cans will be discolored.

Note: Although there are many sizes of canned foods on supermarket shelves, the cans used in putting by food at home are: *No. 2*—which holds about 2½ cups; and *No. 2½*—which holds about 3½ cups. (The differance in measure between No. 2 and pints, and No. 2½ and quarts accounts in part for the difference in processing time given for jars and cans.) A No. 303 can (16 ounces) is mentioned only rarely for home-canning; it's processed like a No. 2.

WHAT FOODS TO CAN IN WHICH

Throughout the individual instructions we've included the type of can—plain, R-enamel or C-enamel—to use if you use cans at all.

However, here's a rule-of-thumb to go by if you're canning a food we don't go into:

R-enamel. Think of "R" as standing for "red" and you'll have the general idea: beets, all red berries and their juices, cherries and grapes and their juices, plums, pumpkins and winter squash, rhubarb and sauerkraut (which is very acid).

C-enamel. Think of "C" as standing for "corn" (which has no acid) and for "cauliflower" (whose typically strong flavor indicates sulfur) and you get: corn—and hominy, very low-acid Lima and other light-colored shell beans (and, combined with corn, succotash); cauliflower—and things with such related taste as plain cabbage, Brussels sprouts, broccoli, turnips and rutabagas; plus onions, seafood and tripe.

Plain. This is the catch-all, and may take these foods, as well as others: most fruits, tomatoes, meats, poultry, greens, peas, and green/snap/string/wax beans, and certain made dishes (like baked beans, etc.).

If you're canning mixed vegetables, use *C-enamel for preference* if one or more of the ingredients would go in C-enamel by itself.

Remember, though, that the heavens won't fall if you mix up *plain and C-enamel* for fruits and vegetables. Nor will "red" acid foods be bad if they're not canned in *R-enamel:* they just won't have their full color.

WHAT'S RE-USABLE?

Answer: Only the sealer.

No damaged, rusted, dented, bent *new* can or lid may ever be used in the first place.

No can or lid may be *re-used* at all.

Preparing cans, lids & sealer

If cans and lids are to be processed in a Boiling-Water Bath (212 F./100 C.) or Pressure-processed, they need not be sterilized before filling.

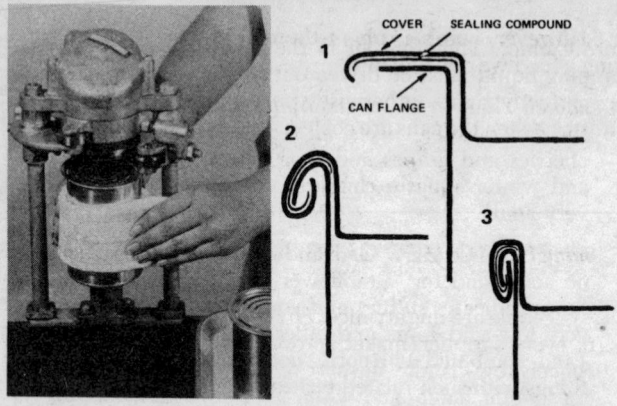

A can of food exhausted to 170 F. (hence the dry cloth pad to protect fingers) will be sealed with the rolling seam shown in stages at right.

Wash cans in clear hot water, scald; drain upside down on sterile cloths so air in the room cannot contaminate the inside of the can. To sterilize, wash, submerge in hot water and bring to boiling, and boil for 15 minutes; remove and drain as above.

The gaskets of some lids could be of a material that *must not be wetted* or it won't seal right; other gaskets are of a rubber-like composition that must be treated as carefully as the compound on non-boilable metal screwband lids. Therefore be sure to ask your supplier for explicit instructions for sterilizing the lids.

Until you know how to sterilize lids, reserve your cans for use only in the B-W Bath and Pressure Canner. Wipe factory dust from the lids with a scrupulously clean damp cloth immediately before sealing them to the can.

Adjust the sealer according to the manufacturer's directions, and test its efficiency by sealing a can that's partly filled with water, then submerge the can in boiling water: if any air bubbles rise from the can after a few seconds, the lid is not seamed tightly enough. Adjust the chucks of your sealer until you have a perfect seam joining the lid to the can.

In and out of the canner

Unlike jars, cans do not vent during processing: they are sealed completely before they go into the B-W Bath or Pressure Canner. Therefore air in the food's tissues and in the canning liquid must be driven out *before* the lid is crimped on perfectly tight—otherwise you wouldn't end up with a vacuum when the cans are cooled.

WHERE TO GET CANS AND CAN-SEALERS

People are mighty nice. *PFB* sent a letter of inquiry to every supplier/maker across the nation who might conceivably handle the sort of cans used for processing food at home, and we heard from them all. Only those named below seem at this writing (1975) to be sources of supply for home-canning equipment, though. (But all of them asked that we say that "tin cans" are made of *steel*, not reclaimed scrap metal, and this we are happy to do.)

For cans and the sealers:

> Freund Can Company
> 155 West 84th Street
> Chicago, Illinois 60620
> (Distributor for American Can and Continental Can)

> Ives-Way Products, Inc.
> 820 Saratoga Lane
> Buffalo Grove, Illinois 60090
> (Formerly Rowe Automatic Can Sealer)

For can-sealers only:

> Dixie Canner Equipment Company
> 786 East Broad Street
> P.O. Box 1348
> Athens, Georgia 30601

In the Dominion, two possible sources are: Continental Can of Canada Ltd., P.O. Box 4021 Terminal A, Toronto, Ontario; and American Can of Canada Ltd., One International Boulevard, Rexdale, Toronto, Ontario.

So *all* cans of food require an extra intermediate step that may be bypassed with most jars (the exceptions being jars of meat, poultry and seafood), and we'll deal with it now before we turn to the Raw and Hot packs in the general procedure for filling any container with food to be canned.

THE EXTRA STEP: "EXHAUSTING"

To drive out enough air to make the desired vacuum, you heat the food to a minimum of 170 F./77 C. at the center of the filled can or jar. You achieve this temperature either by bringing loose food to boiling in a kettle and then ladling it into the container, or by heating the container of raw or cold food until a glass food thermometer thrust into the middle of the contents registers 170 F. (or the 180 F. preferred by the Ohio Extension Service, if you live in the Buckeye State).

Fill the cans according to the instructions for the specific food and put them, still open, on a rack in a large kettle; add hot water to about 2 inches *below* the tops of the cans. Cover the kettle and boil the whole business until your pencil-shaped food thermometer, stuck down in the center of the food until its bulb is halfway to the bottom of the can, registers 170 F. You can get the same result by boiling open cans in a Pressure Canner at Zero pounds pressure (as described for jars in Canning Seafood, q.v.).

The individual directions will repeat the need to "exhaust to 170 F." even with cans just filled with boiling-hot food. This deliberate repetition is our way of emphasizing the importance of checking the contents with your pencil thermometer: it's easy to let food get cool as you turn to sealing the cans, and the given processing times depend on having the food at 170 F. when the sealed cans go into the canner.

COOLING THE CANS

Cans *must be cooled quickly* after their processing time is up, lest their contents cook further (this is why, if you're using cans, you vent your Pressure Canner to hasten its return to Zero pounds).

Fill your sink or a washtub with very cold water and drop the processed cans into it. As each can is removed from the canner its ends should be slightly convex, bulging from the pressure of the hot food inside it; if the ends don't bulge, it means that the can was imperfectly sealed before it went into the processing kettle. The ends will flatten to look slightly concave when the contents have cooled and shrunk, indicating that the desired vacuum has formed inside.

Change the water when it warms, or add ice, to hasten cooling. Remove cans when they're still warm so they'll air-dry quickly. They may be set on a cold surface and they are safe in a draft; stand them six inches apart to let air circulate around them, and wait to stack them until they are completely cold. But if your space is limited and you must stack them to cool, hold the upper tier on wire cake racks laid across the tops of the first layer—and staggered for maximum circulation of air.

General Steps in Canning

The introductions to the sections about fruits, tomatoes, vegetables, meats, poultry, fish and shellfish tell how to select and handle fresh raw materials, and the individual instructions tell you specifically how to can them. Therefore right here it's enough to stress again the importance of refrigeration and sanitation to control the spoilers, and the importance of using the right processing method to ensure the best—and safest—canned food for your family.

Let's see how to fill the jars/cans with food before they go into the Boiling-Water Bath or the Pressure Canner, and how to treat them after they come out.

RAW PACK AND HOT PACK

Many foods may be packed in their containers either *Raw* (in some manuals also designated as "Cold") or *Hot*. The food is trimmed, cleaned, peeled, cut up, etc., in the same manner for both packs. The same amount of canning liquid is added to the container of solid food, regardless of whether it's Hot pack or Raw—roughly ½ to ¾ cup for pints or No. 2 cans, 1 to 1½ cups for quarts or No. 2½ cans; also the liquid should always be very hot. The optional seasoning is added just before processing either pack. And with both packs the containers are handled identically after being removed from the canner and cooled.

Below are the general considerations in choosing which pack to use. However, we do point out that the latest research is prompting independent experts to recommend Hot pack in preference to Raw, as resulting in a more satisfactory product.

RAW PACK

Foods of relatively low density tend to hold their shape better with the Raw pack, simply because they are subjected to less handling physically (this is what we recommend for *whole* tomatoes, for example).

Boiling sirup, juice or water is added to raw foods that require added liquid for processing.

Jars of Raw-packed food must start their processing treatment in hot *but never boiling* water, otherwise they're likely to crack. Even when a jar has been exhausted, the water in the processing kettle should not be boiling.

Raw-packed foods usually take longer than Hot packs to process in a Boiling-Water Bath; this is especially true for the denser foods.

HOT PACK

Food that is precooked a little or almost fully is made more pliable, and so permits a closer pack. (Foods differ in the amount of preheating they need, though: spinach is merely wilted before it's packed Hot, but green string beans boil for 5 minutes.)

Fruit that is canned without sweetening is always packed Hot.

In a Boiling-Water Bath, Hot-packed food requires less processing time than Raw does, because it is thoroughly hot beforehand. However, there generally is no difference in the time required for Pressure Canning either pack: by the time you start counting—the minute pressure reaches 10 pounds (240 F./116 C.)—Raw-packed food has become as hot as if it had been packed Hot to begin with. (One of the interesting exceptions is summer squash, which needs *longer* Pressure-processing for Hot pack, because the precooked squash is much more dense in the container than the crisp raw pieces are.)

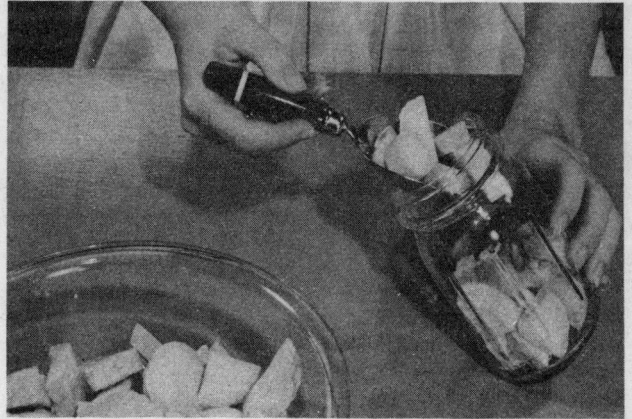

Pressure-processing will condense this Raw pack of summer squash a great deal, even after it's tamped down and given generous headroom.

LEAVING HEADROOM

In packing jars of food that will be processed in a Hot-Water Bath, Boiling-Water Bath or Pressure Canner, there must be some leeway left between the lid and the top of the food or its liquid. This space—called *headroom* in the instructions that follow later—allows for expansion of solids or the bubbling-up of liquid during processing. Without it, some of the contents would be forced out with the air, thus leaving a deposit of food on sealing surfaces, and ruining the seal.

Too much headroom may cause food at the top to discolor—*and* could even prevent a seal, unless processing time were long enough to exhaust all the excess amount of air.

The right headroom for each food and its processing liquid is specified in the individual instructions, with more headroom given to starchy foods—shell beans, green peas, corn, hominy, etc.—because they swell in the canner.

Note: In the older 3-piece closure for mason jars (glass lids + separate rubber ring + metal screwband to hold things in place), the center of the glass lid protrudes a bit down into the jar. With these glass lids, therefore, we recommend that you add from ¼ to ½ inch *more* to the headroom given in the directions, in order to compensate for space lost in the neck of the jar.

Examples of headroom in canning: left, 1 inch for chicken packed Raw in a jar, and exhausted; below, leaving 1 inch of space above boiling water added to Raw-packed summer squash; and, opposite, ¼ inch for whole-kernel corn packed Hot in cans, with boiling precooking liquid added right up to the brim.

Cans generally require no headroom between liquid and lid: all air is driven out in the exhausting step, and the lids are sealed on, before the cans are processed.

FILLING, BUBBLING & CAPPING

Whether it's in cans or jars, you pack most raw food firmly (except for the starchy vegetables that expand, mentioned above), and most hot foods rather loosely.

Have your clean, scalded containers and sealers ready. If you're using old-style masons with porcelain-lined caps, or bailed jars with rounded glass tops, gently fit the wet rubber rings around the necks of the jars, stretching them as little as possible; make sure they rest snugly on the ledge of glass that supports them. (Fittings that seal on the rim of the jar, and of course can lids, are put on after the containers are filled.)

Prepare and fill only one container at a time—don't set them up in a row, standing open to airborne spoilers. Set the jar/can in a clean pie dish, or whatever, to catch spills and overflows.

As you pack Raw, shake the container or tunk it on the bottom to settle the pieces of food; use a slender rubber spatula to make room for chunks you're fitting in; don't compress the food so much that it will spring up again, though, and invade the headroom. Pour in your boiling liquid—sirup, water, juice—to the desired level, keeping track of the headroom you must leave.

Follow directions for headroom in Hot pack. Your wide-mouth funnel is most handy here for controlling dribbles of food that must be wiped away completely.

REMOVING BUBBLES

In either pack in any container, run the blade of a table knife down between the food and the side of the jar/can at several points. There is likely to be more air trapped in the liquid between pieces of food in Raw pack than in Hot; but use your knife with both. Take care not to stir or to fold in more air.

For insurance, use your knife again—but sparingly—when you "top up" a can with boiling liquid after exhausting it to 170 F.

CAPPING

With a scrupulously clean cloth wrung out in boiling water, wipe the sealing rim of cans to remove any liquid, food tissue or fatty substance, because any of these can inter-fere with the seal. Wrap a clean, damp cloth around the body of the hot can to make it comfortable to hold; set the can in the sealer. Wipe the lid (rinse it quickly in hot water if, in some way, it has become dusty—and this is the *first* time it may be wetted), place it on the sealing rim of the can, and crimp it in place according to the directions for making the seal complete. Set the can in the processing kettle, and deal with the next one.

Wipe the sealing edges of the jars—*rim* of the glass, if you're using 2-piece screwband metal lids or the 3-piece glass closure; *rubber ring* around the neck of the jar, if it's a bailed closure or takes a porcelain-lined cap. Remove every vestige of food or other substance that would prevent a perfect seal.

Set the metal lid of your modern masons on the rim of the jar, *sealing compound next to the glass;* screw down the band *firmly tight.* (And see the Postscript, especially pages 531 and 533.)

Set a wet rubber ring on the rim of the jar that takes a 3-piece glass closure; seat the glass lid on the rubber so the center of the lid protrudes down into the headroom; screw the metal band down tight, then give it a ¼-inch counter-turn to allow the jar to vent during processing. (You'll re-tighten it after the jar is processed.)

Screw the porcelain-lined zinc cap down tight on the rubber-padded shoulder of the old-style mason jar, then give it a ¼-inch counterturn to allow for venting. (You'll re-tighten it after the jar is processed.)

Set the rounded glass lid of your bailed jar on the rubber ring, turning the cap so the notch in the top will catch the wire most securely; push the longer hoop up over the top of the lid until it's held firmly in the notch. (But *don't* snap the lower bail down on the shoulder of the jar: you'll do this after the jar is processed.)

POST-PROCESSING JOBS

Your B-W Bath kettle is off the heat, your Pressure Canner has returned to Zero—but not until tomorrow will to-day's canning be finished.

As you remove cans to drop them in cold water, mark the ones with poor seals. Right now, hot from the canner, their ends *should* bulge from internal pressure sealed in; and in the cooling tub, tiny bubbles must not appear around the lids. Therefore flat ends or bubbles mean pinpoint gaps in the sealing seam.

On the other hand, your jars seal themselves as they cool. Of course you never tighten the modern "dome"/"snap"/"self-sealing" 2-piece screwband metal closures. But when you "complete seals if necessary" by flipping down bails or screwing down caps or glass lids that have separate rubber rings, *you* are not actually making the seal: you're merely securing the lids firmly in place so the ensuing vacuum can complete the seal.

For this reason, *don't open the jar* to add liquid if you notice one whose liquid has partly boiled out during process-ing. Simply treat the jar normally and stand it up beside the others. After cooling undisturbed overnight, it might have a good, safe seal.

Two more things. Cans aren't shockproof just because

they're already sealed and are made of metal, so don't bump them around or transport them until they have cooled and rested overnight. And don't invert or shake—or even tilt— jars before they're cold and sealed.

Checking seals

The day after canning is the time for checking the seals of your containers, and preparing them for storage. And *this is your ONLY chance to salvage food* that has failed to seal; once it is stored, a bad seal means that you must destroy it.

CAN SEALS

By now the ends of your cans should be pulled slightly inward, proving that there is a vacuum inside. However, if the end of a can has not collapsed, press it hard: if it stays in, the seal is O.K.

Be right finicky as you check your cans, because at this early stage a poor seal will not have had time to become dramatic. Therefore the ends won't be bulging 'way out (certainly not even so much as they did right after the cans were processed yesterday); nor will the seams be leaking gassy, spoiled food (although traces of food at any seam are an obvious sign, since the cans got well rinsed by the water in which they cooled).

Springy ends and bits of food mean bad seals. Open the cans and refrigerate the contents, to serve or reprocess.

JAR SEALS

First, don't be dumped if there's a haze of dried canning liquid on the outside of some jars. All your jars vented in the processing kettle, remember; and several may have lost liquid, which clouded the water in the canner and so left a slight deposit on the jars.

Food particles lodged around the base of the closure could mean trouble, though. You'll know for sure when you check your seals.

On your modern mason jars, the metal lids of the 2-piece screwband closure will have snapped down, pulled in by the vacuum that means a good seal. If you find a lid that is not concave, press it: if it springs back, the seal is gone; if it stays down, well and good—but set this jar to one side for a tougher test you'll give questionable jars in a minute.

Test bailed jars and the old-style masons with porcelain-lined zinc caps by tilting them far enough so the food presses against the closure. If bubbles start at the lids and rise through the contents, the seal is poor. Moisture appearing at the sealing point is a bad sign too.

Test your flat glass lids by *gently* removing the screw-bands that hold them down on their rubber rings. If the band is reluctant to turn, for heaven's sake don't force it—this could shift the lid enough to break the seal right there. A hot cloth held around the band is usually enough to make it expand and come free. Tilt the jar: any seepage or bubbles that mean a poor seal?

The really tough second test can be applied to all your jars except for old masons with the zinc cap. Take a jar you have doubts about—modern mason with its metal lid, any with a glass lid, any with a bailed closure—and remove the screwband or release the bail, including the longer hoop that holds the rounded glass top in place. Set the jar in a pan that's padded with a perfectly clean towel *and lift the jar by its sealed-on lid*. If the seal is weak, the weight of the jar will break it and the jar will drop and spill, leaving the metal disc or glass lid in your fingers. (The towel protects the fallen jar from cracking; collect the spilled contents to serve or re-process.)

CHECKLIST FOR POOR SEALS

One poor seal out of a full canner batch is a disappointment, but not a worry. Nor is the food in only one container worth reprocessing, which means getting cranked up to start over from scratch with a fresh jar or can and a fresh closure, and repeating, with utmost care this time, every step in the whole canning procedure. Best to eat it right away.

But two poor seals?—not so good. And more than two seal failures in one canner batch will tell you that you're making at least one mistake in your canning technique. For your family's sake, see if it's one of these:

Shortcuts in sanitation or preparing food "because it will sterilize anyway." Manufacturers' directions not followed in preparing closures.

Imperfect or makeshift containers/closures. Sealing edges of
jars, permanent closures (glass tops/lids, porcelain-lined
zinc caps) have nicks/cracks/warps. Jars are "one-trip"
commercial containers, plus commercial lids from peanut
butter, mayonnaise, pickle, etc., or baby-food jars. Rusty
or bent bails on old jars (with extra padding to take up
slack?). Re-used sealers (rubber rings or metal lids with
sealing compound). Dented/bent cans or can lids.

Haphazard filling, exhausting, capping of containers. Packs
too tight or too loose. Too much or too little headroom.
Trapped air not removed; air not exhausted from food
tissues or canning liquid (either by Hot pack or exhausting in the container or boiling the liquid). Sealing rims
not wiped clean after filling. Modern 2-piece screwband
metal lids not firmly tight; other jar lids not vented; can-
sealer not tested and adjusted.

Processing kettle mismanaged. Water in B-W Bath not 2
inches over tops of containers; Pressure Canner not
vented enough to remove air. True boil not reached and
maintained in B-W Bath; called-for pressure not reached
(gauge unchecked?) or maintained. Full processing time
not used. Makers' instructions for handling not followed.

Processing method inadequate. Maverick or makeshift
method used. B-W Bath used for foods that require
Pressure-processing.

Containers mishandled after processing. Modern 2-piece
screwband lids were fiddled with; seals not completed as
as needed for other closures. Jars opened to replace lost
canning liquid. Jars/cans roughly handled. Jars inverted
to cool.

Cleaning & labeling

Wipe all jars carefully with a clean, damp cloth, paying
special attention to the area around the seals.

Some experts say to remove the screwbands from the 2-
piece modern mason closure. Certainly you should take them
off if there are signs of food lodged underneath (but presumably you've done this already with jars you suspected of
having poor seals). But *never force* a reluctant screwband:
hold a hot cloth around it to make it loosen by expanding;

if this doesn't work, mark the jar for special watching while it is stored, and turn to the next jar.

The modern metal closure doesn't need a screwband to ensure the seal, of course—but the band does protect the seal in case you plan to transport the jars or stack them. *Don't replace the screwbands if you take them off,* because you could twist the lid just enough so the torque will break the seal.

Wash any screwbands you remove and dry them thoroughly before storing for future use.

The glass lids of 3-piece closures are especially vulnerable to jostling and bumping, so weigh carefully the possibility of broken seals before removing bands here.

Label each container with the name of the food, the date it was canned, and any special treatment you gave the food (the last information will be mighty handy if the product is notably good or notably bad when you come to serve it).

STORING

Storage for all canned food must be cool and dry and— if the food is in jars—dark. Even when home-canned foods are adequately processed, they will lose Vitamin C, carotene, thiamine, riboflavin and niacin at temperatures above 50 degrees Fahrenheit; and light hastens oxidation of fats and oils, destroys fat-soluble and light-sensitive vitamins, and fades the color of the food.

"Cool" means 32–50 F./0–11 C. Containers must not freeze, lest the food expand and break the seal. Canned food that is held in storage too warm can still spoil because certain thermoduric bacteria (or their spores) can reactivate and grow at room temperature or higher. Therefore keep canned food away from heating pipes or cozy nooks behind furnaces, etc.

Damp or humid storage can corrode or rust the metal of cans and closures, and thereby endanger the seals.

Berries and fruits are especially likely to bleach in jars exposed to light, but other foods, too, can become pale and unappetizing if stored in the light. "Cool" and "dry" have priority over "dark," so you may have to protect jars of food from light by wrapping them individually in paper, or putting them in the cartons the jars came in (be sure to put cardboard dividers between the jars).

Put any containers with suspect seals in a special place

so you will use them first or be able to keep an eye on them easily.

Arrange your food on the storage shelves so "last in" is "last out"—this way you'll keep a good rotation.

And to check your canned food periodically for signs of spoilage that have developed during storage.

SIGNS OF SPOILAGE AND WHAT TO DO

Before a container is opened, you can see signs of spoilage that indicate the food is unfit or actually dangerous to eat:

—Seeping seams, bulging ends on cans.
—Seepage around the seal, even though it seems firmly seated.
—Mold around the seal or visible in the contents.
—Gassiness (small bubbles) in the contents.
—Cloudy or yeasty liquid.
—Shriveled or spongy-looking food.
—Food an unnatural color (often very dark).

When the container is opened, these are additional signs of spoilage:

—Spurting liquid, pressure from inside as the container is opened.
—Fermentation (gassiness).
—Food slimy, or with too soft a texture.
—Musty or disagreeable or downright nasty odor.
—Mold, even a fleck, on the underside of the lid or in the contents.

If any such signs are evident in unopened or opened containers DESTROY THE CONTENTS SO THEY CANNOT BE EATEN BY PEOPLE OR ANIMALS. Burn the food if you can. Otherwise put food and containers and closures in a large enameled kettle, pour in water to cover by several inches, and add ¼ cup of strong detergent and an effective household disinfectant according to directions on the label for sterilizing (do *not* use bleach here, because of its vapor when hot). Bring the kettle to boiling, and boil hard for 20 minutes. Fish out the containers and closures, flush everything else down the toilet. Discard the sterilized metal cans and closures, and all sealers. Sterilized jars and glass lids may be used again if they are perfect and undamaged. In a solution made in the proportion of 1 part household chlorine bleach to each 4 parts water, wash all utensils, cloths and surfaces that might have come in contact with the spoiled food.

A further precaution deals with canned food that has no obvious signs of spoilage either before or after it is opened, and this is: *NEVER taste canned food* without boiling it first for 15 minutes—20 minutes for greens, corn, meats, poultry and seafood. If it has a bad smell or foams unduly during boiling, destroy it so it cannot be eaten by people or animals, and treat the containers as described above.

Canning Fruits

Procedures for canning tomatoes—which are classed botanically as fruit—are given a separate section of their own, following "Canning Fruit Juices" and just preceding "Canning Vegetables."

You simply must follow carefully all the right steps for selecting, preparing and processing fruits: there have been cases reported of botulism poisoning from home-canned apricots, blackberries, figs, huckleberry juice, peaches and pears—even though these are all in the traditional strong-acid grouping. Records of these cases indicate that there was careless handling somewhere along the line. (See "Dealing with the Spoilers" in Chapter 1.)

Use firm, just-ripe fruit and berries. If only a shade over-ripe they have lost a bit of their natural acid content, and they also will be likely to float to the top of the jar; if notably overripe they have lost a critical amount of their acid, and also their flavor and texture will be disappointing. Use absolutely fresh produce and can it as soon as possible. If you have to hold it over for a day, refrigerate it.

The individual instructions that follow use the Boiling-Water Bath at 212 F. (or the pasteurizing Hot-Water Bath at 190 F. for some juices).

In case you find a tempting old recipe that relies on the "open-kettle" procedure now proven to be inadequate, here is how to translate it for the B-W Bath. Prepare and cook the fruit according to the old recipe. When it's *thoroughly* cooked, pack the boiling-hot fruit and its juice in clean jars, leaving ½ inch of headroom. Adjust lids, and process in a B-W Bath (212 F.)—pints for 10 minutes, quarts for 15 minutes.

Complete metric conversion tables are in Chapter 1, but the following apply particularly to this section: 170 Fahrenheit = 77 Celsius / 190 F. = 88 C. / 200 F. = 93 C. / 212 F. = 100 C. / 1 U.S. cup (½ U.S. pint) = .24 Liter / 1 U.S. pint (2 cups) = .48 L. / 1 U.S. quart (4 cups) = .95 L. / 1 U.S. teaspoon = 5 milliliters (5 cubic centimeters) / 1 U.S. tablespoon = 15 ml. (15 cc.) / ½ inch = 1.25 centimeters

---

STEPS IN CANNING ALL FRUITS

Make your sirup ahead of time, in an amount based on the quantity of fruit you intend to can (see the yield table in Chapter 1).

Next, collect your utensils and containers. It's vital to have scrupulously clean utensils, cloths and work surfaces, including cutting-boards and counters, and a good supply of fresh water of drinking quality.

Jars/cans and their sealers must be perfect and perfectly clean. They need not be sterilized, since the adequate Boiling-Water Bath will sterilize the inside of containers during processing. Prepare containers and their sealers as in "About Jars and Cans" earlier.

To prevent canned fruit from discoloring

Work with only one canner batch at a time. Wash the fruit thoroughly in fresh drinking water, but don't let it soak. Lift it from the water to allow sediment to settle at the bottom of the wash water. Be extra gentle with berries. Remove stems, hulls, pits, skins, cores as described in instructions for individual fruits. Cut away all soft or bruised spots and any places where skin is broken: such blemishes can spoil your batch.

SUGAR SIRUPS FOR CANNING FRUITS

Sugar helps to retain the texture and color of canned fruit, so most fruits are canned with a sugar sirup to suit the sweetness of the fruit and the family's taste (see also Sweeteners in Chapter 1).

Commercial canners in the past have leaned toward Heavy sirup except in so-called diet packs. More recently, though, partly because of calorie-counting and partly because of the high cost of sugar, Heavy sirup is not being used so much by any canners. The instructions for individual fruits recommend which sirups to use.

Roughly, estimate ½ to ¾ cup of sirup for each pint or No. 2 can, and increase the allowance proportionately for larger containers.

To make sirup, mix sugar with water in the proportions given below, heat sugar and water together until sugar is dissolved, skimming if necessary.

Juice may be used for part or all the water, as desired. Prepare it by crushing sound, fully ripe and juicy fruit that has been set aside for the purpose. Bring it to simmering over low heat for several minutes, stirring gently; strain the hot pulp through a jelly bag.

Water and/or Juice	Sugar	Yield
4 cups	2 cups	5 cups *thin* sirup
4 cups	3 cups	5½ cups *medium* sirup
4 cups	4¾ cups	6½ cups *heavy* sirup

Don't use *artificial sweeteners* in sirups for canning fruit.

Light corn sirup or mild-flavored *honey* may be substituted for up to ½ the white sugar called for in the table above. Top-grade *maple sirup* is usually too hard to come by to be used for routine canning, but anyway it should take the place of only ¼ the sugar because of its pronounced flavor.

Raw sugar or *brown* (which are semi-refined) may be substituted for white sugar in proportions to taste, but their sweetening power differs slightly from that of white sugar, and they will darken the fruit and tend to change its flavor. Probably they should be reserved for fruits with strong flavor and dark juice.

Sorghum and *molasses* are too overpowering on their own to be satisfactory in canning fruit.

Cut apples, apricots, nectarines, peaches and pears discolor in air. Either coat the cut pieces well as you go along with a solution of 1 teaspoon crystalline ascorbic acid (Vitamin C, the safest and best anti-oxidant) to each 1 cup water; OR drop the pieces in a solution of 2 tablespoons salt and 2 tablespoons vinegar for each 1 gallon of cold water—but not for longer than 20 minutes, lest nutrients leach out too much —then rinse and drain the pieces well before packing the Raw or Hot way. Optional: to prevent their darkening while in the containers, add ¼ teaspoon Vitamin C to each 1 quart during packing—IF they haven't been treated with ascorbic acid as they were being prepared. (See also "Anti-discoloration Treatments" in Chapter 1.)

And remember: Fruit canned with too much headroom or too little liquid will tend to darken at the top of the container.

Packing, processing and all the rest . . .

We refer you to the blow-by-blow General Steps in Canning (immediately preceding this section on fruits) instead of offering a quick paraphrase here. Everything in that detailed account applies to packing, processing, checking and storing jars or cans of fruit.

Canned Fruit Troubles and What to Do

The only time you can tinker SAFELY with a container of canned food is during the 24 hours after it comes from the canner and before it is stored away.

If you find a faulty seal during this lull, repack and reprocess the fruit from scratch according to the original instructions, cutting no corners. There of course will be a loss in quality, especially in texture, from doing it over; and if there's only one poor seal it's probably simpler to eat the fruit right away, or refrigerate it for a couple of days, then serve it.

However, several poor seals warn you that something was dangerously wrong with your packing or processing, and the failure could affect the whole batch.

For example: leaving insufficient headroom in jars, or failing to have the contents of cans exhausted to a minimum of 170 F., plus trapped air left in either type of container—these usually cause bits of food to be forced out during the processing period, with resulting poor seals. Repack in clean containers *with fresh, new sealers,* and reprocess for the full time.

Failing to keep at least 2 inches of boiling water over the tops of the containers, and not keeping the water in the canner at a full boil from beginning to end of the processing period—these are fairly common causes of poor seals. Repack in clean containers *with fresh, new sealers,* and reprocess for the full time.

But, if, after the containers have been stored away, you find any of the following, DESTROY THE CONTENTS SO THAT THEY CANNOT BE EATEN BY PEOPLE OR ANIMALS; then boil the containers and closures in hot soapy water to cover for 15 minutes. Discard the boiled sealers or rubbers; if sound, the sterilized jars may be used again.

—Broken seals, bulging lids on cans.
—Seepage around the seal, even though it seems firmly seated.
—Mold, even a fleck, in the contents or around the seal or on the underside of the lid.
—Gassiness (small bubbles) in the contents.
—Spurting liquid, pressure from inside as the container is opened.
—Spongy or slimy food.
—Cloudy or yeasty liquid.
—Off-color, disagreeable smell, mustiness.

NOT PRIZE-WINNING, BUT EDIBLE

If the containers and contents offer none of the signs of spoilage noted above, and if the storage has been properly cool, you can have less-than-perfect fruits that still are O.K. to eat. Such as:

Floating fruit. The fruit was overripe, or it was packed too loosely, or the sirup was too heavy.

Darker fruit at the top of the container. Too much headroom above the liquid.

Bleached-looking berries. With no signs of spoilage present,
this could mean that jars were exposed to light in the
storage area; wrap the jars in paper or stash them in
closed cartons if the storage wasn't dark.

170 F. = 77 C. / 190 F. = 88 C. / 200 F. = 93 C. / 212 F. = 100 C. /
1 U.S. cup (½ U.S. pint) = .24 L. / 1 U.S. pint (2 cups) = .48 L. / 1
U.S. quart = .95 L. / 1 U.S. tsp. = 5 ml. (5 cc.) / 1 U.S. T. = 15 ml.
(15 cc.) / ½ inch = 1.25 cm.

Apples

Even with root-cellaring and drying (see both), you'll
want some apples put by as sauce, dessert slices or pie
timber. And of these, probably the handiest thing is to can
applesauce and slices done in sirup, and to freeze the slices
you'll use for pies. Canning apple cider is described later on
under Juices. See Recipes.

GENERAL HANDLING

Boiling-Water Bath. Use Hot pack only. Use jars or plain
cans. Process with Thin Sirup, plain water, or with sweetening as desired.

Because apples oxidize in the air, work quickly with only
one canner batch at a time. Wash, peel and core apples (save
peels and cores for jelly, as described in Applesauce, below);
treat prepared pieces with either of the anti-discoloration
solutions described in the general steps, earlier. Drain, and
carry on with the specific handling. (See also Freezing for
pie timber.)

SLICES (HOT PACK ONLY)

Rinse drained, prepared pieces. Cover with hot Thin
Sirup or water, boil gently for 5 minutes. Lift out and drain,
saving cooking sirup or water. Pack hot.

In jars. Fill clean, hot jars, leaving ½ inch of headroom. Add
boiling-hot Thin Sirup or water, leaving ½ inch of headroom; adjust lids. Process in a Boiling-Water Bath (212
F.)—pints for 15 minutes, quarts for 20 minutes. Remove jars; complete seals if necessary.

In plain cans. Fill, leaving only ¼ inch of headroom. Add
boiling-hot Thin Sirup or water to the top of the can.
Exhaust to 170 F. (*c.* 10 minutes); seal. Process in a B-W
Bath (212 F.)—10 minutes for either No. 2 or No. 2½
cans. Remove cans; cool quickly.

APPLESAUCE (HOT PACK ONLY)

Prepare by your favorite rule and according to how
you'll use it—chunky or strained smooth; sweetened or not;
with spices (cinnamon, nutmeg, whatever) or plain. Because
of complete precooking and being packed so hot, processing
time is relatively short and is designed to ensure sterilization
and a good seal.

Pare crisp, red apples, cut in quarters or eighths and remove core parts; drop pieces in anti-discoloration solution.
(*Don't throw away the peels and cores: save them to boil
up for a beautiful juice for jelly.*) Put about 1 inch of water
in a large enameled or stainless-steel kettle, fill with well-
rinsed apple pieces to within 2 inches of the top. Bring to
a boil, stirring now and then to prevent sticking, and cook
until apples are tender. Leave as is for chunky sauce, or put
it through a sieve or food mill for smoothness. Sweeten to
taste if you like (see "Sweeteners" in Chapter 1); bring it
briefly to boiling to dissolve any sweetening. Pack very hot.

In jars. Fill clean, hot jars with piping-hot sauce, leaving ½
inch of headroom; adjust lids. Process in a Boiling-Water
Bath (212 F.)—10 minutes for either pints or quarts.
Remove jars; complete seals if necessary.

In plain cans. Pack to the top with hot sauce. Exhaust to
170 F. (*c.* 10 minutes); seal. Process in B-W Bath
(212 F.)—10 minutes for either No. 2 or No. 2½ cans.
Remove cans; cool quickly.

BAKED APPLES (HOT PACK ONLY)

Sometimes people can baked apples. Prepare them in a
favorite way and bake until *half done;* pack hot in wide-
mouth jars or plain cans as for Apple Slices, adding hot Thin
Sirup. Adjust jar lids or exhaust and seal cans. Process in a
Boiling-Water Bath (212 F.)—20 minutes for either pint or
quart jars, 10 minutes for either No. 2 or 2½ cans (reaching

170 F. by exhausting shortens processing time). Complete seals if necessary for jars; cool cans quickly.

Apricots

Can these exactly as you would Peaches (q.v.), but leave the skins on if you like. Some varieties tend to break up when they're heated, so handle them very gently. Hot pack preferred.

Berries, see grouped handling at the end of Fruits section.

Cherries, Sour (for Pie)

Because these are used primarily as pie timber, they may be canned in water—but they have better flavor in Thin Sirup. Either way, you'll add the extra sweetening at the time you thicken the juice when you're building the pie.

GENERAL HANDLING

Boiling-Water Bath. Use Raw or Hot pack. Use jars or R-enamel cans. Prepare Thin Sirup for Raw pack; heat in their own juice with sugar for Hot pack.

Wash, stem and pit cherries. (Use a small sterilized hairpin or the looped end of a paper clip if you don't have a pitting gadget.) Shake fruit down in the containers for a firm pack.

RAW PACK

In jars. Jog cherries down several times during packing; leave ½ inch of headroom. Add boiling sirup, leaving ½ inch of headroom; adjust lids. Process in a Boiling-Water Bath (212 F.)—pints for 20 minutes, quarts for 25 minutes. Remove jars; complete seals if necessary.

In R-enamel cans. Make a firm pack, leaving only ¼ inch of headroom. Add boiling sirup to top. Exhaust to 170 F. (*c.* 10 minutes); seal. Process in a B-W Bath (212 F.)—

No. 2 cans for 20 minutes, No. 2½ cans for 25 minutes. Remove cans; cool quickly.

PREFERRED: HOT PACK

Measure pitted cherries and put them in a covered kettle with ½ cup of sugar for every 1 quart of fruit. There should be enough juice to keep the cherries from sticking. Set on lowest burner. Cover the kettle, and bring fruit very slowly to a boil to bring out the juice. Be prepared to add a little boiling water to each jar if you haven't enough juice to go around.

In jars. Fill with hot fruit and juice, leaving ½ inch of headroom; adjust lids. Process in a Boiling-Water Bath (212 F.)—pints for 10 minutes, quarts for 15 minutes. Remove jars; complete seals if necessary.

In R-enamel cans. Fill to the top with hot fruit and juice. Exhaust to 170 F. (*c.* 10 minutes); seal. Process in a B-W Bath (212 F.)—No. 2 cans for 15 minutes, No. 2½ for 20 minutes. Remove cans; cool quickly.

Cherries, Sweet

GENERAL HANDLING

Boiling-Water Bath. Use Raw or Hot pack. Use jars or cans (plain cans for light varieties like Royal Ann; R-enamel for dark red or "black" types like Bing).

If you're going to serve these as is, or combined with other fruits in a compôte, you don't pit them (they'll hold their shape better unpitted); but do prick each cherry with a needle to keep it from bursting while it's processed. Use Medium or Heavy sirup for Raw pack; for Hot pack add more sugar than for Sour Cherries.

Wash cherries, checking for blemishes, and discard any that float (they may be wormy); remove stems. Shake down for a firm pack.

RAW PACK

In jars. Fill firmly, leaving ½ inch of headroom. Add boiling sirup, leaving ½ inch of headroom; adjust lids. Process in a Boiling-Water Bath (212 F.)—pints for 20 minutes, quarts for 25 minutes. Remove jars; complete seals if necessary.

In plain or R-enamel cans. Make a firm pack, leaving only ¼ inch of headroom. Fill to top with boiling sirup. Exhaust to 170 F. (*c.* 10 minutes); seal. Process in a B-W Bath (212 F.)—No. 2 cans for 20 minutes, No. 2½ cans for 25 minutes. Remove cans; cool quickly.

PREFERRED: HOT PACK

Measure washed and pricked cherries into a covered kettle, adding ¾ cup of sugar for every 1 quart of fruit. Because there is not much juice in the pan, add a little water to keep fruit from sticking as it heats. Cover and bring very slowly to a boil, shaking the pan gently a few times (instead of stirring, which breaks the fruit). Heat some Medium or Heavy sirup to have on hand in case there's not enough juice to go around when you fill the containers.

In jars. Proceed and process as for Raw pack.

In plain or R-enamel cans. Proceed and process as for Raw pack.

Dried Fruits

FEASIBILITY

Any dried fruit may be freshened and canned. But why do it, when they keep so well as is (see Drying)—unless you foresee a particular need for a few servings of them stewed up ready for the table?

GENERAL HANDLING

Boiling-Water Bath only. Use Raw or Hot pack. Use jars or plain cans.

Freshen by covering with cold water and letting stand overnight. Drain, saving the soaking water (heated to boiling) to use in processing, and proceed with Raw pack.

If you're in a hurry, cover with water, bring to a boil,

and simmer until the fruit is plumped. Drain, saving the cooking water for processing, and proceed with Hot pack.

RAW PACK

In jars. Fill, leaving ½ inch of headroom. Add 2 to 4 table-spoons sugar (depending on sweetness of fruit) to pints, 4 to 6 tablespoons to quarts. Add boiling soaking water, leaving ½ inch of headroom; adjust lids. Process in a Boiling-Water Bath (212 F.)—pints for 20 minutes, quarts for 25 minutes. Remove jars; complete seals if necessary.

In plain cans. Fill, leaving only ¼ inch of headroom. Add 2 to 4 tablespoons sugar to No. 2 cans, 4 to 6 tablespoons to No. 2½ cans. Fill to top with boiling soaking water. Exhaust to 170 F. (*c.* 15 minutes); seal. Process in B-W Bath (212 F.)—No. 2 cans for 15 minutes, No. 2½ for 20 minutes. Remove cans; cool quickly.

PREFERRED: HOT PACK

In jars. Fill with hot fruit, sweeten, and add hot cooking water as for Raw pack. Process in a B-W Bath (212 F.) —pints for 15 minutes, quarts for 20 minutes. Remove jars; complete seals if necessary.

In plain cans. Fill with hot fruit, sweeten, and add hot cooking water as for Raw pack. Process in a B-W Bath (212 F.)—No. 2 cans for 15 minutes, No. 2½ for 20 minutes. Remove cans; cool quickly.

Figs

The green-colored Kadota variety makes a particularly attractive product, but whatever kind you use should be tree-ripened yet still firm.

Some casual old instructions would have you soften (or even remove) fig skins by treating the fruit with a strong soda solution—*but don't do it*. Any such alkaline will counteract some of the acidity upon which we rely to make the stipulated Boiling-Water Bath efficient.

GENERAL HANDLING

Long Boiling-Water Bath. Use Hot pack only. Use jars or plain cans. Prepare Thin Sirup.

Wash ripe, firm figs; do not peel or remove stems. Cover with boiling water and let simmer for 5 minutes. Drain and pack hot.

HOT PACK ONLY

In jars. Fill with hot figs, leaving ½ inch of headroom. Add boiling sirup. Add 2 teaspoons lemon juice to pints, 4 teaspoons lemon juice to quarts (an optional very thin slice of fresh lemon may also be added to each jar for looks). Adjust lids. Process in a Boiling-Water Bath (212 F.)—pints for 85 minutes, quarts for 90 minutes. Remove jars; complete seals if necessary.

In plain cans. Fill with hot fruit, leaving only ¼ inch of headroom. Top off with boiling sirup and 2 teaspoons lemon juice to No. 2 cans, 4 teaspoons lemon juice to No. 2½ cans (an optional very thin slice of fresh lemon may also be added to each can for looks). Exhaust to 170 F. (*c.* 10 minutes); seal. Process in a B-W Bath (212 F.)— No. 2 cans for 85 minutes, No. 2½ cans for 90 minutes. Remove cans; cool quickly.

Canning Frozen Fruits and Berries

For delivery in early fall, large farm-supply chain stores (look under Feeds or Grain, in the Yellow Pages) often have good buys in 4-gallon containers of frozen fruits and berries —good buys because you order them ahead, and they usually cost no more than the going price of store-bought fresh fruit; and all the washing and peeling and slicing is already done, to boot. A list from one such outfit offers sweet and sour cherries, strawberries, halved purple plums, sliced peaches, sliced Spy apples and applesauce, all with sugar; and blueberries, blackberries, red raspberries, rhubarb and crushed pineapple, all without sugar.

The hitch: You must take delivery when they come in, and hence be prepared to can your order immediately (unless you have a freezer big enough to hold the bulk packages). And you should be a serious canner: 4 gallons of prepared fruit is a lot to deal with in one swoop, and requires organizing the time and utensils for the job.

HANDLING

Boiling-Water Bath only. Use Hot pack only. Use jars or cans—plain or R-enamel as recommended for the specific raw food (q.v.).

Defrost fruit slowly in the unopened package. Drain off all juice and measure it: if the fruit was unsugared, add sweetening in proportion to make Thin or Medium sirup, as desired; if it was sugared, add more sweetener to taste if you want to. Bring juice to boiling. Add fruit and boil it gently for 2 or 3 minutes. Proceed with Hot-packing and processing in a Boiling-Water Bath as for the specific fresh fruit (you may need to add some boiling water to each container if there's not enough juice to go around).

Grapes

Tight-skinned seedless grapes are the ones to can if you can any—for fruit cocktail, compôtes, gelatine desserts and salads (but grapes for juice may be any sort you have plenty of).

GENERAL HANDLING

Boiling-Water Bath. Use Raw or Hot pack. Use jars or cans (plain, or R-enamel cans if it's a dark grape).

Sort, wash and stem.

RAW PACK

In jars. Fill tightly but without crushing grapes, leaving ½ inch of headroom. Add boiling Medium Sirup, leaving ½ inch of headroom; adjust lids. Process in a Boiling-Water Bath (212 F.)—pints for 15 minutes, quarts for 20 minutes. Remove jars; complete seals if necessary.

In cans (plain or R-enamel). Fill, leaving ¼ inch of headroom. Add boiling Medium Sirup to top. Exhaust to 170

F. (*c*. 10 minutes); seal. Process in a B-W Bath (212 F.) —No. 2 cans for 20 minutes, No. 2½ for 25 minutes. Remove cans; cool quickly.

PREFERRED: HOT PACK

Prepare as for Raw pack. Bring to a boil in Medium Sirup. Drain, reserving sirup, and pack.

In jars. Pack with hot grapes, leaving ½ inch of headroom. Add boiling sirup, leaving ½ inch of headroom; adjust lids. Process in a Boiling-Water Bath (212 F.)—pints for 15 minutes, quarts for 20 minutes. Remove jars; complete seals if necessary.

In cans (plain or R-enamel). Fill with hot grapes, leaving ¼ inch of headroom. Add boiling sirup to the top. Exhaust to 170 F (*c*. 10 minutes); seal. Process in B-W Bath (212 F.)—No. 2 cans for 20 minutes, No. 2½ for 25 minutes. Remove cans; cool quickly.

Grapefruit (or Orange) Sections

FEASIBILITY

Only if you have a good supply of tree-ripened fruits is canning worth-while—but canning makes an infinitely handier product than freezing does. Don't overlook Mixed Fruit; and don't forget marmalades and conserves (q.v.).

GENERAL HANDLING

Boiling-Water Bath only. Use Raw pack only. Use jars only (cans could give a metallic taste to home-canned citrus).

Wash fruit and pare, removing the white membrane as you go. Slip a very sharp thin-bladed knife between the dividing skin and pulp of each section, and lift out the section without breaking. Remove any seeds from individual sections. Prepare Thin Sirup.

RAW PACK ONLY

In jars only. Fill hot jars with sections, leaving ½ inch of headroom. Add boiling Thin Sirup, leaving ½ inch of headroom; adjust lids. Process in a Boiling-Water Bath (212 F.)—10 minutes for either pints or quarts. Remove jars; complete seals if necessary.

Juices, see grouped handling at the end of Fruits section.

Peaches

FEASIBILITY

The benefits and pleasures of canning are exemplified in peaches: home-canned peaches are full of flavor, are versatile, and are considered by many cooks to be better than frozen ones. See also Drying, Preserves and cobblers in Recipes.

GENERAL HANDLING

Boiling-Water Bath only. Use Raw or Hot pack. Use jars or plain cans.

Wash and scald briefly (by dipping a few peaches at a time in boiling water) and dunk quickly in cold water; slip off loosened skins. Halve, pit; scrape away dark fibers in the pit cavity because they sometimes turn brown in canning. Slice if you like. Treat peeled pieces immediately with an anti-discoloration solution as for Apples, above. Rinse, drain. Pack with Thin or Medium Sirup. If the peaches are especially juicy, make the sirup with their juice instead of water (see "Sirups for Canning Fruits").

RAW PACK

In jars. Pack halves or slices attractively, leaving ½ inch of headroom. Add boiling sirup, leaving ½ inch of headroom; adjust lids. Process in a Boiling-Water Bath (212 F.)—pints for 25 minutes, quarts for 30 minutes. Remove jars; complete seals if necessary.

In plain cans. Fill carefully, leaving only ¼ inch of headroom. Add boiling sirup to the top. Exhaust to 170 F. (c. 10 minutes); seal. Process in a B-W Bath (212 F.)—No. 2 cans for 30 minutes, No. 2½ for 35 minutes. Remove cans; cool quickly.

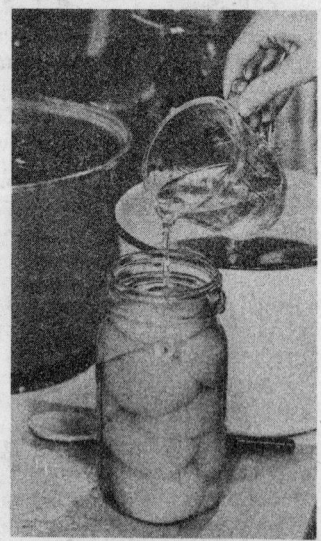

Skins come easily off scalded peaches, above, and cut fruit is held in an anti-darkening solution; right, boiling sirup must cover the beautifully packed peaches with ½ inch of headroom to spare.

PREFERRED: HOT PACK

Simmer prepared peaches in hot sirup for 2 minutes. Drain, reserving sirup.

In jars. Fill with hot peaches, leaving ½ inch of headroom. Add boiling sirup, leaving ½ inch of headroom; adjust lids. Process in a Boiling-Water Bath (212 F.)—pints for 20 minutes, quarts for 25 minutes. Remove jars; complete seals if necessary.

In plain cans. Fill with hot peaches, leaving only ¼ inch of headroom. Add boiling sirup to the top. Exhaust to 170 F. (*c.* 10 minutes); seal. Process in a B-W Bath (212 F.) —No. 2 cans for 25 minutes, No. 2½ for 30 minutes. Remove cans; cool quickly.

Peaches, Brandied

GENERAL HANDLING

Boiling-Water Bath only. Use Hot pack only. Use jars only (because they look so pretty: which is part of their fun).

The peaches should be small to medium in size, firm-ripe, and with attractive color; blemish-free of course. Wash. Using a coarse-textured towel, rub off all their fuzz. Weigh them.

For every 1 pound of peaches, make a Heavy Sirup of 1 cup sugar to 1 cup water. Bring sirup to boiling and, when sugar is dissolved, add the whole peaches and simmer them for 5 minutes. Drain; save the sirup and keep it hot.

HOT PACK ONLY, IN JARS ONLY

Without crushing, fit peaches in hot jars, leaving ½ inch of headroom. Pour 2 tablespoons of brandy over the peaches in each 1-pint jar, using proportionately more brandy for quarts. Fill jars with hot sirup, leaving ½ inch of headroom; adjust lids. Process in a Boiling-Water Bath (212 F.)—pints for 15 minutes, quarts for 20 minutes. Remove jars; complete seals if necessary.

Pears

Bartlett pears are ideal for canning, to serve alone or as
a salad or in a compôte. They will be too soft for successful
canning, though, if they've ripened on the tree: use ones that
were picked green (but full grown) and allowed to ripen in
cool storage, between 60 and 65 F.

Very firm-fleshed varieties like Seckel and Kieffer are
generally spiced or pickled; they make a satisfactory product
if ripened in storage and simmered in water till nearly tender
before packing with sirup. The so-called winter pears—such
as Anjou and Bosc—are usually eaten fresh; they are cold-
stored much the way apples are, but they are not likely to
keep as long (see Root-Cellaring).

GENERAL HANDLING

Use a Boiling-Water Bath. Use Hot pack only. Use jars
or plain cans.

Wash; cut in halves or quarters. Remove stems, core (a
melon-ball scoop is handy for this); pare. Treat pieces
against oxidation with either of the solutions described for
apples.

Make Thin or Medium Sirup.

Small pears may be canned whole: pare them, but leave
the stems on. It takes about 9 small whole pears to fill a pint
jar or a No. 2 can.

RAW PACK

In jars. Pack halves or quarters attractively, leaving ½ inch
of headroom. Add boiling sirup, leaving ½ inch of headroom; adjust lids. Process in a Boiling-Water Bath
(212 F.)—pints for 25 minutes, quarts for 30 minutes.
Remove jars; complete seals if necessary.

In plain cans. Fill carefully, leaving only ¼ inch of headroom. Add boiling sirup to the top. Exhaust to 170 F.
(*c.* 10 minutes); seal. Process in a B-W Bath (212 F.)—
No. 2 cans for 30 minutes, No. 2½ for 35 minutes. Remove cans; cool quickly.

HOT PACK ONLY

Simmer fruit in sirup for 2 minutes; drain, reserving hot sirup.

In jars. Fill with hot pears, leaving ½ inch of headroom. Add boiling sirup, leaving ½ inch of headroom; adjust lids. Process in a Boiling-Water Bath (212 F.)—pints for 20 minutes, quarts for 25 minutes. Remove jars; complete seals if necessary.

In plain cans. Fill with hot pears, leaving only ¼ inch of headroom. Add boiling sirup to the top. Exhaust to 170 F. (*c.* 10 minutes); seal. Process in a B-W Bath (212 F.) —No. 2 cans for 25 minutes, No. 2½ for 30 minutes. Remove cans; cool quickly.

MINT VARIATION

Prepare as above; the pears may be cut up or left whole. To Medium Sirup, add enough natural peppermint extract and green food coloring to give the desired taste and color.

Simmer the pears in this sirup for 5 to 10 minutes, depending on size and firmness of fruit, before packing Hot and processing in a B-W Bath as above.

Pears, Spiced

Seckel, Kieffer and similar hard varieties are best for spicing. Bartletts or other soft pears may be used if they are underripe.

GENERAL HANDLING

Use a Boiling-Water Bath. Use Hot pack only. Use jars only (like Brandied Peaches, these are very attractive to look at; and you could take a ribbon at the fair!).

Wash, peel and core 6 pounds of pears. Gently boil them covered in 3 cups of water until they start to soften.

Make a very heavy sirup of 4 cups sugar and 2 cups white vinegar. Tie in a small cloth bag 3 or 4 3-inch sticks of cinnamon, ¼ cup whole cloves, and 4 teaspoons cracked ginger. Simmer the spice bag in the sirup for 5 minutes.

Add the pears and the water in which they were partially cooked to the spiced sirup, and simmer for 4 minutes. Drain pears, saving the hot sirup and discarding the spice bag.

HOT PACK ONLY, IN JARS ONLY

Pack hot pears attractively in clean hot jars. Add spiced sirup, leaving ½ inch of headroom; adjust lids. Process in a Boiling-Water Bath (212 F.)—pints for 15 minutes, quarts for 20 minutes. Remove jars; complete seals if necessary.

Pineapple

Fresh pineapple is as easy to can as any fruit—and may be packed in any plain or minted and colored sirup (see also "Canning Frozen Fruits").

GENERAL HANDLING

Use a Boiling-Water Bath. Use Hot pack only. Use jars or plain cans.

Scrub firm, ripe pineapples. Cut a thin slice from each end. Cut like a jelly roll in ½-inch slices, or in 8 lengthwise wedges. Remove the skin, the "eyes" and the tough-fiber core from each piece. Leave in slices or wedges, or cut small or chop: let future use guide your hand.

Simmer pineapple gently in Light or Medium Sirup for about 5 minutes. Drain; save the hot sirup for packing.

HOT PACK ONLY

In jars. Fill with fruit, leaving ½ inch of headroom. Add hot sirup, leaving ½ inch of headroom; adjust lids. Process in a Boiling-Water Bath (212 F.)—pints for 15 minutes, quarts for 20 minutes. Remove jars; complete seals if necessary.

In plain cans. Fill with fruit, leaving ¼ inch of headroom. Add hot sirup to the top of the cans. Exhaust to 170 F. (*c.* 10 minutes); seal. Process in a B-W Bath (212 F.)— No. 2 cans for 20 minutes, No. 2½ for 25 minutes. Remove cans; cool quickly.

Plums (and Italian Prunes)

GENERAL HANDLING

Use Boiling-Water Bath. Use Raw or Hot pack. Use jars or cans—R-enamel for red plums, plain for greenish-yellow varieties.

Firm, meaty plums (such as the Greengage) hold their shape better for canning whole than the more juicy types do. Freestone plums and prunes are easily halved and pitted for the tighter pack.

Choose moderately ripe fruit. Wash. If to be canned whole, the skins should be pricked several times with a large needle to prevent the fruit from bursting. Halve and pit the freestone varieties. Prepare Medium or Heavy Sirup, and have it hot.

RAW PACK

In jars. Fill with raw fruit, leaving ½ inch of headroom. Add boiling sirup, leaving ½ inch of headroom; adjust lids. Process in a Boiling-Water Bath (212 F.)—pints for 20 minutes, quarts for 25 minutes. Remove jars; complete seals if necessary.

In cans (R-enamel for red fruit, plain for light-colored). Pack raw fruit, leaving ¼ inch of headroom. Add boiling sirup to the top of the can. Exhaust to 170 F. (*c.* 10 minutes); seal. Process in a B-W Bath (212 F.)—No. 2 cans for 15 minutes, No. 2½ for 20 minutes. Remove cans; cool quickly.

PREFERRED: HOT PACK

Heat prepared plums to boiling in sirup. If they're halved and are very juicy, heat them slowly to bring out the juice; measure the juice, and for each 1 cup juice add ¾ cup sugar —give or take a little, according to your taste—to make a Medium Sirup. Reheat to boiling for just long enough to dissolve the sugar. Drain fruit, saving the hot sirup. Have some hot plain Medium Sirup on hand for eking out sweetened juice.

In jars. Pack hot fruit, leaving ½ inch of headroom. Add
boiling sirup, leaving ½ inch of headroom; adjust lids.
Process in a Boiling-Water Bath (212 F.)—pints for 20
minutes, quarts for 25 minutes. Remove jars; complete
seals if necessary.

In cans (R-enamel for red fruit, plain for light-colored). Pack
hot fruit, leaving ¼ inch of headroom. Fill to top with
boiling sirup. Exhaust to 170 F. (*c.* 10 minutes); seal.
Process in a B-W Bath (212 F.)—No. 2 cans for 15 minutes, No. 2½ for 20 minutes. Remove cans; cool quickly.

Rhubarb (or Pie Plant)

*Never eat rhubarb LEAVES: they are high in oxalic acid,
which is poisonous.*

Safe to eat are the tart, red stalks of this plant, which are
excellent pie timber, make a tangy dessert sauce, are a favorite ingredient in old-time preserves. However, your best use
would be to can sweetened sauce by the method given below, and to freeze the raw pieces for pies. Rhubarb juice
makes a good hot-weather drink. See Recipes.

GENERAL HANDLING AS SAUCE

Use a Boiling-Water Bath. Use Hot pack. Use jars or R-
enamel cans.

For best results, can it the same day you cut it. If the
stalks are young enough, they need not be peeled (their red
color makes an attractive product). *Discard leaves,* trim
away both ends of the stalks, and wash; cut stalks in ½-inch
pieces. Measure. Put rhubarb in an enameled kettle (because
of the tartness), mixing in ½ cup of sugar for each 1 quart
(4 cups) of raw fruit. Let it stand, covered, at room temperature for about 4 hours to draw out the juice. Bring slowly to
a boil; let boil no more than 1 minute (or the pieces will
break up). (*Alternative:* Bake sugared rhubarb in a heavy,
covered pan in a slow oven, *c.* 275 F., for 1 hour.)

HOT PACK ONLY

In jars. Fill with hot fruit and its juice, leaving ½ inch of
headroom; adjust lids. Process in a Boiling-Water Bath

(212 F.)—10 minutes for either pints or quarts. Remove jars; complete seals if necessary.

In R-enamel cans. Pack hot fruit and juice to the top of the cans. Exhaust to 170 F. (*c.* 10 minutes); seal. Process in a B-W Bath (212 F.)—10 minutes for either No. 2 or No. 2½ cans. Remove cans; cool quickly.

Canning Berries

Don't get so taken up with making jellies and jams in berry-time that you forget to can some too: they may be done for serving solo and in compôtes and salads; or, with slightly different handling, for use in cobblers, pies and puddings. See Recipes, and also look at Freezing, Drying.

For purposes of general handling, berries—except for strawberries which are a law unto themselves—are divided in two categories: *soft* (raspberries, blackberries, boysenberries, dewberries, loganberries and youngberries), and *firm* (blueberries, cranberries, currants, elderberries, gooseberries and huckleberries). The texture usually determines which pack to use; but some of the firm ones may be dealt with in more than one way, and such variations are described separately below for the specific berries.

It goes without saying that you'll want to use only perfect berries that are ripe without being at all mushy. Pick them over carefully, wash them gently and drain; stem or hull them as necessary. Work with only a couple of quarts at a time because all berries, particularly the soft ones, break down quickly by being handled.

General Procedure for Most Berries

All berries are acid, so a Boiling-Water Bath for the prescribed length of time is the best process for them. A couple of them, being especially tough-skinned and highly acid, may even be canned by the Cold-Water method (provided that your storage is cold enough and you don't hope to keep them too long).

Use Raw pack generally for *soft* berries, because they break down so much in precooking.

A Hot pack in general makes a better product of most *firm* berries.

Use jars or cans—R-enamel cans for all red berries, but plain cans for gooseberries.

All may be canned either with sugar or without—but just a little sweetening helps hold the flavor even of berries you intend to doll up later for desserts. Thin and Medium sirups are used more often than Heavy, with Medium usually considered as giving a better table-ready product than Thin.

170 F. = 77 C. / 190 F. = 88 C. / 200 F. = 93 C. / 212 F. = 100 C. / 1 U.S. cup (½ U.S. pint) = .24 L. / 1 U.S. pint (2 cups) = .48 L. / 1 U.S. quart = .95 L. / 1 U.S. tsp. = 5 ml. (5 cc.) / 1 U.S. T. = 15 ml. (15 cc.) / ½ inch = 1.25 cm.

RAW PACK (SOFT BERRIES)

In jars. Fill clean, hot jars, shaking to settle the berries for a firm pack; leave ½ inch of headroom. Add boiling Thin or Medium sirup, leaving ½ inch of headroom; adjust lids. Process in a Boiling-Water Bath (212 F.)—pints for 15 minutes, quarts for 20 minutes. Remove jars; complete seals if necessary.

In R-enamel cans. Fill, shaking for a firm pack; leave only ¼ inch of headroom. Add boiling Thin or Medium sirup to the top of the can. Exhaust to 170 F. (*c.* 10 minutes); seal. Process in a B-W Bath (212 F.)—No. 2 cans for 15 minutes, No. 2½ for 20 minutes. Remove cans; cool quickly.

STANDARD HOT PACK (MOST FIRM BERRIES)

Measure berries into a kettle, and add ½ cup of sugar for each 1 quart of berries. On lowest burner, bring very slowly to a boil, shaking the pan to prevent berries from sticking (rather than stirring, which breaks them down). Remove from heat and let them stand, covered, for several hours. *This plumps up the berries and keeps them from floating to the top of the container when they're processed.* For packing, reheat them slowly. As insurance, have some hot Thin or Medium sirup on hand in case you run short of juice when filling the containers.

In jars. Fill with hot berries and juice, leaving ½ inch of headroom. Proceed and process as for Raw pack.

In R-enamel or plain cans. Fill to the top with hot berries and juice, leaving no headroom. Proceed and process as for Raw pack.

UNSWEETENED HOT PACK (MOST FIRM BERRIES)

This is often used for sugar-restricted diets; it is also another way of canning berries intended for pies.

Pour just enough cold water in a kettle to cover the bottom. Add the berries and place over very low heat. Bring to a simmer until they are hot throughout, shaking the pot—not stirring—to keep them from sticking.

Pack hot fruit and its juice, leaving headroom as above; remove any air bubbles by running a knife blade around the inner side of the container. Process as for Raw pack.

Specific Berries—except Strawberries

BLACKBERRIES

Raw pack. Usually considered soft, so for over-all versatility use Raw pack under the General Procedure above. With boiling water or Thin or Medium sirup—but Medium sirup if you want them table-ready. In jars or R-enamel cans.

BLUEBERRIES

Though in the firm category, they actually break down too much in the standard Hot pack (but they make a lovely sauce for ice cream, etc., if you want to can them by Hot pack with a good deal of extra sweetening). Old-timers dried them (q.v.) to use like currants in fruit cake.

Raw pack. With boiling water or sirup (Medium recommended). In jars or R-enamel cans. Proceed and process under General Procedure above.

Raw pack variation. If you want to hold them as much like their original texture and taste as possible when canned (to use like fresh berries in cakes, muffins, pies), you must blanch them. Put no more than 3 quarts of berries in a single layer of cheesecloth about 20 inches square.

Gather and hold the cloth by the corners, and dunk the bundle to cover the berries in boiling water until juice spots show on the cloth—*about 30 seconds*. Dip the bundle immediately in cold water to cool the berries. Drain them.

Fill jars, leaving ½ inch of headroom. Add no water or sweetening; adjust lids. Process as for standard Raw pack under General Procedure above.

BOYSENERRIES

Soft; in Raw pack under General Procedure. Use jars or R-enamel cans.

CRANBERRIES

These hold so well fresh in proper cold storage (see Root-Cellaring) or in the refrigerator, and they also freeze (q.v.), so they probably make the most sense canned if they're done as whole or jellied sauce. Whole raw, unsweetened berries may also be canned by the Cold-Water method, but their storage life of course is shorter.

Use jars only.

For about 6 pints Whole Sauce. Boil together 4 cups sugar and 2 cups water for 5 minutes. Add 8 cups (about 2 pounds) of washed, stemmed cranberries, and boil without stirring until the skins burst. Pour boiling hot into clean *hot jars*, leaving ½ inch of headroom, and run a sterile knife or spatula around the inner side of the jar to remove trapped air.

Boiling-Water Bath. After filling with ½ inch of headroom, *adjust* lids, and process pints in a B-W Bath (212 F.) for 10 minutes. Remove jars; complete seals if necessary.

For 4 pints Jellied Sauce. Boil 2 pounds of washed stemmed berries with 1 quart of water until the skins burst. Push berries and juice through a food mill or strainer. Add 4 cups sugar to the resulting purée, return to heat, and boil almost to the jelly stage (see "Testing for Doneness" in Jelly section). Pour hot into sterilized straight-sided jars (so it will slip out easily); seal with sterilized lids as for any jelly.

Cold-Water Canning for raw cranberries. Wash and stem cranberries. Sterilize jars, lids and gaskets. Scrub and scald a deep pail. Fill each jar with cranberries, set it in the pail, and let cold drinking water fill the jar and keep running into the pail until the water comes 6 inches over the top of the jar. Put on the cover and complete the seal under water.

Store the sealed jars in a place that's dry, dark and cold, but where they won't freeze.

CURRANTS

Currants are a novelty in certain sections of the United States where, during the 1920's, the bushes were uprooted because they harbored an insect destructive to the white pine. If you are fortunate enough to have some, by all means make jelly with them. Turn extra ones into dessert sauce; dry some; and they freeze.

Classed as firm berries, they can by Hot pack under General Procedure; in jars or R-enamel cans.

DEWBERRIES

Soft; in Raw pack under General Procedure. Use jars or R-enamel cans.

ELDERBERRIES

Best use is for jelly or wine.

If you do can them, use a Hot pack under General Procedure—and add 1 tablespoon lemon juice for each 1 quart of berries, to improve their flavor. In jars or R-enamel cans.

On the outside chance that you'd want to use them in muffins and cakes, experiment with blanching and the Raw pack variation described under Blueberries.

GOOSEBERRIES

Another scarce fruit in many sections of the country because, like the currant, they harbored the white-pine blister rust insect, and were eradicated as a conservation measure. But they make such heavenly old-fashioned pies, tarts and preserves!

Although they're firm, they do well in Raw pack with very sweet sirup. Or they may be done with Hot pack (where they hold their shape less well). Or the green (slightly underripe) ones may be Cold-Water Canned.

Wash; pick them over, pinch off stem ends and tails. Some cooks prick each berry with a sterile needle to promote a better blending of sweetening and juice (but we can't imagine a quicker way to drive ourselves up the wall; we'll leave it to osmosis).

Raw pack. Heavy Sirup is recommended for these very tart berries. And they'll probably pack better if you put ½ cup of hot sirup in the bottom of the container before you start filling. Use jars or plain cans. Process as in General Procedure above.

Hot pack. Follow the General Procedure for a standard Hot pack—but you may want to increase the sugar to ¾ cup for each 1 quart of berries. Process.

Cold-Water Canning for raw green gooseberries. Follow the method given above for Cranberries. And here you certainly do *not* prick the berries beforehand.

HUCKLEBERRIES

Being cousins of the blueberry, these firm berries are handled like Blueberries (q.v.).

LOGANBERRIES

They're soft, so pack Raw under General Procedure.

RASPBERRIES

Tenderest of the soft berries, these really do better if they're frozen. Put by some as jam or jelly, of course; try canning some sauce or juice. And if you have them in your garden, please don't forget to take a basket of fresh-picked raspberries to some older person who doesn't have a way to get them any more.

If you can them, use Raw pack and Medium Sirup. Use jars or R-enamel cans. Follow the General Procedure above.

YOUNGBERRIES

Another softie. Raw pack, as under General Procedure.

Strawberries

The most popular berry in the United States, these never-theless are often a disappointment when canned, because they fade and float if they are handled in the standard way recommended for most other soft berries. Here's how to have a blue-ribbon product—and no short cuts, please.

GENERAL HANDLING

Use a Boiling-Water Bath. Use Hot pack only (even though they're soft). Use jars or R-enamel cans.

Wash and hull perfect berries that are red-ripe, firm, and without white or hollow centers. Measure berries. Using ½ to 1 cup sugar for each 4 cups of berries, spread the berries and sugar in shallow pans in thin alternating layers.

Cover with waxed paper or foil if necessary as a protection against insects, and let stand at room temperature for 2 to 4 hours. Then turn into a kettle and simmer for 5 minutes in their own juice. Have some boiling Thin Sirup on hand if there's not enough juice for packing.

HOT PACK ONLY

In jars. Fill, leaving ½ inch of headroom (adding a bit of hot sirup if needed); adjust lids. Process in a Boiling-Water Bath (212 F.)—pints for 10 minutes, quarts for 15 minutes. Remove jars; complete seals if necessary.

In R-enamel cans. Fill to the top with hot berries and juice. Exhaust to 170 F. (*c.* 10 minutes); seal. Process in a B-W Bath (212 F.)—No. 2 cans for 15 minutes, No. 2½ for 20 minutes. Remove cans; cool quickly.

FOR A NICE SAUCE

Some ½-pint jars of strawberries will be welcome for toppings on ice cream or puddings.

Prepare the berries as above, but use 1 to 1¼ cups sugar to each 4 cups of berries. Process the ½-pints in a B-W Bath (212 F.) for 10 minutes. Remove jars; complete seals if necessary.

Canning Fruit Juices

It might seem a marginal use of time and material to can fruit juices, but they may be used to advantage in several ways: as beverages at breakfast or in place of commercial soft drinks for the family, or as the base for punches served by a country caterer (a Grange group, for instance); or the juices may be put by for future jelly-making.

Beverage juices will have better flavor if they are pre-sweetened at least partially. Use sugar, or the sweetener of your choice. (But for sugar-restricted diets, *use only the non-nutritive artificial sweetener approved by your doctor.* And unless you have had success in cooking with it, postpone adding the sweetener until serving time—some of these sugar substitutes leave an aftertaste if heated.)

Juices intended for jelly are not sweetened until you are making your jelly.

Fruit (and berry) sirups, which are concentrated, are better made from regular canned juice at the time you'll be wanting to use them, because you could run into a problem with pectin in some cases and end up with something too gooey for your purpose.

CONTAINERS FOR CANNED JUICES

Modern glass home-canning jars with fresh sealers are your best choice for juices or nectars because they're made to withstand heat-processing and years of re-use, and they provide a good seal *if* filling and processing are done conscientiously.

Cans also are suitable for juices. One manufacturer recommends R-enamel cans for all juices, even though the fruits from which they are made do not lose color in plain cans. Juices in cans, which are light-proof, need not be stored in the dark as do glass jars.

The manufacturers of commercial containers for several name-brands of mayonnaise, peanut butter and salad dressing have told *PFB* that they regard such jars and bottles (with their closures) as "one-trip containers," and therefore *do not recommend them for re-use* in home-canning.

Also not suitable for re-use in heat-processing at home are soft-drink bottles; nor would crimped-on caps be satisfactory.

General Procedure for Most Juices

Because boiling temperature (212 F. at sea level) can impair the fresh flavor of almost all fruit juices, these are usually processed by the *Hot*-Water Bath (given as 190 F.), which is pasteurization. Be sure there is at least 1 to 2 inches of definitely simmering water above the tops of the containers throughout the processing time, as for the B-W Bath.

The various nectars are processed in a Boiling-Water Bath (212 F.) because of their greater density.

Use Hot pack only. Use jars or cans (R-enamel suggested).

Choose firm-ripe, blemish-free fruit or berries; wash carefully, lifting the fruit from the water to let any sediment settle, and to avoid bruising. Then stem, hull, pit, core, slice —whatever is needed for preparing the particular fruit. *Simmer* the fruit until soft; strain through a jelly bag to extract clear juice (see "Equipment for Jellies, Jams, Etc." in the Preserving Kettle section).

170 F. = 77 C. / 190 F. = 88 C. / 200 F. = 93 C. / 212 F. = 100 C. / 1 U.S. cup (½ U.S. pint) = .24 L. / 1 U.S. pint (2 cups) = .48 L. / 1 U.S. quart = .95 L. / 1 U.S. tsp. = 5 ml. (5 cc.) / 1 U.S. T. = 15 ml. (15 cc.) / ½ inch = 1.25 cm.

APPLE CIDER (BEVERAGE)

Get cider fresh from the mill and process it without delay (though it can be held in a refrigerator in sterilized covered containers for 12 hours, if necessary). To prepare, strain it through a clean, dampened jelly bag, and in a large kettle bring it to a good simmer at 200 F., *but do not boil.* Pack hot.

Hot pack only, in jars. Pour strained fresh cider into hot sterilized jars, leaving ¼ inch of headroom; adjust lids. Process in a Hot-Water Bath at 190 F.—30 minutes for either pints or quarts. Remove jars; complete seals if necessary.

Hot pack only, in cans (R-enamel suggested). Fill to the top with strained fresh cider, leaving no headroom. Exhaust to 170 F. (*c.* 15 minutes); seal. Process in a H-W Bath at 190 F.—30 minutes for either No. 2 or No. 2½ cans. Remove cans; cool quickly.

APPLE JUICE (FOR JELLY LATER)

Add some underripe apples to the batch for more pectin. Wash and cut up apples, discarding stem and blossom ends. *Do not peel or core* (you may even use the peels left over from making Applesauce). Barely cover with cold water and bring to a boil over moderate heat, and simmer until apples are quite soft—about 30 minutes. Strain hot through a dampened jelly bag.

Reheat to 200 F. and pack *hot;* process as for Apple Cider, above.

APRICOT NECTAR

Nectars—most often made from apricots, peaches and pears—are simply juices thickened with *finely* sieved pulp of the fruit; usually they are "let down" with ice water when served. For sweetening, which is optional, honey or corn sirup may be substituted (see "Sweeteners" in Chapter 1 for proportions); artificial non-nutritive sweetening, if wanted, should be added at serving time to avoid a flavor change caused by heat-processing.

Use a Boiling-Water Bath. Use Hot pack only. Use ½-pint or pint jars or No. 2 cans (R-enamel suggested).

Wash, drain; pit and slice. Measure, and treat with an anti-oxidant if desired (see "General Handling for All Fruits" earlier). In a large enameled kettle add 1 cup boiling water to each 4 cups of prepared fruit, bring to simmering, and cook gently until fruit is soft. Put through a fine sieve or food mill. Measure again, and to each 2 cups of fruit juice-plus-pulp add 1 tablespoon lemon juice and about ½ cup sugar, or sweetening to taste. Reheat and simmer until sugar is dissolved. Pack hot.

Hot pack only, in jars. Pour hot nectar into ½-pint or pint jars, leaving ½ inch of headroom; adjust lids. Process in a Boiling-Water Bath (212 F.)—15 minutes for either ½-pints or pints. Remove jars; complete seals if necessary.

Hot pack only, in No. 2 cans (R-enamel suggested). Fill No. 2 cans to the top with simmering nectar, leaving no headroom; seal. Process in a Boiling-Water Bath (212 F.)—for 15 minutes. Remove cans; cool quickly.

BERRY JUICES

Crush and simmer berries in their own juice until soft; strain through a jelly bag—allow several hours for draining. If you twist the bag for a greater yield, the juice should be strained again through clean cloth to make it clear.

Measure, and to each 4 quarts of strained juice add 4 tablespoons lemon juice, plus sugar to taste—usually 1 to

2 cups. (If the juice is for jelly later, omit lemon juice and sugar at this time.) Reheat juice to a 200 F. simmer. Pack.

Hot pack only, in jars. Pour simmering juice into hot scalded jars, leaving ¼ inch of headroom; adjust lids. Process in a Hot-Water Bath at 190 F.—30 minutes for either pints or quarts. Remove jars; complete seals if necessary.

Hot pack only, in R-enamel cans. Fill cans to the top with hot juice, leaving no headroom; seal (at simmering stage it will already be more than 170 F., so exhausting is not necessary). Process in a H-W Bath at 190 F.—30 minutes for either No. 2 or No. 2½ cans. Remove cans; cool quickly.

CHERRY JUICE

Prepare as for Berry Juices. To each 4 quarts of strained juice add 2 tablespoons lemon juice, but adjust sweetening to the tartness of the cherries. (If the juice is for jelly later, omit lemon juice and sugar at this time.) Reheat juice to a 200 F. simmer. Pack hot. Process in a Hot-Water Bath at 190 F.—30 minutes for either pint or quart jars or for No. 2 or No. 2½ cans.

CRANBERRY JUICE

Boiling-Water Bath only. Hot pack only. Use jars only.

Pick over the berries and wash. Measure, and add an equal amount of water. Bring to boiling in an enameled kettle and cook until berries burst. Strain through a jelly bag (squeezing the bag adds to the yield: re-strain if you want beautifully clear juice). Add sugar to taste, and bring just to boiling. Pack hot.

If you're canning this juice only for special-diet reasons, omit sugar. Add the artificial non-nutritive sweetener *prescribed by your doctor* just before serving, lest heat-processing give the sweetener an unwanted aftertaste.

Hot pack only, in jars only. Pour boiling juice into clean hot jars, leaving ¼ inch of headroom; adjust lids. Process in a B-W Bath (212 F.)—10 minutes for either pints or quarts. Remove jars; complete seals if necessary.

CURRANT JUICE

Prepare and process as for Berry Juices, above.

GRAPE JUICE

Hot-Water Bath only. Use Hot pack only. Use jars or R-enamel cans.

The extra intermediate step of refrigerating the juice will prevent crystals of tartaric acid (harmless, but not beautiful) in the finished product. It's easier to work with not more than 1 gallon of grapes at a time.

Select firm-ripe grapes; wash, stem. Crush and measure into an enameled or stainless-steel kettle; add 1 cup water for each 4 quarts of crushed grapes. Cook gently *without boiling* until fruit is very soft—about 10 minutes. Strain through a jelly bag, squeezing it for a greater yield.

Refrigerate the juice for 24 hours. Then strain again for perfect clearness, being mighty careful to hold back the sediment of tartaric acid crystals in the bottom of the container.

Add ½ cup sugar for each 1 quart of juice (or omit sweetening), and heat to a 200 F. simmer.

Hot pack only, in jars. Pour simmering juice into hot scalded jars, leaving ¼ inch of headroom; adjust lids. Process in a Hot-Water Bath at 190 F.—30 minutes for either pints or quarts. Remove jars; complete seals if necessary.

Hot pack only, in R-enamel cans. Fill cans to the top with simmering juice, leaving no headroom; seal (the step of exhausting to 170 F. is not necessary when juice is simmering-hot). Process in a H-W Bath at 190 F.—30 minutes for either No. 2 or No. 2½ cans. Remove cans; cool quickly.

PEACH NECTAR

Prepare and process as for Apricot Nectar, above.

PEAR NECTAR

Prepare and process as for Apricot Nectar, above.

PLUM JUICE (AND FRESH PRUNE)

Hot-Water Bath only. Use Hot pack only. Use jars or R-enamel cans.

Choose firm-ripe plums with attractive red skins. Wash; stem; cut in small pieces. Measure. Put in an enameled or stainless-steel kettle, add 1 cup water for each 1 cup prepared fruit. Bring slowly to simmering, and cook gently until fruit is soft—about 15 minutes. Strain through a jelly bag. Add 1/4 cup sugar to each 2 cups juice, or to taste. Reheat just to a 200 F. simmer.

Pack and process as for Berry Juices, above.

RHUBARB JUICE

This makes good sense if you have extra rhubarb, because it can be used for a delicious quencher, and was the main ingredient of a hill-country wedding punch in olden days (see Recipe). And rhubarb is said to be good for our teeth.

Boiling-Water Bath only. Use Hot pack only. Use jars or R-enamel cans.

Wash and trim fresh young red rhubarb, but *do not peel*. Cover the bottom of the kettle with 1/2 inch of water, add rhubarb cut in 1/2-inch pieces. Bring to simmering; and cook gently until soft—about 10 minutes. Strain through a jelly bag. Reheat juice, adding 1/4 cup sugar to each 4 cups of juice to hold the flavor, and simmer at 200 F. until sugar is dissolved.

Hot pack, in jars. Pour simmering juice into hot scalded jars, leaving 1/4 inch of headroom; adjust lids. Process in a B-W Bath (212 F.)—10 minutes for either pints or quarts. Remove jars; complete seals if necessary.

Hot pack, in R-enamel cans. Fill cans to the top with simmering juice; seal (exhausting is not necessary if juice is simmering-hot). Process in a B-W Bath—10 minutes for either No. 2 or No. 2 1/2 cans. Remove cans; cool quickly.

Canning Fruit for Special Diets

For canning large pieces of fruit without sugar or other sweetener, follow individual instructions given earlier, but in addition to omitting sugar *use Hot pack only,* and eke out the natural liquid with extra unsweetened boiling juice (not water) if necessary to fill containers.

Pint jars (½-pints for the person with a very small appetite) are usually the best size to use unless you're canning sugarless fruit for several people in the family. Processing time is the same for pint and ½-pint jars.

Fruit Purées

Infants and those on low-residue diets require fruit whose natural fiber has been reduced to tiny particles in a sieve, food mill or blender. These purées are generally processed without sweetening—certainly the strained fruits for babies are unsweetened.

Any favorite fruit may be canned as a purée, because the only limiting factor would be an above-average amount of fiber to deal with. Apples, apricots, peaches and pears are the most popular for purées.

Use Boiling-Water Bath. Use Hot pack only. Use standard ½-pint canning jars with appropriate closures (*not* commercial babyfood jars, whose sealers are not re-usable).

APPLE PURÉE

Follow directions for Applesauce earlier, but omit sweetening; sieve, pack and process as for Apricot Purée, below.

APRICOT PURÉE (OR PEACH OR PEAR)

Use perfectly sound, ripe fruit. Wash, drain; pit and slice. In a large kettle crush a 1-inch layer of fruit to start the juice, then add the rest of the prepared fruit; if there seems not to be enough juice to keep the fruit from sticking or scorching, add no more than ¼ cup water for every firmly packed 1 cup of fruit. Simmer over medium heat, stirring as needed, until the fruit is soft—about 20 minutes. Push the cooked fruit through a sieve or food mill; or whir briefly in a blender and strain. Measure, and add 1 tablespoon fresh lemon juice for each 2 cups of pulp. Reheat to a 200 F. simmer. Pack hot.

Hot pack only, in ½-pint standard canning jars. Pour hot purée into clean, scalded ½-pint jars, leaving ½ inch of headroom. Adjust lids. Process in a Boiling-Water Bath (212 F.) for 10 minutes. Remove jars, complete seals if necessary.

Canning Tomatoes

Tomatoes are by far and away the most popular food for canning at home. One of the reasons for their popularity is that they are so versatile, since canned tomatoes in various forms can be served plain, or titivated, or used as the base for any number of nutritious, family-pleasing and inexpensive main dishes. Another attribute is their abundance, because they are grown at home in every likely corner of North America, and they're sold in vast quantities by professional truck gardeners from uncountable roadside stands.

Their third great virtue has been that they are easy to can—so easy that many a householder considered them to be just about foolproof. And this presumed dependability, regardless of handling or canning method, was ascribed to the fact that traditionally they had always been grouped with fruits on the *pH* scale: the average homemaker believed, therefore, that canned tomatoes were too strong-acid ever to permit the growth of certain bacteria, including the dread toxin-producing spores of *Cl. botulinum*.

Midsummer of 1974 saw a record number of backyard vegetable gardens and a record run on canning jars. It also saw the first general, widespread publication of two happenings that should clobber any idea that home-processed tomatoes cannot harbor dangerous spoilers: *In the first half of 1974, two verified cases of botulism from eating home-canned tomatoes or tomato juice had been reported by the Center for Disease Control of the U.S. Public Health Service.*

As the initial publicity died down, a prime fact emerged: In the preceding decade and a half, successful hybridizing had created a number of popular strains of so-called "table" tomatoes that are less acid than the old-time varieties. This lower-acid content means that some tomatoes should be regarded as having a higher number in the *pH* ratings—which in turn calls for re-thinking a standard method of packing and processing them. (The *pH* scale rates foods according to

Complete metric conversion tables are in Chapter 1, but the following apply particularly to this section: 170 Fahrenheit = 77 Celsius / 212 F. = 100 C. / 240 F. = 116 C. / 1 U.S. cup (½ U.S. pint) = .24 Liter / 1 U.S. pint (2 cups) = .48 L. / 1 U.S. quart (4 cups) = .95 L. / 1 U.S. teaspoon = 5 milliliters (5 cubic centimeters) / 1 U.S. tablespoon = 15 ml. (15 cc.) / ½ inch = 1.25 centimeters / 1 inch = 2.5 cm.

natural acid content, with a low *pH* meaning strongly acid and a higher number meaning lower acid: please see in Chapter 1 the full discussion of the effect of acidity on spoilers and on canning procedures earlier in this chapter.)

THE VARIETY FACTOR

Many of the new hybrid tomatoes have fewer seeds in order to be meatier and more solid, and thus stand up better during handling and shipping. These qualities also make them more desirable for slicing and serving raw; and it also often makes them *less* strong-acid, since much of the acid in a tomato is carried in the jelly-like substance that surrounds the seeds. It's impossible—and unfair—to cite the resulting lower-acid tomatoes by trade name, since even a lower-acid strain can produce strong-acid fruit in some microclimates and some soil conditions. But the present consensus allows the generalization that the strains specially recommended for table use, and the yellow and the "white" varieties, and the so-called Italian "paste" tomatoes may all be treated as being lower-acid than the stand-bys grown a generation ago.

How important is geography?

Not so water-tight is the idea that all tomatoes of the Eastern Seaboard are likely to be stronger-acid than all those of the West. A look at the states that have gone on record as often having lower-acid strains shows the geographic spread: California on the Pacific Coast, Minnesota and Ohio in the Middle West, and New Jersey in the East. (For a number of years the Co-operative Extension Service bulletins from California, Minnesota and Ohio have recommended a handling slightly different from the standard USDA treatment.)

Finally, rainfall, soil chemistry and length of growing season can add or subtract tiny, but important, degrees of acidity—and thus the East/West rule-of-thumb is not something to bank on.

Choosing a kind for canning

It's next to impossible for anyone but the expert plant scientist to recognize as a lower-acid strain the tomatoes bought by the bushel from a roadside stand. *Therefore when these are canned at home, the specific procedures described below must be followed faithfully.*

Growing your own? Although seed cataloguers certainly cannot be expected to cite the *pH* of every tomato they sell (considering the many variables), occasionally they will describe a named strain as "sweeter" or "less acid." Sometimes they may say that a certain named variety is a "good canner," a notation generally understood to mean that fruit of that particular strain holds its shape, color and texture when processed—in short, that it makes a handsome canned product but not that it is necessarily strong-acid.

Incidentally, a hormone substance applied to tomato blossoms to improve the yield also is credited with shortening the growing season and producing meatier fruit with fewer seeds—all of which help to reduce the tomatoes' natural acidity.

So the sensible thing to do if you're planning to grow your own is to *ask the agricultural agent of your county Extension Service which varieties are most suitable for your locality and for your purpose.*

...BUT THE CARE NEVER VARIES

No matter which strain of tomato you can in which form —whole, stewed or puréed; as juice, chili sauce or ketchup— any sloppiness, any cutting of corners will result in tomato products that are disappointing or even nasty or possibly downright dangerous.

The federal, state and non-government experts whom *PFB* consulted about the "great tomato revolution" of '74 agreed 100 percent that clean, careful handling, and due respect for the *Why's* of good packing and processing, are the primary safeguards in canning tomatoes of all varieties.

Selecting the fruit. Use only *firm*-ripe, unblemished tomatoes, ones that have not quite reached the table-ready stage wanted for slicing and serving raw.

Discard any that have rotten spots or mold: the regulations governing commercial canning regard *just one decomposed tomato per 100 sound fruit* as reason enough to condemn the entire lot as unfit for human consumption.

Discard any that have open lesions.

Washing. Wash the fruit carefully in fresh water of drinking quality. If many are spattered with field dirt, or have not been staked or mulched in your own garden, add a little mild detergent and 4 teaspoons of 5 percent chlorine bleach to each 1 gallon of wash water; rinse well in fresh water. (This thoroughness cuts down bacterial load.)

Peeling and cutting. You will be working with clean, sterilized equipment and cutting surfaces—just as you do when handling any food you're putting by. Peel tomatoes by dipping a few at a time in briskly boiling water, then dunking immediately in cold, clean water: the skins will strip off.

Without cutting into the seed cavity, ream out the stem end and core (the point of an apple-corer does a good job). Cut off the blossom end. Cut off any green shoulders to ensure a product of uniform tenderness and flavor. Cut out any bruises, no matter how small.

Cut/chop as individual instructions say to.

Packing. Pack in clean, scalded containers, leaving appropriate headroom. Remove air bubbles. Cap with clean, scalded closures that have been treated according to the maker's instructions (see "About Jars and Cans" starting page 55).

Add boiling juice to Raw-packed whole tomatoes. All other tomato products are packed Hot, with the contents of cans exhausted to a minimum of 170 F. if the tomatoes have cooled after precooking.

Leave the right amount of headroom.

Processing. Old-style open-kettle canning—with its opportunities for wicked airborne spoilers to contaminate food and the interiors of containers and lids, and with its un-reliable temperature control—has been considered the reason for some cases of botulism in supposedly strong-acid foods in years past: DO NOT USE THE OLD OPEN-KETTLE METHOD.

Whether you process in a Boiling-Water Bath or in a Pressure Canner, *time the processing accurately:* from return to the full boil in a B-W Bath; or after 10 pounds is reached, following a 7-to-10-minute strong flow of steam from the vent (depending on the size of the can-ner) to ensure adequate pressure inside.

Removing and cooling containers. Follow instructions given at the beginning of this Chapter 2. Complete the seals on bail-type jar lids as you take them from the canner. NEVER RE-TIGHTEN 2-PIECE SCREWBAND LIDS. AT ANY TIME. EVER.

Remember that *hastening* the cooling of jars can cause them to break; *retarding* the natural cooling of jars can cause flat-sour to develop in the contents. But cool *cans* quickly.

Check, clean, label and store containers according to "Han-dling after Processing" and "Storing All Canned Foods," given in the introduction to Canning Fruits, and at the start of this Chapter 2.

THE CHOICE BETWEEN PROCESSING METHODS

Until the extensive research on lower-acid tomatoes is completed and the reports are correlated, there can be no consensus that says tomatoes must be Pressure-processed to ensure a safe home-canned food. In the meantime each householder must make a judgment call.

If you have reason to think your tomatoes are not quite within the acid range for the B-W Bath (a *pH* rating of be-low 4.5)—because of the variety and the way they were grown, or if you're stuck with fruit that's past its ideal condi-tion—then just let informed good sense choose between:

Either (1) increasing the acid content of the pack yourself (how, is told below), and use a proper B-W Bath (212 F.) for the time specified.

Or (2) processing at 10 pounds pressure (240 F.) for the length of time given for the individual tomato products.

(And there's always *freezing*, q.v., in a space-saving form like sauce.)

GOOD COMPANY

Adding acid to tomatoes is not new. Nor is it gastronomical vandalism in the eyes of French cooks, who are famed for snubbing all but the naturally best ingredients for their dishes. Here's something from one of our favorite older cookbooks, *Nos amis les légumes, recettes de Marie* (Édition Valmorin-Andrieux, no date). "Choisissez des tomates moyennes bien rouges et fermes. Épluchez-les après les avoir passées quelques secondes dans de l'eau bouillante. Mettez-les dans les bocaux, remplissez avec de l'eau salée dans la proportion de 1 cuillerée à café par litre d'eau, *ajoutez le jus d'un citron*. Remplissez les bocaux jusqu' à 3 centimètres du bord. Faites stériliser . . ." etc., etc.

Added acid + B-W Bath

Adding acid is NOT A CRUTCH. Increasing the acidity DOES NOT MEAN THAT YOU CAN SHORT-CUT ANY STEP in safe canning procedure.

Remember that the temperature of your kitchen during canning season allows most bacteria to *double* their populations *every 15 to 30 minutes*. So if your tomatoes are not carefully selected and washed, and your work surfaces and utensils are not sanitary, the result can be a bacterial load that your processing method can't deal with completely.

As for how much of what acid to add in order to bring the *pH* rating of your tomatoes within the safety range for the B-W Bath, in late April 1975 the head nutritionists of the USDA Extension Service issued their recommendation: ¼ teaspoon of pure crystalline citric acid (U.S.P.) to each pint or No. 2 can, ½ teaspoon of the powdered citric acid to each quart or No. 2½ can—right on top of the tomatoes before the containers are capped/sealed (just as we add the salt for seasoning, which is optional). We also give a workable equivalent in white (distilled) vinegar. (See "Acids" in Chapter 1 for more information.)

Adjust lids on jars; exhaust the contents of cans to 170 F. if the tomatoes have cooled below that temperature in packing. In the B-W canner the jars/cans are covered with a *minimum* of 1 to 2 inches of water; the water is brought to a brisk boil—*212 F.* at sea level, with compensation for high altitudes (according to the table in Chapter 1)—and boiled continuously *by the clock for the full time required.* Remove containers: with jars, complete seals if necessary and cool naturally out of drafts; pop cans in cold water to cool quickly.

After 24 hours, when containers are checked, cleaned and labeled, store your canned tomatoes in a cool, dark, dry place.

Enjoy.

Pressure-processing for tomatoes

It is easy to cite Pressure-processing as the alternative method for canning tomatoes safely at home. However, a sound and well-tested timetable—a timetable of the reasoned sort we all rely on now for canning other foods—will take a good while to establish.

Meanwhile, for the householder who feels secure only with tomatoes done in a Pressure Canner, we offer the following stopgap. First, though, five things:

(1) Pressure-processing is NOT A CRUTCH. IT DOES NOT MEAN THAT YOU CAN SHORT-CUT ANY STEP of good canning procedure—careful selection, sanitation, correct packing, maintaining pressure, accurate timing, ensuring seals, proper storage.

(2) Nutrients in some degree and of course texture to a greater extent will suffer more than they do in a Boiling-Water Bath. The tender flesh of tomatoes will disintegrate more (unless a firming additive is included, q.v. "Calcium chloride," page 28), and an excessive amount of juice is likely to separate from the tissues.

(3) Independent food scientists around the country agree that 5 *pounds pressure is TOO LOW* to get the result desired from Pressure-processing plain tomatoes.

(4) The processing times given below are for *cut-up plain tomatoes ONLY*. The times are not long enough to deal safely with a mixture of tomatoes and lower-acid vegetables like onions, celery, green peppers, or whatever (see Stewed Tomatoes).

(5) The processing vessel used is a conventional Pressure *Canner—NOT a pressure saucepan*, even though it might hold several pint jars. The much smaller size of the saucepan plays hob with any pressure timetable (for why, see under "The Canning Methods" early in this Chapter 2). And anyway such a little saucepan-size batch of cut-up plain tomatoes would be better converted to Plain Sauce (q.v.) and done in the proper Boiling-Water Bath.

These points made, for Pressure Canning cut-up plain tomatoes: Use Hot pack only. Use jars or plain cans (if necessary, exhausting the contents to a minimum of 170 F. after packing). Vent the heated canner 7 minutes for medium-size kettles, 10 minutes for large ones. Time the processing after internal pressure of the canner has reached *10 pounds* (240 F.)—15 minutes for pint jars, 20 minutes for quarts, 15 minutes for No. 2 cans, 20 minutes for No. 2½ cans. Complete jar seals if necessary, cool naturally; remove cans, cool quickly in cold water.

CANNED TOMATO TROUBLES AND WHAT TO DO

New developments bring new strictness, so canned tomatoes in your storeroom should be examined for the same signs that mean vegetables are unfit, or dangerous, to eat:

—Broken seals.
—Bulging cans.
—Seepage around the seal.
—Mold, the tiniest spot, around the seal or on the underside of the lid or in the contents.
—Gassiness in the contents, spurting liquid from pressure inside any container when it is opened.
—Cloudy or yeasty liquid.
—Unnatural color.
—Unnatural or unpleasant odor.

But even if the seals appear intact and the contents look and smell all right, *do not even taste the tomatoes before boiling them for 15 minutes to destroy hidden toxins. Then, during boiling, if they foam or smell bad, destroy them so they cannot be eaten by people or animals.* Wash the containers and sealers in hot soapy water, then cover with fresh water and boil hard for 15 minutes; salvage only sterilized jars—destroy cans and all closures, etc.

170 F. = 77 C. / 212 F. = 100 C. / 240 F. = 116 C. / 1 U.S. cup (½ U.S. pint) = .24 L. / 1 U.S. pint (2 cups) = .48 L. / 1 U.S. quart = .95 L. / 1 U.S. tsp. = 5 ml. (5 cc.) / 1 U.S. T. = 15 ml. (15 cc.) / ½ inch = 1.25 cm. / 1 inch = 2.5 cm.

Troubles galore after six months in storage. There may have been too much headroom; or the sealing rim wasn't wiped clean. Whatever the cause, the seal was poor. So the tomatoes fermented, forcing material to ooze out under the lid (and turning the contents an unnatural brown). Destroy.

SPECIFIC TOMATO PRODUCTS

Unless you can pick with finicky selectiveness from a well-managed small garden of your own, there is bound to be a range of ripeness, size and condition in any good-sized batch of tomatoes you're getting ready to can. So pick them over carefully, of course discarding any rotten or banged-up ones, and let size and degree of acceptable ripeness dictate the form you'll can them in.

Perfect, *firm-ripe*, uniform and small enough to slip easily down into the jar—these are the ones for canning whole to use in salads. Misshapen or overly large fruit are cut to stewing size—quarters, eighths or chunks—and are canned plain or with added vegetables for flavor, they also go for sauce or juice.

Regardless of the form or whether they're to be processed in a Boiling-Water Bath or a Pressure Canner, prepare the tomatoes according to the general directions given above for the B-W Bath.

Whole Tomatoes (Salad Style)

Serve these filled with a salad mixture of chopped vegetables or tuna fish or chicken. (Or try this: In a glass or china bowl combine with each 1 pint of tomatoes-plus-juice ¼ to ⅓ cup thinly sliced red onion, ¼ cup vinaigrette French dressing, perhaps 1 teaspoon Worcestershire sauce, maybe ¼ teaspoon dried basil or oregano; turn gently to mix; cover tightly and refrigerate for several hours or overnight, then serve chilled as a side dish instead of a conventional salad.)

HANDLING

Use a Boiling-Water Bath. Use Raw pack (and this is the only style we prefer Raw-packed). Use jars only.

In advance prepare enough Tomato Juice (q.v. below) to be the canning liquid for the batch—figure on ½ to ¾ cup of hot juice for each pint jar, 1 to 1½ cups of juice for each quart. Hold the juice in a sterilized, covered container until you're ready to heat it to fill the jars. (*Caution:* Don't dilute the acidity of the pack by eking out the tomato juice with boiling water—be ready to use canned juice if your planning went wrong.)

Peel select, thoroughly washed tomatoes by dunking them in briskly boiling water for about 30 seconds, then in cold water; handle gently as you strip off the skins and core the fruit.

Raw pack, in jars only. Fit whole tomatoes snugly—but without pressing so much that you break them—into clean scalded jars, leaving ½ inch of headroom. Add ¼ teaspoon fine crystalline citric acid (or 1 tablespoon white vinegar), and ½ teaspoon salt to pints; add ½ teaspoon citric acid (or 2 tablespoons white vinegar),

and 1 teaspoon salt to quarts. Add boiling juice, leaving
½ inch of headroom for both pints and quarts (hold
the tomatoes away from the jar's side with the blade
of a table knife to let the hot liquid fill all gaps). Adjust
lids. Process in a Boiling-Water Bath (212 F.)—40
minutes for pints, 50 minutes for quarts. Remove jars,
complete seals if necessary.

Cut-up Plain Tomatoes

Boiling-Water Bath preferred (but for Pressure Canning see times, etc., in the general introduction for Canning
Tomatoes). Use Hot pack only. Use jars or plain cans.

Select, wash, peel, according to general handling earlier;
cut in quarters or eighths, saving all juice possible. In a
large enameled kettle bring cut tomatoes to a boil in their
own juice, and cook gently for 5 minutes, stirring so they
don't stick. Pack.

HOT PACK ONLY

B-W Bath, in jars. Fill clean scalded jars with boiling-hot
tomatoes and their juice, leaving ½ inch of headroom.
Add ¼ teaspoon crystalline citric acid (or 1 tablespoon
white vinegar), and ½ teaspoon salt to pints; add ½
teaspoon citric acid (or 2 tablespoons white vinegar),
and 1 teaspoon salt to quarts. Adjust lids. Process in a
Boiling-Water Bath (212 F.)—15 minutes for pints,
20 minutes for quarts. Remove jars; complete seals if
necessary.

B-W Bath, in plain cans. Fill with boiling-hot tomatoes
and juice, leaving ¼ inch of headroom. Add ¼ teaspoon citric acid (or 1 tablespoon white vinegar), and
½ teaspoon salt to No. 2 cans; add ½ teaspoon citric
acid (or 2 tablespoons white vinegar), and 1 teaspoon
salt to No. 2½ cans. If the tomatoes have cooled unavoidably, exhaust to 170 F. (*c.* 10 minutes); seal.
Process in a B-W Bath (212 F.)—15 minutes for No. 2
cans, 20 minutes for No. 2½ cans. Remove cans; cool
quickly.

Stewed Tomatoes with Added Vegetables

The addition of lower-acid vegetables to tomatoes means that you must Pressure-process the mixture *according to the specific rule for the lowest-acid vegetable in the combination.*

How much of which of the usual vegetables is added for interest of course depends on the family's taste. However, *density* of the pack is an important factor in any timetable for processing. Therefore we say that the total amount of several added vegetables *should not exceed one-fourth the volume of tomatoes in the mixture.* For example, to 8 cups of prepared cut tomatoes we would add no more than 1 cup chopped celery, ½ cup chopped onion and ½ cup chopped green pepper. Incidentally, this balance of added vegetables also makes for good flavor.

(Although tomatoes with zucchini squash are a popular side dish, the amount of squash added is generally so large that this mixture is better if you combine canned Squash (q.v.) with as much canned Stewed Tomatoes as you like, and the two are heated together just before serving.)

GENERAL HANDLING

Pressure Canning only. Use Hot pack only. Use jars or plain cans.

To avoid diluting acidity or flavor, it's a good idea to prepare 3 or 4 cups of Tomato Juice (q.v.) to have ready in case you need extra hot liquid when filling the containers; or use canned juice, heated.

Wash, peel, core and cut the tomatoes in quarters or smaller, saving the juice; measure. Add the desired proportion of well-washed coarsely chopped celery, finely chopped onions, or chopped seeded green peppers. Combine the vegetables in a large enameled kettle and boil them gently in their own juice *without added water* for 10 minutes, stirring to prevent sticking.

HOT PACK ONLY, IN JARS

Ladle boiling hot into clean hot jars, leaving ½ inch of headroom. Add ¼ teaspoon citric acid (or 1 tablespoon white vinegar), and ½ teaspoon salt to pints; add ½ teaspoon citric acid (or 2 tablespoons vinegar), and 1 teaspoon salt to quarts. (If there is too little free liquid, make up the difference with boiling tomato juice, *not water*.) Adjust lids; process. After processing, remove jars; complete seals if necessary.

With only onion added, Pressure-process at 10 pounds (240 F.)—25 minutes for pints, 30 minutes for quarts.

With celery added, Pressure-process at 10 pounds (240 F.)—30 minutes for pints, 35 minutes for quarts.

With green peppers added, Pressure-process at 10 pounds (240 F.)—35 minutes for pints, 45 minutes for quarts.

HOT PACK ONLY, IN PLAIN CANS

Ladle boiling hot into cans, leaving ½ inch of headroom. Add ¼ teaspoon citric acid (or 1 tablespoon white vinegar), and ½ teaspoon salt to No. 2 cans; add ½ teaspoon citric acid (or 2 tablespoons white vinegar), and 1 teaspoon salt to No. 2½ cans. Fill to the top with boiling-hot cooking liquid or tomato juice, *not water;* seal. (Exhausting to 170 F. is not needed unless the food has cooled unavoidably before packing.) Process. After processing, remove cans; cool quickly.

Pressure-process No. 2 cans at 10 pounds (240 F.)—25 minutes with *only onions* added, 30 minutes with celery added, 35 minutes with green peppers added.

Pressure-process No. 2½ cans at 10 pounds (240 F.)—30 minutes with *only onions* added, 35 minutes with celery added, 45 minutes with green peppers added.

Plain Tomato Sauce (Purée)

This is a handy way indeed to can tomatoes, and it makes a better base for red Italian-style pasta sauces than does Tomato Paste (which Americans often tend to use too much of in such sauces anyway). The texture should fall about halfway between juice and paste.

Do not add onions or celery, etc., now *or you must* Pressure-process the sauce (see Stewed Tomatoes).

HANDLING

Use a Boiling-Water Bath. Use Hot pack only. Use
½-pint or pint jars only.

Prepare and sieve the fruit as for Tomato Juice
(below). In a large enameled kettle bring the juice to
boiling, and boil gently until thickened but not so stiff as
Tomato Paste—about 1 hour or a little longer. Stir often
so it doesn't stick.

Hot pack only, in ½-pint or pint jars. Pour into clean
scalded jars, leaving ¼ inch of headroom in ½-pints,
½ inch in pints. Add ⅛ teaspoon citric acid (or 1½
teaspoons white vinegar), and ½ teaspoon salt to ½-
pints; add ¼ teaspoon citric acid (or 1 tablespoon
vinegar), and ½ teaspoon salt to pints. Adjust lids.
Process in a B-W Bath (212 F.)—30 minutes for either
½-pints or pints. Remove jars; complete seals if necessary.

Tomato Paste

If you're canning many tomatoes, you'll surely want a
few little jars of paste put by too. Work with small batches,
because it scorches easily during the last half of cooking.
And forgo onions, garlic, celery, etc., because such flavors
may not be wanted in delicate sauces you merely want to
color with the paste.

HANDLING

Use a Boiling-Water Bath. Use Hot pack only. Use
½-pint jars only.

Carefully wash, peel, trim and chop the tomatoes
saving all the juice possible (4 to 4½ pounds of tomatoes
will make about 4 ½-pint jars of paste). In an enameled
kettle bring the chopped tomatoes to a boil, then reduce
heat and simmer for 1 hour, stirring to prevent sticking.
Remove from heat and put the cooked pulp and juice
through a fine sieve or food mill. Measure, return to the
kettle, and for every 2 cups of sieved tomatoes add ¼ teaspoon citric acid (or 1 tablespoon white vinegar), and
½ teaspoon salt. Reheat, and continue cooking very slowly,

stirring frequently, until the paste holds its shape on the spoon—about 2 hours more.

Hot pack only, in ½-pint jars. Ladle hot paste into clean hot jars, leaving ¼ inch of headroom. Adjust lids, and process in a B-W Bath (212 F.) for 35 minutes. Remove jars; complete seals if necessary.

Chili Sauce and **Tomato Ketchup**, see the Preserving Kettle chapter.

Tomato Juice

Canned tomato juice is noted for encouraging growth of the highly heat-resistant bacillus that causes *flat-sour* spoilage, a sneaky and nasty-tasting condition indeed. However, even though the organism is very hard to destroy, it can be avoided quite easily: just follow carefully all the requirements for handling food in a sanitary way.

This care extends from every piece of sterilized equipment to the tomatoes themselves. Choose only firm-ripe, red, perfect tomatoes—no injured ones, none with soft spots or broken skins. Wash them thoroughly. With a stainless-steel knife cut away stem and blossom ends and cores. Cut the fruit in eighths and put it in an enameled kettle (which won't react with the acid of the tomatoes), and simmer it, stirring often, until soft. Put the tomatoes through a fine sieve or food mill: the finer the pulp, the less likely that the juice will separate during storage. Measure the juice into the kettle, and for each 4 cups of juice add ½ teaspoon citric acid (or see "Acids," Chapter 1), and 1 teaspoon salt. Reheat all to simmering. Pack hot.

Hot pack only, in jars. Fill clean, hot jars with the very hot juice, leaving ½ inch of headroom; adjust lids. Process in a Boiling-Water Bath—15 minutes for either pints or quarts. Remove jars, complete seals if necessary.

Hot pack only, in plain cans. Fill to the top with boiling juice, leaving no headroom: seal (juice already 170 F. or over doesn't need further exhausting). Process in a B-W Bath (212 F.)—20 minutes for either No. 2 or No. 2½ cans. Remove cans; cool quickly.

Annette's Tomato Cocktail

This is delicious as an appetizer or for aspic: the seasonings of course may be varied—but being low-acid vegetables, they all must be thoroughly precooked before being added to the juice (otherwise the mixture would have to be processed at 10 pounds' pressure like Stewed Tomatoes, below). Annette Pestle says she has never had this cocktail separate after it sits in properly cool storage; but in case yours does, just give it a good shake.

Use a Boiling-Water Bath. Use Hot pack only. Use jars or plain cans.

To make about 7 quarts of cocktail, you'll need 8 quarts of cut-up tomatoes. Wash thoroughly the firm-ripe unblemished tomatoes; remove stems, blossom ends and cores; cut in small pieces. In a large enameled kettle simmer the tomatoes over low heat until soft; put through a fine sieve or food mill to remove skins and seeds, and set the strained juice aside. Rinse the kettle, and into it measure 2 cups of the tomato juice, add 2 diced medium onions, 1¼ cups diced celery (including a few leaves), 1 large seeded and chopped green pepper, 3 bay leaves, 8 or 10 fresh basil leaves (or 2 teaspoons dried basil), 1½ tablespoons salt, ½ teaspoon ground pepper, 3 tablespoons sugar, and 2 teaspoons Worcestershire sauce. Boil over medium heat—stirring, and adding extra juice as needed to keep the mixture from sticking—until soft, about 30 minutes. Pick out the bay leaves, then press the vegetables through a fine sieve or food mill. Add 3½ teaspoons crystalline citric acid (or ⅔ cup bottled lemon juice OR ¾ cup white vinegar) and the rest of the tomato juice. Bring to simmering. Pack hot.

Hot pack only, in jars. Fill jars with hot juice, leaving ½ inch of headroom; adjust lids. Process in a Boiling-Water Bath (212 F.)—15 minutes for pints, 20 minutes for quarts. Remove jars, complete seals if necessary.

Hot pack only, in plain cans. Fill cans with hot juice to the tops, leaving no headroom; seal immediately (juice heated to 170 F. or over does not need further exhausting). Process in a B-W Bath (212 F.)—15 minutes for No. 2 cans, 20 minutes for No. 2½ cans. Remove cans; cool quickly.

Country Tomato Soup

This plain and good soup is not diluted for serving. The amounts given make about 4½ quarts.

Pressure Canning only (onions and peppers are low-acid). Use Hot Pack only. Use jars only.

Wash 1 peck (8 quarts) of ripe red tomatoes; remove stem ends and cores and cut in pieces. In a large kettle, cook and stir the tomatoes until soft—about 15 minutes. Push the pulp and juice through a wire strainer or food mill to remove skins and seeds; return the purée to the kettle but do not reheat yet.

Cook together until soft (in enough water just to cover) 3 large onions and 2 green peppers—all finely chopped. Sieve, and add to the puréed tomatoes in the kettle. Mix together ¾ cup of sugar, 3 tablespoons salt and 8 tablespoons cornstarch; blend in 3 tablespoons white vinegar and just enough more water or cool tomato juice to make a smooth paste. Pour slowly into the sieved tomatoes, stirring all the while. Heat to boiling and stir until the liquid clears. Pack hot.

Hot pack only, in jars. Pour boiling-hot soup into clean, hot jars, leaving ½ inch of headroom for pints, 1 inch for quarts. Adjust the lids. Pressure-process at 10 pounds (240 F.)—pints for 20 minutes, quarts for 30 minutes. Remove jars; complete seals if necessary.

Canning Vegetables

All fresh natural vegetables are low-acid, and therefore MUST be processed in a regular Pressure Canner.

Because Pressure-processing softens them, vegetables for canning may be a little more mature—but no less *fresh* —than those for freezing.

The Hot pack helps to ensure the seal by driving air from the tissues of the food before it goes into the containers. Therefore when both Raw and Hot pack are given in the following instructions, we say "Preferred: Hot pack" (except with vegetables that may be Hot-packed only). Hot pack markedly increases the density of foods like squash and celery, etc., and hence requires longer processing than Raw pack does.

Note: Salt added to vegetables in canning is merely a seasoning and therefore of course is optional. Pure canning salt is ideal, but the amounts called for are so small that the fillers, etc., in your regular table salt won't cloud the canning liquid. However, salt substitutes should *not* be added to the pack before processing, lest the finished product have an unwanted aftertaste; add your salt substitute to the heated vegetable just before serving.

EQUIPMENT FOR CANNING VEGETABLES

All the utensils and standard kitchen furnishings that you used for fruits (see the list at the beginning of this Chapter 2) will do, *with this exception: use only a standard Pressure Canner for processing* (your B-W Bath kettle will make a good vessel for blanching). Do not use a pressure *saucepan* even though you're doing only a couple of small jars: such a saucepan heats and cools so quickly that it violates the carefully researched timetable for Pressure-processing canned foods.

Complete metric conversion tables are in Chapter 1, but the following apply particularly to this section: 170 Fahrenheit = 77 Celsius / 228 F. = 109 C. / 240 F. = 116 C. / 1 U.S. cup (½ U.S. pint) = .24 Liter / 1 U.S. pint (2 cups) = .48 L. / 1 U.S. quart (4 cups) = .95 L. / 1 U.S. teaspoon = 5 milliliters (5 cubic centimeters) / 1 U.S. tablespoon = 15 ml. (15 cc.) / ½ inch = 1.25 centimeters / 1 inch = 2.5 cm.

You will need a blanching basket or some other means of holding many prepared vegetables loosely in the boiling water, because using the precooking water as the canning liquid helps to save nutrients.

CANNED VEGETABLE TROUBLES
AND WHAT TO DO

It goes without saying that every conscientious canner will select only prime vegetables; will prepare them care-

HOW ALL THE FAMILY CAN PLAY RUSSIAN ROULETTE WITHOUT ACTUALLY TRYING

Anybody who has the get-up-and-get to can vegetables at home also has—or should have—the desire to can them safely: and this means to process them *only in a Pressure Canner*. With no shortcuts, no scamping.

Why does *PFB* harp on this? Because in 1974, when the new surge of home-canning was just getting underway, there was the largest number of food-borne botulism outbreaks reported by state health departments since 1935, in the depth of the Great Depression. Home-canned food was associated with 94 percent of the outbreaks in which the vehicle for the deadly toxin was known; nearly two-thirds of the guilty home-canned foods were vegetables or vegetable combinations.

Maybe you can harp too, and help people to realize that it would take *22 hours of continuous boiling* in a B-W Bath to have the same sterilizing effect on low-acid foods as does Pressure-processing between 240 and 250 F. for about 45 minutes.

When people say that "canning vegetables in the old wash boiler was good enough for Great-aunt Gussie, so what's the big deal?" you can give them the answer. The big deal is a matter of life and death.

fully, pack them and Pressure-process them right; will check seals, and will store them in a cool, dark, dry place 24 hours after they have been canned.

But if, after they have been in storage, you find any of the following, *destroy the contents so that they cannot be eaten by people or animals.* Wash containers and closures (and rubbers, if used) in hot soapy water, and boil them hard for 15 minutes in clean water to cover. Then throw away the sterilized cans, lids, sealing discs and rubbers; if sound, the sterilized jars may be used again.

—Broken seal.

—Bulging cans.

—Seepage around the seal.

—Mold, even a speck, in the contents or around the seal or on the underside of the lid.

—Gassiness (small bubbles) in the contents.

—Spurting liquid, pressure from inside the container as it is opened.

—Cloudy or yeasty liquid.

—Unnatural or unpleasant odor.

Even innocent-looking containers with apparently good seals can contain hidden toxins. Therefore, before tasting any home-canned vegetables, boil them hard for 15 minutes (corn and greens require 20 minutes, and should be stirred a little to distribute the heat). If they foam or smell bad during boiling, destroy them completely so they can't be eaten by people or animals, and salvage only the sterilized jars, as above.

Asparagus

Asparagus keeps more spring flavor if you freeze it; but it cans easily—whole or cut up. See Recipes.

GENERAL HANDLING

Only Pressure Canning for asparagus: it has even less acid than string beans. Use Raw or Hot pack. Use jars or plain cans.

Wash; remove large scales that may have sand behind them; break off tough ends; wash again. If you're canning it whole, sort spears for length and thickness, because you'll pack them upright; otherwise cut spears in 1-inch pieces.

RAW PACK

In jars. Whether asparagus is whole (spears packed upright) or cut up, leave ½ inch of headroom. Add ½ teaspoon salt to a pint, 1 teaspoon to a quart. Add

170 F. = 77 C. / 228 F. = 109 C. / 240 F. = 116 C. / 1 U.S. cup (½ U.S.
pint) = .24 L. / 1 U.S. pint (2 cups) = .48 L. / 1 U.S. quart = .95 L. / 1
U.S. tsp. = 5 ml. (5 cc.) / 1 U.S. T. = 15 ml. (15 cc.) / ½ inch = 1.25
cm. / 1 inch = 2.5 cm.

boiling water, leaving ½ inch of headroom; adjust
lids. Pressure-process at 10 pounds (240 F.)—pints for
25 minutes, quarts for 30 minutes. Remove jars; complete seals if necessary.

In plain cans. Pack as for jars, leaving only ¼ inch of
headroom. Add ½ teaspoon salt to No. 2 cans, 1 teaspoon to No. 2½. Fill to top with boiling water. Exhaust to 170 F. (*c.* 10 minutes); seal. Pressure-process
at 10 pounds (240 F.)—20 minutes for either No. 2
or No. 2½. Remove cans; cool quickly.

PREFERRED: HOT PACK

Whole spears—stand upright in a wire blanching basket and dunk it for 3 minutes in boiling water up to *but
not covering* the tips; drain and pack upright (tight but
not squdged). Cut-up—cover clean 1-inch pieces with
boiling water for 2 to 3 minutes; drain and pack.

The added processing liquid can be the boiling-hot
blanching water—if it's free of grit—instead of fresh
boiling water.

In jars. Complete the pack and Pressure-process as for Raw
pack, above.

In plain cans. Complete the pack and Pressure-process as
for Raw pack, above.

Beans, "Butter," see Beans, Lima (fresh)

Beans (dried), Baked

Pressure Canning only. Hot pack only—either hot from
the oven or reheated. Use jars or plain cans. If the cooked
beans have got cold or too dry, add a bit of hot water as
you reheat them, so they'll be juicy enough again.

HOT PACK ONLY

In jars. Pack beans hot, leaving 1 inch of headroom. Pressure-process at 10 pounds (240 F.)—pints for 80 minutes, quarts for 100 minutes. Remove jars; complete seals if necessary.

In plain cans. Pack beans hot, leaving only ¼ inch of headroom. Exhaust to 170 F. (*c.* 15 minutes); seal. Pressure-process at 10 pounds (240 F.)—No. 2 cans for 95 minutes, No. 2½ for 115 minutes. Remove cans; cool quickly.

Beans (dried), in Sauce

Only Pressure Canning. Only Hot pack. Use either jars or plain cans.

Use clean beans—kidney, navy, pea or yellow-eye as you prefer—that are free of any musty odor. Cover with boiling water, boil 2 minutes. Remove from heat and let soak 1 hour in their cooking water. Yield: 1 cup dry beans will make about 2 or 2½ cups after soaking.

To pack, reheat to boiling, drain (saving water to use in the sauce), and put hot beans in containers.

To make about 4 cups of sauce, enough to do 4 to 5 pints of beans:

Tomato—4 cups tomato juice, 3 tablespoons sugar, 2 teaspoons salt, 1 tablespoon onion pulp, a pinch of whatever spices you like, a few grains of cayenne. Heat to boiling. (Blah but O.K. in a bind: 1 cup tomato catsup, 3 cups of the drained-off soaking water, and 1 teaspoon salt, or to taste; boil.)

Molasses—4 cups of drained-off soaking water, 3 tablespoons dark molasses, 1 tablespoon vinegar, 2 teaspoons salt, ½ teaspoon dry mustard, ¼ teaspoon ginger; boil. Maple sirup is great in beans; so is some onion pulp. Experiment!

HOT PACK ONLY

In jars. Fill only ¾ full with drained hot beans. Lay on top of the beans a 1-inch-square slice of salt pork or bacon end (more for quarts). Add boiling sauce, leaving 1 inch of headroom. Adjust lids, Pressure-process at 10 pounds (240 F.)—pints for 65 minutes, quarts for 75 minutes. Remove jars; complete seals if necessary.

In plain cans. Fill only ¾ full with drained hot beans. Lay on top of the beans a 1-inch-square slice of salt pork or bacon end (more for No. 2½). Add boiling sauce, leaving only ¼ inch of headroom. Exhaust to 170 F. (*c.* 20 minutes); seal. Pressure-prcess at 10 pounds (240 F.)—No. 2 cans for 65 minutes, No. 2½ for 75 minutes. Remove cans; cool quickly.

Beans—Green/Italian/Snap/String/Wax

These beans are among the most popular vegetables to put by, so don't forgo canning them because they're the leading vehicle for botulism among foods *processed carelessly at home.* Do read about "mushy" canned beans in the Postscript on page 535—and *never* under-process them.

They freeze well; maybe you'd like to freeze the very young, just-from-the-garden ones, and can the rest. And old-timers used to dry them (see "Leather Britches").

GENERAL HANDLING

Pressure Canning only. Use Raw or Hot pack (Hot makes them supple and permits a more solid pack). Use jars or plain cans.

The name "snap" comes from the crisp way the young ones break when they're fresh-picked; if you must hold them overnight before canning, refrigerate them in bags. Wash, trim ends. Sort roughly for size: you may want some whole in a fancy pack (upright like asparagus spears), or others frenched or cut on a slant in 1-inch pieces. If you're stuck with doing some bigger, older ones, though, break off tips and tails, unzip their strings along their length, cut them in small pieces, and pack them by themselves. There's plenty of use for all types.

RAW PACK

In jars. Fill as tightly as you can, leaving ½ inch of headroom. Add ½ teaspoon salt to pints, 1 teaspoon to quarts. Add boiling water, leaving ½ inch headroom. Adjust lids. Pressure-process at 10 pounds (240 F.)—pints for 20 minutes, quarts for 25 minutes. Remove jars; complete seal if necessary.

In plain cans. Pack as tightly as you can, leaving only ¼ inch of headroom. Add ½ teaspoon salt to No. 2 cans, 1 teaspoon to No. 2½. Fill to top with boiling water. Exhaust to 170 F. (c. 10 minutes); seal. Pressure-process at 10 pounds (240 F.)—No. 2 cans for 25 minutes, No. 2½ for 30 minutes. Remove cans; cool quickly.

PREFERRED: HOT PACK

Cover clean, trimmed beans with boiling water and boil 5 minutes. Drain, keeping the hot cooking water. Pack whole beans upright; use a wide-mouth funnel to pack the cut ones.

In jars. Fill with hot beans, leaving ½ inch of headroom. Add ½ teaspoon salt to pints, 1 teaspoon to quarts. Add boiling-hot cooking water, leaving ½ inch of headroom. Adjust lids. Pressure-process at 10 pounds (240 F.)—pints for 20 minutes, quarts for 25 minutes. Remove jars; complete seals if necessary.

In plain cans. Fill loosely with hot beans, leaving only ¼ inch of headroom. Add ½ teaspoon salt to No. 2 cans, 1 teaspoon to No. 2½. Fill to top with boiling-hot cooking water. Exhaust to 170 F. (c. 10 minutes); seal. Pressure-process at 10 pounds (240 F.)—No. 2 cans for 25 minutes, No. 2½ for 30 minutes. Remove cans; cool quickly.

Beans, fresh Lima (Shell beans)

Pressure Canning only. Use Raw or Hot pack. Use jars or C-enamel cans.

It's good but not vital to deal with one variety at a time—if only because sometimes different-sized types require different amounts of headroom.

Shell the beans and wash them before packing. They must be packed loosely because, like all starchy legumes, they swell.

RAW PACK

In jars. If it's a small variety, leave 1 inch of headroom for pints, 1½ inches of headroom for quarts; if it's a large variety, leave ¾ inch of headroom for pints, 1¼ inches of headroom for quarts. *Do not press or shake the beans down.* Add ½ teaspoon salt to pints, 1 teaspoon salt to quarts. Add boiling water, leaving ½ inch of headroom (the water will be well over the top of the beans); adjust lids. Pressure-process at 10 pounds (240 F.)—pints for 40 minutes, quarts for 50 minutes. Remove jars; complete seals if necessary.

In C-enamel cans. Fill with beans, leaving ¾ inch of headroom for either No. 2 or No. 2½ cans; *don't press or shake down.* Add ½ teaspoon salt to No. 2 cans, 1 teaspoon salt to No. 2½. Fill cans to top with boiling water. Exhaust to 170 F. (*c.* 10 minutes); seal. Pressure-process at 10 pounds (240 F.)—40 minutes for either No. 2 or No. 2½. Remove cans; cool quickly.

PREFERRED: HOT PACK

Cover, shelled, washed beans with boiling water and cook 1 minute after water returns to boiling. Drain, saving the hot cooking water.

In jars. Fill loosely with drained hot beans, leaving 1 inch of headroom for either pints or quarts. Add ½ teaspoon salt to pints, 1 teaspoon salt to quarts. Add boiling-hot cooking water, leaving 1 inch of headroom; adjust lids. Pressure-process at 10 pounds (240 F.)—pints for 40 minutes, quarts for 50 minutes. Remove jars; complete seals if necessary.

In C-enamel cans. Fill loosely with drained hot beans, leaving only ½ inch of headroom. Add ½ teaspoon salt to No. 2 cans, 1 teaspoon salt to No. 2½. Fill to the top

with boiling-hot cooking water. Exhaust to 170 F.
(*c.* 10 minutes); seal. Pressure-process at 10 pounds
(240 F.)—40 minutes for either No. 2 or 2½. Remove
cans; cool quickly.

Beets

Beets keep well in a root cellar (q.v.). Between canning and freezing, can them: they can beautifully. Use
canned beets plain, titivated as a relish, in hash (see
Recipes).

GENERAL HANDLING

Only Pressure Canning for beets: they rank with home-
canned string beans as carriers of *Cl. botulinum* toxin.
Because they're firm-fleshed, use Hot pack only. Use jars
or R-enamel cans.

Sort for size; leave on tap root and 2 inches of stem
(otherwise they bleed out their juice before they get in
the containers). Wash carefully. Cover with boiling water
and boil until skins slip off easily (15 to 25 minutes,
depending on size/age). Drop them in cold water for just
long enough to be able to slip off skins; skin, trim away
roots, stems, any blemishes. Leave tiny beets whole; cut
larger ones in slices or dice. Now they are ready to pickle
or can.

HOT PACK ONLY

In jars. Fill with hot beets, leaving ½ inch of headroom.
Add ½ teaspoon salt to pints, 1 teaspoon salt to quarts.
Add fresh boiling water, leaving ½ inch of headroom;
adjust lids. Pressure-process at 10 pounds (240 F.)—
pints for 30 minutes, quarts for 35 minutes. Remove
jars; complete seals if necessary.

In R-enamel cans. Fill with hot beets, leaving only ¼ inch
of headroom. Add ½ teaspoon salt to No. 2 cans, 1
teaspoon salt to No. 2½. Cover to top with boiling
water. Exhaust to 170 F. (*c.* 10 minutes); seal. Pressure-process at 10 pounds (240 F.)—30 minutes for

either No. 2 or No. 2½ cans. Remove cans; cool quickly.

Beets, Pickled

Boiling-Water Bath (vinegar makes them so acid that a B-W Bath is quite O.K.). Use Hot pack only. Use jars only.

Wash; leave tap root and bit of stem. Boil until tender (how long, depends on size). Dunk in cold water to handle; trim, strip off skins, slice. While beets are cooking, make a Pickling Sirup of equal parts of vinegar and sugar (but ¼ of the vinegar may be replaced by water if you like a less-sharp pickle), and bring it to a boil.

Fill clean, hot jars with hot beet slices, leaving ½ inch of headroom. Add ½ teaspoon salt to pints, 1 teaspoon to quarts. Add boiling Pickling Sirup, leaving ½ inch of headroom; adjust lids. Process in a Boiling-Water Bath (212 F.)—30 minutes for either pint or quart jars. Remove jars; complete seals if necessary.

Broccoli

Broccoli and similarly strong-flavored vegetables—Brussels sprouts, cabbage (unless it's done as sauerkraut, and see also Root-Cellaring) and cauliflower—usually discolor when canned and grow even stronger in flavor. But here's how you can broccoli and the others if you have to.

GENERAL HANDLING

Pressure Canning only. Use Hot pack only. Use jars or C-enamel cans.

Wash all-green spears, trimming off leaves, any old blossoms, and woody parts of the stems. Soak in cold salt water (1 tablespoon salt to 1 quart water) for 10 minutes to drive out bugs, etc. Drain and wash in fresh water. Cut in 2-inch pieces, splitting thick stalks; or size as you like.

HOT PACK ONLY

Cover trimmed, clean, cut broccoli with boiling water and boil 3 minutes; drain.

In jars. Pack, leaving 1 inch of headroom. Add ½ teaspoon salt to pints, 1 teaspoon to quarts. Add boiling water, leaving 1 inch of headroom; adjust lids. Pressure-process at 10 pounds (240 F.)—pints for 30 minutes, quarts for 35 minutes. Remove jars; complete seals if necessary.

In C-enamel cans. Pack, leaving only ½ inch of headroom. Add ½ teaspoon salt to No. 2 cans, 1 teaspoon salt to No. 2½. Fill to top with boiling water. Exhaust to 170 F. (*c.* 10 minutes); seal. Pressure-process at 10 pounds (240 F.)—30 minutes for either No. 2 or No. 2½ cans. Remove cans; cool quickly.

Brussels Sprouts

These really should be frozen (they're hateful soggy and watery). But if canning, treat them like Broccoli, making a special point of the salt-water soak because they may have worms.

Cabbage

This really should be root-cellared. Or put by as Sauerkraut (q.v.).

If canned as is, see Broccoli, but pressure-process cans at 10 pounds (240 F.) for *40* minutes.

Carrots

Like beets, carrots can be harvested late in the season to can when weather is cooler. Don't bother with overlarge, woody ones. Fancy tiny carrots should be harvested early.

GENERAL HANDLING

Only Pressure Canning. Raw or Hot pack. In jars or C-enamel cans.

Sort for size. Wash, scrubbing well, and scrape. (An energetic scrub with your stiffest brush often will do for the very small ones; or parboil them just enough to loosen the skins, dunk them in cold water, slip off their skins, then use Hot pack.) Slice, dice, cut in strips—whatever.

RAW PACK

In jars. Fill tightly, leaving 1 inch of headroom. Add ½ teaspoon salt to pints, 1 teaspoon to quarts. Add boiling water, leaving ½ inch of headroom (water comes above the carrots); adjust lids. Pressure-process at 10 pounds (240 F.)—pints for 25 minutes, quarts for 30 minutes. Remove jars; complete seal if necessary.

In C-enamel cans. Fill tightly, leaving ½ inch of headroom. Add ½ teaspoon salt to No. 2 cans, 1 teaspoon to No. 2½. Add boiling water to top. Exhaust to 170 F. (*c.* 10 minutes); seal. Pressure-process at 10 pounds (240 F.)—No. 2 cans for 25 minutes, No. 2½ for 30 minutes. Remove cans; cool quickly.

PREFERRED: HOT PACK

Cover clean, scraped, cut or whole carrots with boiling water, bring again to a full boil; drain, but save the water to put in the jars for processing.

In jars. Pack hot carrots, leaving just ½ inch of headroom. Proceed as for Raw pack, using the cooking water for the added processing liquid.

In C-enamel cans. Pack, leaving only ¼ inch of headroom. Proceed as for Raw pack, using cooking water as the processing liquid, and reducing Pressure-process time— No. 2 cans for 20 minutes, No. 2½ for 25 minutes.

Cauliflower

Immeasurably better frozen; but can it like Broccoli.

Celery

Pressure Canning only. Use Hot pack only. Use jars or plain cans.

Wash thoroughly, trim off leaves (but if it's destined for stew, a few bits of chopped leaf are good flavor), cut stalks in 1-inch pieces. Cover with boiling water, boil 3 minutes. Drain, saving the cooking water.

HOT PACK ONLY

In jars. Fill with hot celery, leaving 1 inch of headroom. Add
$\frac{1}{2}$ teaspoon salt to pints, 1 teaspoon to quarts. Add boiling-hot cooking water, leaving 1 inch of headroom; adjust lids. Pressure-process at 10 pounds (240 F.)—pints
for 30 minutes, quarts for 35 minutes. Remove jars; complete seals if necessary.

In plain cans. Fill with hot celery, leaving $\frac{1}{2}$ inch of headroom. Add $\frac{1}{2}$ teaspoon salt to No. 2 cans, 1 teaspoon to
No. $2\frac{1}{2}$. Fill to top with boiling cooking water. Exhaust
to 170 F. (*c.* 10 minutes); seal. Pressure-process at 10
pounds (240 F.)—30 minutes for either No. 2 or No.
$2\frac{1}{2}$ cans. Remove cans; cool quickly.

Corn, Cream Style

Canning is the better, and certainly handier, way of putting by cream-style corn. Its density demands that it be
home-canned *only in pint jars or No. 2 cans:* an extremely
low-acid vegetable, it would be pressure-cooked to death for
the much longer time needed to process the interior of larger
containers.

GENERAL HANDLING

Pressure Canning only. Use Raw or Hot pack. Use pint
jars or No. 2 C-enamel cans.

Get it ready by husking, de-silking and washing the ears.
Slice the corn from the cob *halfway through the kernels,* then
scrap the milky juice that's left on the cob in with the cut
corn (this is where the "cream" comes in).

RAW PACK

In pint jars only. Fill with corn-cream mixture, leaving $1\frac{1}{2}$
inches of headroom (more space than usual is needed for
expansion). Add $\frac{1}{2}$ teaspoon salt. Add boiling water,
leaving $\frac{1}{2}$ inch of headroom (water will be well over
the top of the corn); adjust lids. Pressure-process at 10
pounds (240 F.) for 95 minutes. Remove jars; complete
seals if necessary.

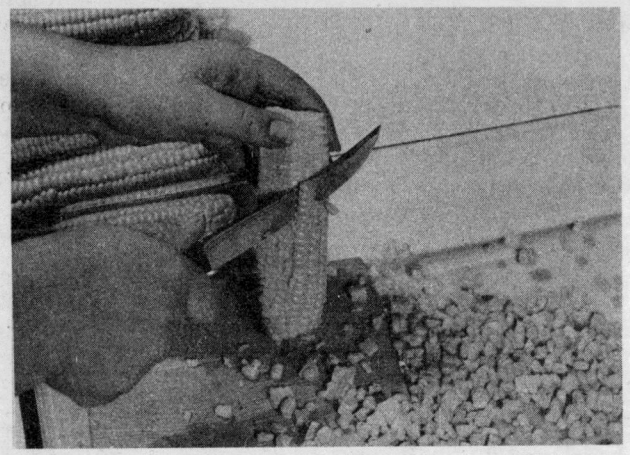

In No. 2 C-enamel cans only. Fill without shaking or pressing down, leaving ½ inch of headroom. Add ½ teaspoon salt. Fill to top with boiling water. Exhaust to 170
F. (*c.* 25 minutes); seal. Pressure-process at 10 pounds
(240 F.) for 105 minutes. Remove cans; cool quickly.

PREFERRED: HOT PACK

Prepare as for Raw pack. To each 4 cups of corn-cream
mixture, add 2 cups boiling water. Heat to boiling, stirring,
over medium heat (it scorches easily).

In pint jars only. Fill with boiling corn and liquid, leaving
1 inch of headroom. Add ½ teaspoon salt (but you don't
need more water); adjust lids. Pressure-process at 10
pounds (240 F.) for 85 minutes. Remove jars; complete
seals if necessary.

In No. 2 C-enamel cans only. Fill to the top with boiling corn
and liquid. Add ½ teaspoon salt (but you don't need
more water). Exhaust to 170 F. (*c.* 10 minutes); seal.
Pressure-process at 10 pounds (240 F.) for 105 minutes.
Remove cans; cool quickly.

Corn, Whole Kernel

Pressure Canning only. Use Raw or Hot pack. Use jars
or C-enamel cans. Less dense than cream-style corn, it may

be canned in quarts and No. 2½ cans just as well as in pints and No. 2 cans.

Husk, de-silk and wash fresh-picked ears. Cut from the cob at about ⅔ the depth of the kernels (this is deeper than for cream-style, but still avoids getting bits of cob).

RAW PACK

In jars. Fill, leaving 1 inch of headroom—and don't shake or press down. Add ½ teaspoon salt to pints, 1 teaspoon salt to quarts. Add boiling water, leaving ½ inch of headroom (water will come well over top of corn); adjust lids. Pressure-process at 10 pounds (240 F.)—pints for 55 minutes, quarts for 85 minutes. Remove jars; complete seals if necessary.

In C-enamel cans. Fill, leaving ½ inch of headroom—and don't shake or press down. Add ½ teaspoon salt to No. 2 cans, 1 teaspoon salt to No. 2½. Add boiling water to top. Exhaust to 170 F. (*c.* 10 minutes); seal. Pressure-process at 10 pounds (240 F.)—60 minutes for either No. 2 or No. 2½ cans. Remove cans; cool quickly.

PREFERRED: HOT PACK

Prepare as for Raw pack. To each 4 cups of kernels, add 2 cups boiling water. Bring to boiling over medium heat, stirring so it won't scorch. Drain, saving the hot liquid.

In jars. Fill with kernels, leaving 1 inch of headroom. Add ½ teaspoon salt to pints, 1 teaspoon salt to quarts. Add boiling-hot cooking liquid, leaving 1 inch of headroom; adjust lids. Pressure-process at 10 pounds (240 F.)—pints for 55 minutes, quarts for 85 minutes. Remove jars; complete seals if necessary.

In C-enamel cans. Fill with hot kernels, leaving ½ inch of headroom. Add ½ teaspoon salt to No. 2 cans, 1 teaspoon salt to No. 2½. Add boiling-hot cooking liquid to top (water will come well over top of corn). Exhaust to 170 F. (*c.* 10 minutes); seal. Pressure-process at 10 pounds (240 F.)—60 minutes for either No. 2 or No. 2½ cans. Remove cans; cool quickly.

Before tasting canned low-acid food, boil it hard for 15 minutes to destroy any hidden toxins (corn and greens require 20 minutes). If it foams unduly or smells bad during boiling, destroy it completely so it can't be eaten by people or animals.

Eggplant

This loses its looks when it's canned, but some cooks feel that it makes an unhandy product if frozen alone (better to precook it as a favorite casserole, etc., they say, and freeze it as a convenience food).

GENERAL HANDLING

Pressure Canning only. Use Hot pack only. Use jars or plain cans.

Wash, pare and slice or cube eggplant. Sprinkle lightly with salt and cover with cold water (to help draw out its juice). Let soak 45 minutes; drain. In fresh water, boil for 5 minutes. Drain, and pack hot without adding salt.

In jars. Fill clean hot jars, leaving 1 inch of headroom. Add boiling water, leaving 1 inch of headroom; adjust lids. Pressure-process at 10 pounds (240 F.)—pints for 30 minutes, quarts for 40 minutes. Remove jars; complete seals if necessary.

In plain cans. Pack, leaving only ¼ inch of headroom. Fill to top with boiling water. Exhaust to 170 F. (*c.* 10 minutes); seal. Pressure-process at 10 pounds (240 F.)—No. 2 cans for 35 minutes, No. 2½ for 40 minutes. Remove cans; cool quickly.

Greens—Spinach, etc., and Wild

All garden greens—spinach, chard, turnip or beet tops —can nicely; so do wild ones like dandelions and milkweed (fiddleheads and cowslips are usually such treats that they're eaten as they come in). Greens freeze well, and with more garden freshness.

GENERAL HANDLING

Pressure Canning only. Use Hot pack only (to make greens solid enough in the container). Use jars or plain cans.

Using spinach as the example *for garden greens:* Remove bits of grass, poor leaves, etc., from just-picked leaves; cut out tough stems and coarse midribs. Wash thoroughly, lifting from the water to let any sediment settle. Put about 2½ pounds of clean, wet leaves in a large cheesecloth, tie the top, and steam the spinach for about 10 minutes—or until well wilted, and pack.

Prepare *wild greens*—dandelions, milkweed and fiddleheads—according to directions given for the individual greens in Recipes.

HOT PACK ONLY

In jars. Fill with greens, leaving ½ inch of headroom. Add only ¼ teaspoon salt to pints, ½ teaspoon salt to quarts. Add boiling water, leaving ½ inch of headroom; adjust lids. Pressure-process at 10 pounds (240 F.)—pints for 70 minutes, quarts for 90 minutes. Remove jars; complete seals if necessary.

In plain cans. Fill with greens, leaving only ¼ inch of headroom. Add only ¼ teaspoon of salt to No. 2 cans, ½ teaspoon salt to No. 2½. Cover to top with boiling water. Exhaust to 170 F. (*c.* 10 minutes); seal. Pressure-process at 10 pounds (240 F.)—No. 2 cans for 65 minutes, No. 2½ for 75 minutes. Remove cans; cool quickly.

Hominy (Lye-hulled Corn)

This traditional Southern vegetable is made from dried whole-kernel field corn, not from precious home-parched sweet corn (see Drying), after the hulls are removed by long cooking in a weak lye solution.

Alkali—lye, and washing or baking sodas, its much milder cousins—is a quick destroyer of Vitamin C and many of the B vitamins.

GENERAL HANDLING

Pressure Canning only. Use Hot pack only. Use jars or C-enamel cans. Plan on the rough ratio of 1 cup of dried corn making up to 4 cups of canned hominy; this swelling must take place during precooking, before it is packed.

WARNINGS ABOUT LYE

The stuff called "lye" may be any one of several highly caustic alkaline compounds that, in the presence of only the moisture in the air on a muggy day, can become activated, burning and eating deeply into animal or other organic tissue—including human skin.

ANTIDOTE FOR SEARING CONTACT: slosh immediately with cold water, follow with boric-acid solution (eyes) or vinegar.

If you buy household lye/caustic soda for hulling corn, *make sure it's suitable for use with food,* and is designated as "lye" or "lycons" on the can and contains no aluminum, nitrates or stabilizers. Above all, avoid commercial drain-openers, either crystalline or liquid. (See Soapmaking for more details.)

Use only enameled- or granite-ware pots or kettles —*never use utensils of aluminum,* which reacts violently with lye in water.

Hulling and precooking. In an enameled kettle, dissolve 4 tablespoons of suitable household lye in 8 quarts of water; add 8 cups of dried field corn. Boil 30 minutes; let stand off heat 20 minutes more. Drain; wash off lye with several hot-water rinses; cool by rinsing in cold water. Work off the hulls and dark tips of the kernels by rubbing the hominy or washing it vigorously in a colander. When hulls and tips are removed, boil it in fresh water to cover for 5 minutes, drain; *repeat four times* (totaling 25 minutes of boiling in five fresh waters). After last repeat, cook in fresh water until kernels are soft—about 45 minutes. Drain and pack hot.

Hulling with homemade lye. When the liquid that has leached through a barrel of ashes (see Soapmaking) is strong enough to float an egg, put it in an enameled kettle with dried corn in the proportion of 4 parts liquid to 1 part corn. Boil until the hulls can be worked and rinsed off. Proceed with rinses, etc., as above.

HOT PACK ONLY

In jars. Fill, leaving ½ inch of headroom. Add ½ teaspoon salt to pints, 1 teaspoon salt to quarts. Add boiling water, leaving ½ inch of headroom; adjust lids. Pressure-process at 10 pounds (240 F.)—pints for 60 minutes, quarts for 70 minutes. Remove jars; complete seals if necessary.

In C-enamel cans. Fill, leaving only ¼ inch of headroom. Add ½ teaspoon salt to No. 2 cans, 1 teaspoon salt to No. 2½. Fill to top with boiling water. Exhaust to 170 F. (c. 10 minutes); seal. Pressure-process at 10 pounds (240 F.)—No. 2 cans for 60 minutes, No. 2½ for 70 minutes. Remove cans; cool quickly.

Mixed Vegetables (in General)

Pressure Canning only. Hot packs only. Use jars or plain or C-enamel cans (for which type of can, see "About Jars and Cans" at the beginning of this Chapter 2).

Rule of thumb for processing: Choose the time required by the single ingredient requiring the longest processing (viz. Stewed Tomatoes with Added Vegetables).

Wash, trim vegetables, peeling if necessary. Cut to uniform size. Cover with boiling water and boil 10 minutes. Drain; save the cooking water to use for processing.

HOT PACK ONLY

In jars. Fill with hot mixed vegetables, leaving ½ inch of headroom. Add ½ teaspoon salt to pints, 1 teaspoon salt to quarts. Add boiling water (fresh, or the cooking water), leaving ½ inch of headroom; adjust lids. Pressure-process at 10 pounds (240 F.)—pints for 60 minutes, quarts for 70 minutes. Remove jars; complete seals if necessary.

In plain or C-enamel cans. Fill with hot mixed vegetables, leaving ½ inch of headroom. Add ½ teaspoon of salt to No. 2 cans, 1 teaspoon of salt to No. 2½ cans. Fill to the top with boiling water (fresh, or the cooking water). Exhaust to 170 F. (c. 10 minutes); seal. Pressure-process at 10 pounds (240 F.)—No. 2 cans for 60 minutes, No. 2½ for 70 minutes. Remove cans; cool quickly.

Mixed Corn and Beans (Succotash)

Pressure Canning only. Use Hot pack only. Use jars or C-enamel cans.

Boil freshly picked ears of corn for 5 minutes; cut kernels from cobs (as for whole-kernel corn, *without* scraping in the milk). Prepare fresh lima beans or green/snap beans, and boil by themselves for 3 minutes. Measure and mix hot corn with ½ to an equal amount of beans.

HOT PACK ONLY

In jars. Fill with hot corn-and-bean mixture, leaving 1 inch of headroom. Add ½ teaspoon salt to pints, 1 teaspoon salt to quarts. Add boiling water, leaving 1 inch of headroom; adjust lids. Pressure-process at 10 pounds (240 F.) —pints for 60 minutes, quarts for 85 minutes. Remove jars; complete seals if necessary.

In C-enamel cans. Fill with hot mixture, leaving ½ inch of headroom. Add ½ teaspoon salt to No. 2 cans, 1 teaspoon salt to No. 2½ cans. Fill to the top with boiling water. Exhaust to 170 F. (*c.* 10 minutes); seal. Pressure-process at 10 pounds (240 F.)—No. 2 cans for 70 minutes, No. 2½ for 95 minutes. Remove cans; cool quickly.

Mushrooms

Canned mushrooms have a bad track record as carriers of *Cl. botulinum* toxin, so use only fresh edible mushrooms— preferably those grown in pre-sterilized soil.

GENERAL HANDLING

Pressure Canning only. Use Hot pack only. Use ½-pint or pint jars, plain No. 2 cans.

Soak them in cold water for 10 minutes to loosen field dirt, then wash well. Trim blemishes from caps and stems.

Leave small buttons whole; cut larger ones in button-size pieces. In a covered saucepan simmer them gently for 15 minutes.

HOT PACK ONLY

In jars. Fill with hot mushrooms, leaving ½ inch of headroom. Add ¼ teaspoon of salt to ½-pint jars, ½ teaspoon salt to pints. To prevent color change, add 1/16 teaspoon of crystalline ascorbic acid to ½-pint jars, 1/8 teaspoon to pints. Add boiling water, leaving ½ inch of headroom; adjust lids. Pressure-process at 10 pounds (240 F.)— 30 minutes for either ½-pints or pints. Remove jars; complete seals if necessary.

In plain No. 2 cans. Fill with hot mushrooms, leaving ½ inch of headroom. Add ½ teaspoon salt and 1/8 teaspoon crystalline ascorbic acid. Fill to top with boiling water. Exhaust to 170 F. (*c.* 10 minutes); seal. Pressure-process at 10 pounds (240 F.) for 30 minutes. Remove cans; cool quickly.

Nut Meats

Nuts are an important source of protein for vegetarian main dishes, and are a popular touch in baked goods, candies, salads, etc. All nuts are rather fatty, and it's this fat that turns rancid and spoils the meats (even nuts in the shell can spoil after a while). Freeze them too.

GENERAL HANDLING

Pressure Canning or Boiling-Water Bath. Use *dry* Hot pack only (nut meats are oven-dried before canning). Use dry, sterilized jars no larger than pints, and with *self-sealing lids* (see "About Jars and Cans").

Spread a shallow layer of nut meats in baking pans, and bake in a very slow oven—not more than 275 F.—watching the nuts carefully and stirring once in a while, until they are dry but not browned: they must not scorch. Keep hot for packing.

HOT PACK ONLY

In self-sealing jars. For pints (or ½-pints), fill dry, sterilized jars, leaving ½ inch of headroom; adjust lids. Pressure-process at 5 pounds (228 F.)—10 minutes for ½-pints or pints. OR process in a Boiling-Water Bath—*but with the water level well below the tops of the jars*—for 20 minutes. Remove jars; remove bands from self-sealing lids after jars have cooled 12 hours.

Okra (Gumbo)

This vegetable popular in Southern, Creole and West Indian cooking cans and freezes equally well: if you plan to use it cut up in soups and stews, it's probably handier canned.

GENERAL HANDLING

Pressure Canning only. Use Hot pack only. Use jars or plain cans.

Wash tender young pods; trim stems but don't cut off caps. Cover with boiling water and boil 1 minute; drain. Leave whole with cap, or cut in 1-inch pieces, discarding cap.

HOT PACK ONLY

In jars. Fill with hot okra, leaving ½ inch of headroom. Add ½ teaspoon salt to pints, 1 teaspoon salt to quarts. Add boiling water, leaving ½ inch of headroom; adjust lids. Pressure-process at 10 pounds (240 F.)—pints for 25 minutes, quarts for 40 minutes. Remove jars; complete seals if necessary.

In plain cans. Fill with hot okra, leaving only ¼ inch of headroom. Add ½ teaspoon salt to No. 2 cans, 1 teaspoon salt to No. 2½ cans. Fill to top with boiling water. Exhaust to 170 F. (*c.* 10 minutes); seal. Pressure-process at 10 pounds (240 F.)—No. 2 cans for 25 minutes, No. 2½ for 35 minutes. Remove cans; cool quickly.

Onions, White

Onions that are properly cured and stored (see Drying) carry over so well that many cooks don't bother to can them —on top of which home-canned onions are apt to be dark in color and soft in texture. They are often canned in combinations (see Mixed Vegetables for rule-of-thumb).

GENERAL HANDLING

Pressure Canning only. Use Hot pack only. Use jars or C-enamel cans.

Sort for uniform size—1 inch in diameter is ideal—and wash. Peel, trimming off roots and stalks. (If you push a hole downward through the middle with a slender finishing nail, their centers will cook with less chance of shucking off outer layers.) Cover with boiling water, parboil gently for 5 minutes. Drain, saving the cooking water for processing.

HOT PACK ONLY

In jars. Fit whole onions in closely, leaving ½ inch of headroom. Add ½ teaspoon salt to pints, 1 teaspoon salt to quarts. Add boiling cooking liquid, leaving ½ inch of headroom; adjust lids. Pressure-process at 10 pounds (240 F.)—pints for 25 minutes, quarts for 30 minutes. Remove jars; complete seals if necessary.

In C-enamel cans. Fit whole onions closely, leaving ¼ inch of headroom. Add ½ teaspoon salt to No. 2 cans, 1 teaspoon to No. 2½ cans. Fill to brim with boiling cooking water. Exhaust to 170 F. (*c.* 10 minutes); seal. Pressure-process at 10 pounds (240 F.)—No. 2 cans for 25 minutes, No. 2½ for 30 minutes. Remove cans; cool quickly.

Parsnips

This is probably the only vegetable that actually improves by wintering over in frozen ground—so why take the shine off it as the "first of spring" treat?

But if you can't keep them in a garden or a root cellar: wash, trim, scrape, and cut them in pieces, then proceed as for Broccoli (q.v.).

Peas, Black-eyed (Cowpeas, Black-eyed Beans)

Pressure Canning only. Use Raw or Hot pack. Use jars or C-enamel cans.

Shell and wash before packing. Take care *not* to shake or press down the peas when you pack the containers: they swell in the containers.

RAW PACK

In jars. Fill with raw peas, leaving 1½ inches of headroom in pints, 2 inches of headroom in quarts. Add ½ teaspoon salt to pints, 1 teaspoon salt to quarts. Add boiling water, leaving ½ inch of headroom (water will come well over the top of the peas); adjust lids. Pressure-process at 10 pounds (240 F.)—pints for 35 minutes, quarts for 40 minutes. Remove jars; complete seals if necessary.

In C-enamel cans. Fill with raw peas, leaving ¾ inch of headroom. Add ½ teaspoon salt to No. 2 cans, 1 teaspoon to No. 2½ cans. Add boiling water, leaving ¼ inch of headroom. Exhaust to 170 F. (*c.* 10 minutes); seal. Pressure-process at 10 pounds (240 F.)—No. 2 cans for 35 minutes, No. 2½ for 40 minutes. Remove cans; cool quickly.

PREFERRED: HOT PACK

After shelling and washing, cover with boiling water, bring to a full, high boil. Drain, saving the blanching water for processing.

In jars. Pack hot, leaving 1¼ inches of headroom in pints, 1½ inches of headroom in quarts. Add ½ teaspoon salt to pints, 1 teaspoon to quarts. Add boiling water (or blanching liquid), leaving ½ inch of headroom in either size of jar; adjust lids. Pressure-process at 10 pounds (240 F.)—pints for 35 minutes, quarts for 40 minutes. Remove jars; complete seals if necessary.

In C-enamel cans. Pack hot, leaving ½ inch of headroom for either size of can. Add ½ teaspoon salt to No. 2 cans, 1 teaspoon salt to No. 2½ cans. Add boiling water, leaving ¼ inch of headroom. Exhaust to 170 F. (*c.* 10 min-

utes); seal. Pressure-process at 10 pounds (240 F.)—
No. 2 cans for 30 minutes, No. 2½ for 35 minutes. Remove cans; cool quickly.

Peas, Green

Pressure Canning only. Use Raw or Hot pack. Use jars
or plain cans.

Shell and wash peas. In packing, take care not to shake
or press the peas down: they swell.

RAW PACK

In jars. Fill, leaving 1 inch of headroom. Add ½ teaspoon
salt to pints, 1 teaspoon to quarts. Add boiling water to
within 1½ inches of the top of the jar (water will come
well below the top of the peas); adjust lids. Pressure-
process at 10 pounds (240 F.)—40 minutes for both
pints and quarts. Remove jars; complete seal if necessary.

In plain cans. Fill, leaving only ¼ inch of headroom. Add ½
teaspoon salt to No. 2 cans, 1 teaspoon to No. 2½ cans.
Fill to top with boiling water. Exhaust to 170 F. (*c.* 10
minutes); seal. Pressure-process at 10 pounds (240 F.)
—No. 2 cans for 30 minutes, No. 2½ for 35 minutes.
Remove cans; cool quickly.

PREFERRED: HOT PACK

Cover washed peas with boiling water, bring to a full
boil. Drain, saving the blanching water to process with if
you like.

In jars. Fill loosely with hot peas, leaving 1 inch of headroom.
Add ½ teaspoon salt to pints. 1 teaspoon to quarts. Add
boiling water, leaving 1 inch of headroom; adjust lids.
Pressure-process at 10 pounds (240 F.)—40 minutes for
either pints or quarts. Remove jars; complete seals if
necessary.

In plain cans. Fill loosely with hot peas, leaving only ¼ inch
of headroom. Add ½ teaspoon salt to No. 2 cans, 1 teaspoon to No. 2½. Fill to the brim with boiling water.
Exhaust to 170 F. (*c.* 10 minutes); seal. Pressure-process

at 10 pounds (240 F.)—No. 2 cans for 30 minutes, No. 2½ for 35 minutes. Remove cans; cool quickly.

Peppers, Green (Bell, Sweet)

Pressure Canning only. Hot pack only. Use jars or plain cans.

Wash, remove stems, cores and seeds. Cut in large pieces or leave whole. Put in boiling water and boil 3 minutes. Drain and pack. (If you like them peeled, take them from the boiling water, dunk in cold water to cool just enough for handling, and strip off the skins; pack.)

Acid is added to green peppers in order to can them *safely* within a reasonable processing time (as in Tomatoes, and see Chapter 1).

HOT PACK ONLY

In jars. Pack hot peppers flat, leaving 1 inch of headroom. Add ½ teaspoon salt and 1 tablespoon white vinegar to pints, 1 teaspoon salt and 2 tablespoons vinegar to quarts. Add boiling water, leaving ½ inch of headroom (water comes over the top of the peppers); adjust lids. Pressure-process at 10 pounds (240 F.)—pints for 35 minutes, quarts for 45 minutes. Remove jars; complete seals if necessary.

In plain cans. Pack flat leaving ½ inch of headroom. Add ½ teaspoon salt and 1 tablespoon white vinegar to No. 2 cans, 1 teaspoon salt and 2 tablespoons vinegar to No. 2½ cans. Fill to top with boiling water. Exhaust to 170 F. (*c.* 10 minutes); seal. Pressure-process at 10 pounds (240 F.)—No. 2 cans for 20 minutes, No. 2½ for 25 minutes. Remove cans; cool quickly.

Pimientos

Pressure Canning only. Use Hot pack only. Use ½-pint or pint jars, or No. 2 plain cans.

Wash, cover with boiling water, and simmer until skins can be peeled off—4 to 5 minutes. Dunk in cold water so they can be handled, trim stems, blossom ends, and skin them as for Green Peppers, above; pack hot, adding acid for safety as indicated.

HOT PACK ONLY

In jars. Pack flat in clean, hot ½-pint or pint jars, leaving ½ inch of headroom. Add ¼ teaspoon salt and 1½ teaspoons white vinegar to ½-pints, ½ teaspoon salt and 1 tablespoon vinegar to pints. Add *no water;* adjust lids. Pressure-process at 10 pounds (240 F.)—20 minutes for either ½-pints or pints. Remove jars; complete seals if necessary.

In No. 2 plain cans. Pack flat, leaving only 1/8 inch of headroom. Add ½ teaspoon salt and 1 tablespoon vinegar, but *no water.* Exhaust to 170 F. (c. 10 minutes); seal. Pressure-process at 10 pounds (240 F.) for 20 minutes. Remove cans; cool quickly.

Potatoes, Sweet (and Yams)

Sweet potatoes don't keep as well in open storage as white potatoes do, and it makes sense to can some of them to serve hot with butter, or mashed, or glazed.

GENERAL HANDLING

Pressure Canning only. Hot pack only—but either *wet* or *dry* (dry-packed glaze well). Use jars or plain cans.

Sort for size, wash; boil or steam until only half cooked and skins come off easily—20 minutes or so. Dunk in cold water so they can be handled, slip off skins and trim away any blemishes. If they're large, cut them in pieces.

HOT PACK—DRY

In jars. Fill tightly, leaving 1 inch of headroom—pack more or less upright, pressing gently to fill spaces (but not cramming them in a shapeless mush). *Add nothing to them,* not even salt; adjust lids. Pressure-process at 10 pounds (240 F.)—pints for 65 minutes, quarts for 95 minutes. Remove jars; complete seals if necessary.

In plain cans. Fill tightly as for jars, but to the top, leaving no
headroom. *Add nothing,* not even salt. Exhaust to 170
F. (*c.* 10 minutes); seal. Pressure-process at 10 pounds
(240 F.)—No. 2 cans for 80 minutes, No. 2½ for 95
minutes. Remove cans; cool quickly.

HOT PACK—WET

Prepare as in General Handling (parboiling, skinning,
etc.).

In jars. Fill more loosely than for dry-packed, leaving 1 inch
of headroom. Add ½ teaspoon salt to pints, 1 teaspoon
salt to quarts. Add boiling water or Medium Sirup (see
"Sirups for Canning Fruits"); adjust lids. Pressure-
process at 10 pounds (240 F.)—pints for 55 minutes,
quarts for 90 minutes. Remove jars; complete seals if
necessary.

In plain cans. Fill more loosely than for dry-packed, leaving
only ¼ inch of headroom. Add ½ teaspoon salt to No. 2
cans, 1 teaspoon to No. 2½. Fill to top with boiling water
or Medium Sirup. Exhaust to 170 F. (*c.* 10 minutes);
seal. Pressure-process at 10 pounds (240 F.)—No. 2
cans for 70 minutes, No. 2½ cans for 90 minutes. Remove cans; cool quickly.

Potatoes, White ("Irish")

These potatoes don't home-freeze at all well (unless
they're partially precooked in a combination dish, q.v.)—
so cold-store them (see Root-Cellaring). But it's possible for
them to be too immature to store without spoiling—so you
can them. Delicate tiny, new potatoes can well, and are
good served hot with parsley butter or creamed.

GENERAL HANDLING

Pressure Canning only. Use Raw or Hot pack. Use jars
or plain cans. If around 1 to 1½ inches in diameter, they may
be canned whole; dice the larger ones.

Wash and scrape just-dug new potatoes, removing all
blemishes. (If you're dicing them, prevent darkening during

preparation by dropping the dice in a solution of 1 teaspoon salt for each 1 quart of cold water.) Drain before packing by either method.

RAW PACK

In this pack, *diced get 5 minutes longer processing time,* because they're more dense in the containers than whole ones are.

In jars. Fill with whole or diced potatoes, leaving ½ inch of headroom; add ½ teaspoon salt to pints, 1 teaspoon salt to quarts. Add boiling water, leaving ½ inch of headroom; adjust lids. Pressure-process at 10 pounds (240 F.)—*whole,* pints or quarts for 40 minutes; *diced,* pints or quarts for 45 minutes. Remove jars; complete seals if necessary.

In plain cans. Fill, leaving only ½ inch of headroom. Add ½ teaspoon salt to No. 2 cans, 1 teaspoon salt to No. 2½. Fill to the top with boiling water. Exhaust to 170 F. (*c.* 10 minutes); seal. Pressure-process at 10 pounds (240 F.)—*whole,* No. 2 cans for 40 minutes, No. 2½ for 45 minutes; *diced,* No. 2 for 45 minutes, No. 2½ for 50 minutes. Remove cans; cool quickly.

PREFERRED: HOT PACK

Cover clean, scraped *whole* potatoes with boiling water, boil 10 minutes; drain. Drain anti-discoloration solution off *diced* potatoes; cover them with boiling fresh water, boil 2 minutes; drain.

In jars. With whole or diced potatoes, leave ½ inch of headroom. Add ½ teaspoon salt to pints, 1 teaspoon to quarts. Add boiling water, leaving ½ inch of headroom; adjust lids. Pressure-process at 10 pounds (240 F.)—*whole,* pints for 30 minutes, quarts for 40 minutes; *diced,* pints for 35 minutes, quarts for 40 minutes. Remove jars; complete seals if necessary.

In plain cans. Fill with whole or diced potatoes, leaving only ¼ inch of headroom. Add ½ teaspoon salt to No. 2 cans, 1 teaspoon to No. 2½ cans. Fill to top with boiling water. Exhaust to 170 F. (*c.* 10 minutes); seal. Pressure-process at 10 pounds (240 F.)—*whole* or *diced*—No. 2 cans for

35 minutes, No. 2½ for 40 minutes. Remove cans; cool quickly.

Pumpkin (and Winter Squash)

Pressure Canning only. Hot pack only. Use jars or R-enamel cans. Cube it or strain it (strained takes longer to process).

Dry-fleshed pumpkin (or winter squash) is best for canning: test it with your thumbnail—it's dry enough if your nail won't cut the surface skin easily.

HOT PACK—CUBED

Wash, cut in manageable hunks, pare and remove seeds; cut in 1-inch cubes. Cover with water and bring to boiling. Drain, reserving hot liquid, and pack hot.

In jars. Fill with hot cubes, leaving ½ inch of headroom. Add ½ teaspoon salt to pints, 1 teaspoon to quarts. Add boiling cooking water, leaving ½ inch of headroom; adjust lids. Pressure-process at 10 pounds (240 F.)—pints for 55 minutes, quarts for 90 minutes. Remove jars; complete seals if necessary.

In R-enamel cans. Fill, leaving only ¼ inch of headroom. Add ½ teaspoon salt to No. 2 cans, 1 teaspoon to No. 2½. Fill to brim with boiling cooking water. Exhaust to 170 F. (*c.* 10 minutes); seal. Pressure-process at 10 pounds (240 F.)—No. 2 cans for 50 minutes, No. 2½ for 75 minutes. Remove cans; cool quickly.

HOT PACK—STRAINED

Prepare as for cubes, but leave in larger pieces if more convenient. Steam or bake until tender; put through a strainer or food mill. Over low heat, simmer until it's heated through, stirring steadily so it won't scorch. Pack hot *with no added liquid or salt.*

In jars. Fill, leaving ½ inch of headroom; adjust lids. Pressure-process at 10 pounds (240 F.)—pints for 65 minutes, quarts for 80 minutes. Remove jars; complete seals if necessary.

In R-enamel cans. Fill, leaving only 1/8 inch of headroom.
Exhaust to 170 F. (*c.* 10 minutes); seal. Pressure-process
at 10 pounds (240 F.)—No. 2 cans for 75 minutes, No.
2½ for 90 minutes. Remove cans; cool quickly.

Rutabagas (and White Turnips)

Root-cellaring is best (q.v.), but rutabagas are good to
ferment like Sauerkraut (see how, in the Curing chapter)—
in which case they are canned like Sauerkraut, below.

However, if you do want *to can them fresh:* Wash, peel,
cube; pack in jars/C-enamel cans, and Pressure-process as
for Broccoli (q.v.).

Salsify (Oyster Plant)

Like parsnips and horseradish, this delicately flavored,
old-fashioned vegetable is able to winter in the ground, and
it may be root-cellared. With a little extra attention so it
won't discolor, it cans well. See Recipes.

GENERAL HANDLING

Pressure Canning only. Use Hot pack only. Use jars or
C-enamel cans.

Its milky juice turns rather rusty when it hits the air, and
you may prevent discoloration of the vegetable by either one
of two ways: (1) Scrub roots well, scrape as for carrots, and
slice, dropping each slice immediately in a solution of 2
tablespoons vinegar and 2 tablespoons salt for each 1 gallon
of cold water; rinse well, cover quickly with boiling water,
boil for 2 minutes; pack hot. Or (2) scrub roots; in a solution
of 1 tablespoon vinegar to each 1 quart of water, boil whole
until skins come off easily—10 to 15 minutes; rinse well in
cold water, skin, leave whole or slice; pack hot.

HOT PACK ONLY

In jars. Fill, leaving 1 inch of headroom. Add fresh boiling
water, leaving 1 inch of headroom; adjust lids. Pressure-
process at 10 pounds (240 F).—pints for 30 minutes,

quarts for 35 minutes. Remove jars; complete seals if necessary.

In C-enamel cans. Fill, leaving ½ inch of headroom; add fresh boiling water to brim. Exhaust to 170 F. (*c.* 10 minutes); seal. Pressure-process at 10 pounds (240 F.) —30 minutes for both No. 2 and No. 2½ cans. Remove cans; cool quickly.

Sauerkraut (Fermented Cabbage)

By all means can your sauerkraut—unless you are able to guarantee cool enough storage for the crock after it is fermented.

GENERAL HANDLING

Boiling-Water Bath processing is adequate for a food as acid as this. Hot pack only. Use jars or R-enamel cans.

HOT PACK ONLY

Heat to simmering—180 to 210 F.—*do not boil,* and pack as directed below. If you don't have enough sauerkraut juice, eke it out with a brine made of 1½ tablespoons salt for each 1 quart of water.

In jars. Fill clean, hot jars, leaving ½ inch of headroom. Add hot juice (or hot brine, above), leaving ½ inch of headroom; adjust lids. Process in a Boiling-Water Bath—pints for 15 minutes, quarts for 20 minutes. Remove jars; complete seals if necessary.

In R-enamel cans. Pack hot, leaving only ½ inch of headroom. Fill to the brim with hot juice or brine (again, as above). Exhaust to 170 F. (*c.* 10 minutes); seal. Process in a Boiling-Water Bath—25 minutes for either No. 2 or No. 2½ can. Remove cans; cool quickly.

Soybeans

Although not particularly interesting by themselves, soybeans are a splended *natural* high-protein addition—aside from being an economical "stretcher"—for ground meats, stews, chowders, casseroles, etc. They are good keepers when dried and, with the right handling, they produce wonderfully nutritious sprouts to eat (see "Seeds and Sprouting" and the end of the Drying section), and they also freeze well. Even with these options, though, it could make sense to can some soybeans, provided that they're available or a good buy.

GENERAL HANDLING

Use shelled, fully developed but still tender, *green* soybeans (old, light-colored ones are better dried, after which they're handled like any dried bean for baking, or whatever).

Pressure Canning only. Hot pack only. Use jars or C-enamel cans. Pack loosely, allowing the extra amount of headroom because they swell.

HOT PACK ONLY

Prepare and process like Lima Beans (q.v.), except *to increase the processing time* at 10 pounds (240 F.) to—*55 minutes* for pint jars, *65 minutes* for quart jars; *50 minutes* for No. 2 cans, *60 minutes* for No. 2½ cans.

Spinach, see Greens

Sprouts

Since the reasons-for-being of sprouts are their freshness, their availability on short notice throughout the year, and their wide variety of uses, it's unlikely that you'd have any around to can. Still, in case crop mismanagement or a friend's largess has given you a surplus that can't be handled any other way, here's what to do.

Sprouts of legumes—peas and beans—seem to lend themselves better than other kinds to the following procedure, which we offer for relative safety: the sprouts will be oven-dried (roasted, actually) and then canned dry. We chose this method because research has not yet established the pressure-processing times needed for canning any raw sprouts at home, or whether we should add acid as commercial canners do: *and this is no area for the layperson to improvise in.*

GENERAL HANDLING

Pick over the sprouts, discarding damaged or spoiled ones; save the husks. Wash thoroughly, being finicky as all get-out if the sprouts have been grown on soil; drain and pat dry between layers of clean paper toweling. Oven-dry in a single layer on a cookie sheet for 1 to 3 hours at 325 F., stirring occasionally. Keep hot.

Pressure Canning only. Hot pack only. Only pint or ½-pint jars with 2-piece self-sealing screwband lids.

Hot pack only, in jars. Fill clean, hot jars with hot, dried sprouts, leaving ½ inch of headroom; *add no liquid.* Put on lids, screwing bands firmly tight. Process at *5 pounds* (228 F.)—10 minutes for either pints or ½-pints. Remove jars; cool slowly.

Squash—Chayote, Summer, Zucchini

Pressure Canning only. Use Raw or Hot pack. Use jars or plain cans.

Wash, trim ends but do not peel. Cut in ½-inch slices; halve or quarter the slices to make the pieces uniform.

RAW PACK

In jars. Pack tightly in clean, hot jars, leaving 1 inch of headroom. Add ½ teaspoon salt to pints, 1 teaspoon to quarts. Add boiling water, leaving ½ inch of headroom (water will come over the top of the squash); adjust lids. Pressure-process at 10 pounds (240 F.)—pints for 25 minutes, quarts for 30 minutes. Remove jars; complete seals if necessary.

In plain cans. Pack tightly, leaving ½ inch of headroom. Add ½ teaspoon salt to No. 2 cans, 1 teaspoon to No. 2½. Fill to the brim with boiling water. Exhaust to 170 F. (*c.* 10 minutes); seal. Pressure-process at 10 pounds (240 F.)—20 minutes for either No. 2 or No. 2½ cans. Remove cans; cool quickly.

PREFERRED: HOT PACK

Prepare as for Raw pack. Cover with boiling water, bring to a boil. Drain, saving the hot cooking liquid for processing.

In jars. Pack hot squash loosely, leaving ½ inch of headroom. Proceed as for Raw pack, but Pressure-process pints for 30 minutes, quarts for 40 minutes. (Hot squash is more dense than raw, so requires longer processing.) Remove jars; complete seals if necessary.

In plain cans. Pack hot squash loosely, leaving ¼ inch of headroom. Proceed as for Raw pack, Pressure-processing both No. 2 and No. 2½ cans for 20 minutes (exhausting to 170 F. gives several minutes' advantage here). Remove cans; cool quickly.

Squash, Winter, see Pumpkin

Tomatoes, see the separate section, before Vegetables

Turnips, White, see Rutabagas

170 F. = 77 C. / 228 F. = 109 C. / 240 F. = 116 C. / 1 U.S. cup (½ U.S. pint) = .24 L. / 1 U.S. pint (2 cups) = .48 L. / 1 U.S. quart = .95 L. / 1 U.S. tsp. = 5 ml. (5 cc.) / 1 U.S. T. = 15 ml. (15 cc.) / ½ inch = 1.25 cm. / 1 inch = 2.5 cm.

Canning Meat and Poultry

Unless you raise food animals yourself (thus taking care that they are housed, fed and slaughtered in the best possible way); or unless you get whole or part carcasses from a local small-scale raiser (whose methods you've investigated carefully beforehand)—your safest buy is from a meat-seller whose goods are purchased under interstate trade regulations. Through improved animal-raising and meat-handling regulations and inspections, most parasites and diseases have been eliminated in meat that passes federal inspection and is allowed to travel in interstate trade. Most notable is the reduction of the feared *Trichina spiralis* in pork, the cause of trichinosis in people.

WARNING ABOUT GAME

Any wild game may be diseased or carry parasites (bear, for example, often have trichinosis). So, if you are a successful hunter or if you've been given a present of venison or other wild meat, *do not eat it cooked Rare.*

Pressure Canning at 10 pounds (240 F.)—for the length of time required for large pieces, small pieces or ground meat, respectively—is the only safe way to put game by in jars or cans.

The Biologist for your state Fish and Game Department in the state capital will be glad to tell you of any disease problems with game in your area.

Although freezing is easier, any fresh, wholesome meats or poultry—beef, pork, lamb and chicken are the most popular—may be canned safely.

Domestic rabbits and small-game animals are canned like Poultry.

Choose only good meat for canning, and handle it with expediency and total cleanliness, because bacteria grow at a frightening rate in meats and poultry if given half a chance.

Note: You must work quickly. Any meat picks up bacteria so don't keep it waiting at room temperature until you can handle it. If you have a large amount to do, temporarily store the part you're not working on in the refrigerator or a meat cooler (32 to 38 F.). Can first thing *tomorrow* what you could not can today: keep right at it until the job is done.

Process all canned animals and birds only in a Pressure Canner at 10 pounds (240 F.) to destroy bacteria—including the spores of *Cl. botulinum.*

Equipment and its care

A Pressure Canner is essential. And you'll need a pencil-shaped glass food thermometer for exhausting jars as well as cans.

Containers may be jars—we recommend the modern straight-sided ones, because they're easier to get the meat out of—or plain cans (meat sometimes makes the coating flake off the interiors of enameled cans: harmless, but unattractive).

Use good-sized sharp knives, including a 3- to 4-inch boning knife.

Wooden cutting-boards and surfaces are best for working on: they are less likely than harder surfaces to dull the knives, and yet will withstand the necessary scrubbing and disinfecting.

If you plan to do large quantities of meat, you should have a high-sided roasting pan and a very big kettle (at least as large as your Boiling-Water Bath kettle).

All tools and utensils must be scrubbed in hot soapy water and rinsed well with fresh boiling water before each use.

To control bacteria, cutting-boards and wooden working surfaces must be scrubbed hard in hot soapy water *both before and after* you handle meat on them, and must be disinfected with a solution of ¼ cup chlorine bleach to 4 cups of water; leave it on for 15 minutes, then rinse off with fresh boiling water.

Signs of spoilage and what to do

The fats in meat, plus its high protein content, make it more susceptible to spoilage than vegetables are. Follow the procedures for examining vegetables ("Canned Vegetable Troubles and What to Do"). In addition, take this warning to heart:

Before tasting home-canned meat and poultry—and their broths—boil the food hard for 20 minutes to destroy hidden toxins (stirring to distribute the heat, and adding water if necessary). If it foams unduly or has a bad odor during boiling, destroy it completely so it can't be eaten by people or animals.

Canning Meats

Salt is merely an optional seasoning in canned meat and poultry; it may be omitted.

If you use salt, your regular table salt will do.

Salt substitutes for special diets can leave an unwanted aftertaste when used in canning: wait to add these seasonings when the food is heated for serving.

CUTS OF MEAT FROM LARGE ANIMALS

Know what a whole, half or quarter of an animal will yield in the way of cuts before you buy one. There is 20 to 25 percent waste to start with, and usually there will be more pounds of stewing and/or ground meat than pounds of steaks and roasts and chops. Various USDA publications, and an article in the September 1974 issue of *Consumer Reports* are mighty helpful about actual yield.

Cutting the meat

Have the carcass cut in serving pieces by a professional meat-cutter to get the most out of your investment.

If you wish to tackle the job yourself, ask your County Extension Service office for a copy of *Farmers Bulletin No. 2209, Slaughtering, Cutting and Processing Beef on the Farm,* and for a copy of *Farmers Bulletin No. 2152, Slaughtering, Cutting and Processing Pork on the Farm.*

Audrey Alley Gorton's *The Venison Book, How to Dress, Cut up and Cook Your Deer* is a helpful guide to field-dressing and butchering large game.

CANNING LARGE PIECES OF MEAT

Pressure Canning only. Use Raw or Hot pack (we prefer Hot pack). Use straight-sided jars or plain cans.

Unlike the canning procedures for Vegetables, in certain

instances the air in food in *jars*—as well as cans—is exhausted before processing.

Prime cuts of beef, pork, lamb, veal and large game are best canned in large pieces. The less choice parts are good for stews and ground meats.

MAJOR CUTS OF BEEF (or ELK or MOOSE)

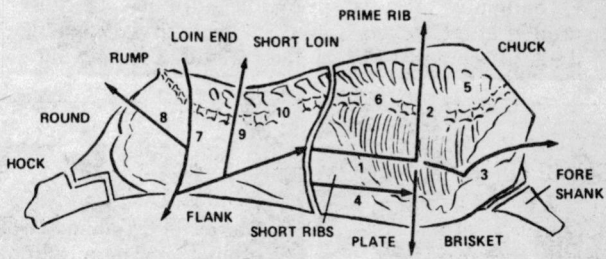

The following USDA pictures show the efficient sequence—and with a minimum of waste—for breaking down a side of beef into major anatomical cuts. From these large basic pieces the householder will cut the pot roasts, oven roasts, steaks, pieces for stewing and grinding—some to be served soon, the rest to put by. Use a meat saw for bones, wickedly sharp knives for cutting tissue.

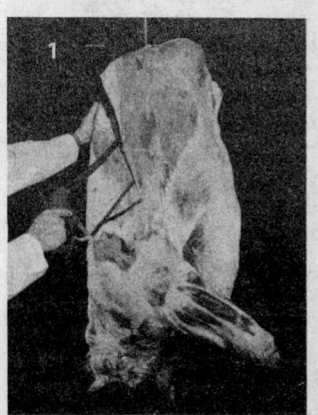

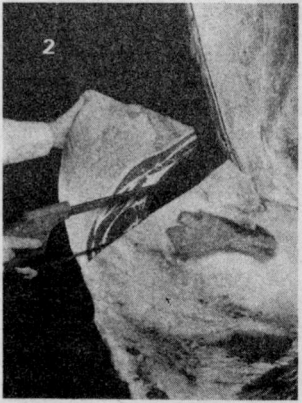

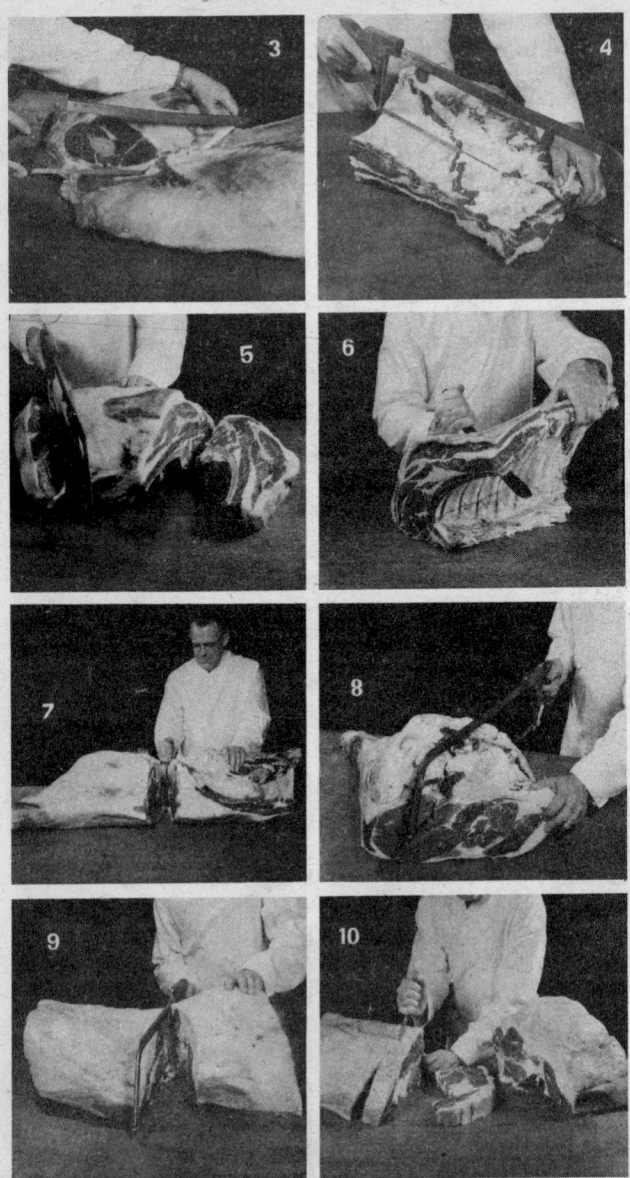

First, the fore- and hindquarter are separated just behind the last rib in the manner indicated by the gap in the diagram. Photos 1–6 deal with the forequarter. *Photos 1 and 2:* removing the prime-rib section. *Photo 3:* separating the plate-and-brisket section from the chuck (foreshank already removed). *Photo 4:* cutting short ribs from the plate. *Photo 5:* cutting chuck steaks. *Photo 6:* dividing the prime-rib section into standing rib roasts.

Turning to the hindquarter, *Photo 7:* having marked the cutting line to end at a point four vertebrae forward from the root of the tail, separate round-and-rump section from the loin. *Photo 8:* separating rump from round. *Photo 9:* having removed the flank (which will go into flank steaks or ground meat), divide the loin into short loin and loin end (which makes eventual sirloin roasts or steaks). *Photo 10:* cutting Porterhouse steaks from the short loin.

Complete metric conversion tables are in Chapter 1, but the following apply particularly to this section: 170 Fahrenheit = 77 Celsius / 240 F. = 116 C. / 1 U.S. cup (½ U.S. pint) = .24 Liter / 1 U.S. pint (2 cups) = .48 L. / 1 U.S. quart (4 cups) = .95 L. / 1 U.S. teaspoon = 5 milliliters (5 cubic centimeters) / 1 U.S. tablespoon = 15 ml. (15 cc.) / ½ inch = 1.25 centimeters / 1 inch = 2.5 cm.

The USDA pictures on page 176 show the large basic cuts from a side of pork; they will be broken down to suit the family's purpose. Use a meat saw on bones, use long *sharp* knives for cutting tissue.

Photo 1: having removed the head (for head cheese or sausage), cut down between the second and third ribs to separate shoulder section from loin-and-bacon midsection. *Photo 2:* remove the square-top ham with a cut at right angles to the shank. *Photo 3:* separating picnic shoulder and butt (both usually cured and smoked). *Photo 4:* trimming the ham. *Photo 5:* separating the loin from the bacon strip and leaf fat. *Photo 6:* cutting spareribs from the bacon strip.

To cut up lamb or goat. The carcass is handled whole, not split lengthwise like the larger beef, pork, elk, etc. Also, the USDA photos show finished cuts (except for chops).

Photo 1: removing breast and flank (both for stew) and foreleg; for rib chops with shorter bone, cut at the dotted line. *Photo 2:* cuts for *A, B, C* and *D* will give,

MAJOR CUTS OF PORK

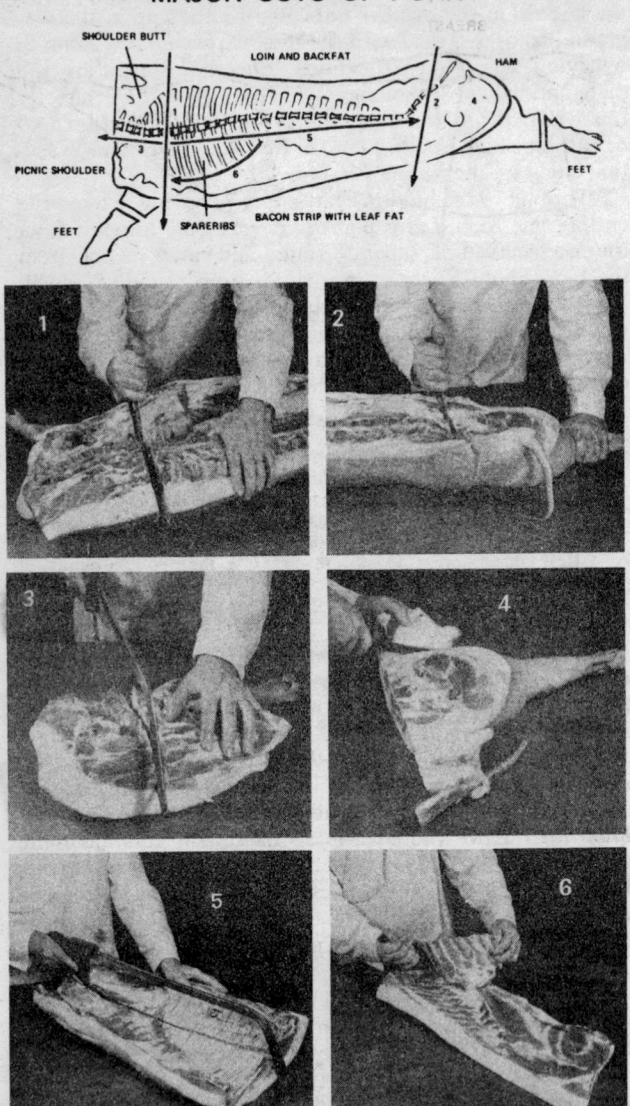

MAJOR CUTS OF LAMB (or GOAT)

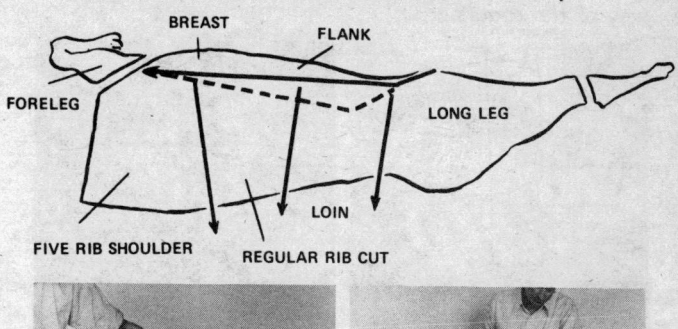

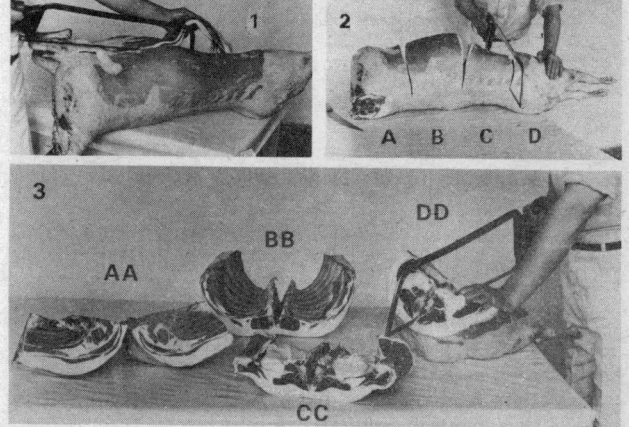

left to right, a five-rib shoulder, regular rib cut, loin and long leg. *Photo 3:* splitting the preceding cuts along the backbone gives *AA*—two shoulder roasts (or to become chops); *BB*—for rib chops; *CC*—for loin chops; *DD*—two leg roasts.

Prepare and Pack Raw

Wipe raw meat with a clean damp cloth. Remove bones *and all surface fat* (fat in canned meat is likely to shorten its storage life, and fat is a Number One seal-spoiler).

Cut in jar/can lengths, with the grain running the long way of the container.

A jar-sized chunk of meat needs no liquid packed Raw, but it must be exhausted (right) even though it is processed in glass, not a can.

IN STRAIGHT-SIDED JARS

Push the long pieces into the jars, leaving 1 inch of headroom. *Add no liquid:* there will be enough juice. Exhaust jars in a slow-boil bath to 170 F. (*c.* 70 minutes). Remove jars from exhaust bath. Add ½ teaspoon salt to pints, 1 teaspoon salt to quarts. Adjust lids.

Pressure-process at 10 pounds (240 F.)—pints for 75 minutes, quarts for 90 minutes. Remove jars; complete seals if necessary.

IN PLAIN CANS

Push long pieces into the cans, leaving *no* headroom. *Add no liquid:* there will be enough juice. Exhaust cans in a slow-boil bath to 170 F. (*c.* 70 minutes). Remove from exhaust bath. Add ½ teaspoon salt to No. 2 cans, ¾ teaspoon salt to No. 2½ cans; seal.

Pressure-process at 10 pounds (240 F.)—No. 2 cans for 65

minutes, No. 2½ cans for 90 minutes. Remove cans; cool quickly.

Prepare and Pack Hot (Precooked)

Put large cut-to-measure pieces of boned, de-fatted meat (see Raw pack preparation, above) in a large, shallow pan. Add just enough water to keep meat from sticking; cover, and cook slowly on top of the stove or in a 350 F. oven until the meat is Medium done, turning it now and then so it precooks evenly.

IN STRAIGHT-SIDED JARS

Pack hot meat loosely, leaving 1 inch of headroom. Add ½ teaspoon salt to pints, 1 teaspoon salt to quarts. Add boiling meat juice (extended with boiling water if necessary), leaving 1 inch of headroom. Wipe jar rims carefully to remove any fat. Adjust lids.

Pressure-process at 10 pounds (240 F.)—pints for 75 minutes, quarts for 90 minutes. Remove jars; complete seals if necessary.

IN PLAIN CANS

Pack hot meat loosely, leaving ½ inch of headroom. Add ½ teaspoon salt to No. 2 cans, ¾ teaspoon salt to No. 2½ cans. Fill cans to the top with boiling meat juice (extended with boiling water if necessary), leaving *no* headroom. Wipe can rims carefully to remove any fat; seal (no exhausting is necessary, because long precooking has driven air from the meat).

Pressure-process at 10 pounds (240 F.)—No. 2 cans for 65 minutes, No. 2½ cans for 90 minutes. Remove cans; cool quickly.

Note: You may also roast large pieces of meat as for the table until it is Medium done; pack as above (extending pan juices with boiling broth or water if necessary), and pressure-process at 10 pounds (240 F.) for the full time required above.

CANNING SMALL PIECES OF MEAT

Use the less choice parts of the animal for future use in stews, main-dish pies, etc.

Pressure Canning only. Use Raw or Hot (precooked) pack. Use straight-sided jars or plain cans.

Prepare and Pack Raw

Remove all surface fat from clean meat. Cut the meat off any bones. As you cut in stewing-size pieces, remove any interior bits of fat; cut away any tough muscle-sheath.

IN STRAIGHT-SIDED JARS

Pack, leaving 1 inch of headroom. *Add no liquid:* there will be enough juice. Exhaust jars in a slow-boil bath to 170 F. (*c.* 70 minutes). Remove from exhaust bath. Add ½ teaspoon salt to pints, 1 teaspoon salt to quarts. Wipe jar rims carefully to remove any fat. Adjust lids.

Pressure-process at 10 pounds (240 F.)—pints for 75 minutes, quarts for 90 minutes. Remove jars; complete seals if necessary.

IN PLAIN CANS

Pack cut-up meat, leaving no headroom. *Add no liquid.* Exhaust cans in a slow-boil bath to 170 F. (*c.* 70 minutes). Remove from exhaust bath. Add ½ teaspoon salt to No. 2 cans, ¾ teaspoon to No. 2½ cans. Wipe can rims carefully to remove any fat. Seal.

Pressure-process at 10 pounds (240 F.)—No. 2 cans for 65 minutes, No. 2½ cans for 90 minutes. Remove cans; cool quickly.

Prepare and Pack Hot (Precooked)

Follow Raw pack preparation. Put meat in a large, shallow pan, with just enough water to keep it from sticking; cover. Precook until Medium done. Stewing-size pieces take less tending if you do them in a 350 F. oven; but you can also precook them to Medium on top of the stove, turning or stirring them from time to time.

If you want to brown the meat before canning it, you don't need to dredge it in flour first: just put it under a hot broiler long enough to brown it on all sides; or brown it in a hot skillet, using an absolute minimum of oil to keep it from sticking. Slosh a little water around the pan to pick up any juice, and save the water to use in precooking the meat, as above.

IN STRAIGHT-SIDED JARS

Pack hot meat loosely, leaving 1 inch of headroom. Add ½ teaspoon salt to pints, 1 teaspoon salt to quarts. Add boiling meat juice (extended with boiling water if necessary), leaving 1 inch of headroom. Wipe jar rims carefully to remove any fat. Adjust lids.

Pressure-process at 10 pounds (240 F.)—pints for 75 minutes, quarts for 90 minutes. Remove jars; complete seals if necessary.

IN PLAIN CANS

Pack hot meat loosely, leaving ½ inch of headroom. Add ½ teaspoon salt to No. 2 cans, ¾ teaspoon salt to No. 2½ cans. Fill cans to the top with boiling meat juice (extended if necessary), leaving *no* headroom. Wipe can rims carefully to remove any fat. Seal.

Pressure-process at 10 pounds (240 F.)—No. 2 cans for 65 minutes, No. 2½ cans for 90 minutes. Remove cans; cool quickly.

CANNING GROUND MEAT

Freezing ground meat gives you a much better product than canning does. But if you must can it, don't can it bulk in a solid mass: make it up in meatballs by your favorite recipe (draining all browning fat off them before packing); or shape it in thin patties—which you may always break up to use later in casseroles, spaghetti sauce or whatever.

Pressure Canning only. Use Raw or Hot pack. Use straight-sided jars or plain cans.

Prepare and Pack Raw

Grind less tender cuts of *lean* meat. Work with fresh, clean cold meat: cold meat is easier to grind than meat at room temperature is. *Don't add any fat*. Make meatballs and brown them quickly on their surface, draining them well; or shape meat in thin patties slightly smaller in diameter than the containers.

IN STRAIGHT-SIDED JARS

Fill with raw patties packed in layers, or with browned raw meatballs, leaving 1 inch of headroom. Add ½ teaspoon salt to pints, 1 teaspoon salt to quarts. Exhaust jars in a slow-boil bath to 170 F. (*c.* 75 minutes). Remove from exhaust bath. Wipe jar rims carefully to remove any fat. Adjust lids.

Pressure-process at 10 pounds (240 F.)—pints for 75 minutes, quarts for 90 minutes. Remove jars; complete seals if necessary.

IN PLAIN CANS

Pack raw patties or browned raw meatballs firmly to the top of the cans. Add ½ teaspoon salt to No. 2 cans, ¾ teaspoon salt to No. 2½ cans. Exhaust in a slow-boil bath to 170 F. (*c.* 75 minutes). Remove from exhaust bath and press meat down in the cans so as to leave ½ inch of headroom. Wipe can rims carefully to remove any fat. Seal.

Pressure-process at 10 pounds (240 F.)—No. 2 cans for 100 minutes, No. 2½ cans for 135 minutes. Remove cans; cool quickly.

Prepare and Pack Hot (Precooked)

Trim and grind *lean* meat as in Raw pack, above. Make meatballs, browning their surface and draining well;

or shape thin patties, slightly smaller in diameter than the containers. In a slow oven (325 F.) precook meatballs or patties until Medium done. Skim off all fat from the drippings in the pan, saving the pan juices.

IN STRAIGHT-SIDED JARS

Pack hot patties (in layers) or hot precooked meatballs, leaving 1 inch of headroom. Cover with boiling *fat-free* pan juices (extended with boiling meat broth if necessary), leaving 1 inch of headroom. Wipe jar rims carefully to remove any fat. Adjust lids.

Pressure-process at 10 pounds (240 F.)—pints for 75 minutes, quarts for 90 minutes. Remove jars; complete seals if necessary.

IN PLAIN CANS

Pack hot patties (in layers) or hot precooked meatballs, leaving ½ inch of headroom. Cover with boiling *fat-free* pan juices (extended with boiling meat broth if necessary), leaving *no* headroom. Wipe can rims carefully to remove any fat. Seal.

Pressure-process at 10 pounds (240 F.)—No. 2 cans for 65 minutes, No. 2½ cans for 90 minutes. Remove cans; cool quickly.

CANNING PORK SAUSAGE

Freezing is better for pork sausage, especially in view of the large amount of fat.

Make your sausage by any tested recipe (see how in the Roundup section), *but use your seasonings lightly* because such flavorings change during canning and storage; and *omit sage*—it makes canned pork sausage bitter.

Pressure Canning only. Use Hot pack only. Use straight-sided jars or plain cans.

Prepare and Pack Hot (Precooked)

Shape raw sausage in thin patties, slightly smaller in diameter than the containers. In a slow oven (325 F.)

precook patties until Medium done. Skim off all fat from the drippings in the pan, saving pan juices.

IN STRAIGHT-SIDED JARS

Pack hot sausage patties in layers, leaving 1 inch of headroom. Cover with boiling *fat-free* pan juices (extended with boiling meat broth if necessary), leaving 1 inch of headroom. Wipe jar rims carefully to remove any fat. Adjust lids.

Pressure-process at 10 pounds (240 F.)—pints for 75 minutes, quarts for 90 minutes. Remove jars; complete seals if necessary.

IN PLAIN CANS

Pack hot patties in layers, leaving ½ inch of headroom. Cover with boiling *fat-free* pan juices (extending with boiling meat broth if necessary), leaving *no* headroom. Wipe can rims carefully to remove any fat. Seal.

Pressure-process at 10 pounds (240 F.)—No. 2 cans for 65 minutes, No. 2½ cans for 90 minutes. Remove cans; cool quickly.

CANNING BOLOGNA-STYLE SAUSAGE

If you don't have adequate cold, dry storage for Bologna-style sausage (see under Roundup for recipe and storing), you may can it.

Pressure Canning only—*and at 15 pounds.* Use Hot pack only. Use straight-sided jars or plain cans.

Prepare and Pack Hot (Cooked Completely)

When the sausage has been simmered until it is completely cooked and floats (see recipe under Smoking), remove it from the kettle, saving the cooking water. Cut the hot sausage in lengths the height of your containers for packing.

If the sausage is not fresh from cooking and therefore

is cold, heat it through by simmering for 10 to 20 minutes, depending on thickness, in a bland broth (bland, so as not to change the flavor of the sausage). Save the broth, cut the hot sausage to length, and pack.

IN STRAIGHT-SIDED JARS

Fit pieces of hot sausage lengthwise in the jars, leaving 1 inch of headroom. Add boiling broth, leaving 1 inch of headroom. Wipe the jar rims carefully to remove any fat. Adjust lids.

Pressure-process at 15 pounds (250 F.)—pints for 50 minutes, quarts for 60 minutes. Remove jars; complete seals if necessary.

IN PLAIN CANS

Fit hot sausage lengthwise in cans, leaving ½ inch of headroom. Add boiling broth, leaving ½ inch of headroom. Wipe the can rims carefully to remove any fat. Seal.

Pressure-process at 15 pounds (250 F.)—No. 2 cans for 45 minutes, No. 2½ cans for 60 minutes. Remove cans; cool quickly.

CANNING CORNED BEEF

Corn the beef (see under Brining in the section on Curing).

Pressure Canning only. Use Hot pack only (but in this case the meat is not precooked as for fresh meat). Use straight-sided jars or plain cans.

Prepare and Pack Hot (Freshened)

Wash the corned beef and cut it in chunks or thick strips to fit your containers, removing all fat. Put the pieces of meat in cold water and bring to boiling. Taste the broth in the kettle: if it's unpleasantly salty, drain the meat, cover it with fresh cold water, and bring again to boiling. This boiling merely freshens (removes salt), *it does not cook the corned beef.*

IN STRAIGHT-SIDED JARS

Fit hot freshened meat in jars, leaving 1 inch of headroom. Add boiling broth in which the meat was freshened, leaving 1 inch of headroom. Wipe jar rims carefully to remove any fat. Adjust lids.

Pressure-process at 10 pounds (240 F.)—pints for 75 minutes, quarts for 90 minutes. Remove jars; complete seals if necessary.

IN PLAIN CANS

Fit hot freshened meat in cans, leaving ½ inch of headroom. Fill cans to the top with boiling broth in which the meat was freshened, leaving *no* headroom. Wipe can rims carefully to remove any fat. Seal.

Pressure-process at 10 pounds (240 F.)—No. 2 cans for 65 minutes, No. 2½ cans for 90 minutes. Remove cans; cool quickly.

CANNING VARIETY MEATS

Most of the variety meats—liver, heart, tongue (see Recipes) and sweetbreads and brains—are best cooked and eaten right away. Certainly sweetbreads and brains, the most delicate foods of the lot, should be served when they are fresh. For liver and kidneys, freezing is recommended. Tongue may be canned satisfactorily, as well as frozen.

Canning Beef Tongue

The following procedure of course may be used for smaller tongues.

Soak the tongue in cold water for several hours, scrubbing it thoroughly and changing the water twice. Put it in a deep kettle, cover with fresh water, and bring to boiling. Skim off the foam well, then salt the water lightly; cover, and cook slowly until Medium done—*not tender* in the thickest part. Remove from kettle and plunge into cold water for a moment; peel off skin and trim off remaining gristle, etc., from the root.

Cut in container-size pieces, and pack Hot as for Large Pieces of Meat, or slice evenly and pack Hot as for Small Pieces of Meat. Pressure-process at 10 pounds (240 F.) for the times required (q.v.).

CANNING FROZEN MEAT

If you're ever faced with a freezing emergency, you may salvage frozen meat by canning it—provided it is good quality to start with, was correctly frozen and stored (see Freezing, and the table "Freezer Storage Life of Various Foods").

First, thaw it slowly in the refrigerator below 40 F. Then handle it as if it were fresh, using Pressure Canning only, Hot pack only, and the processing times that apply for Canning Large Pieces of Meat/Canning Small Pieces of Meat, above.

CANNING SOUP STOCK (MEAT BROTH)

Bony pieces of meat make good soup stock, and it may be worth your while to can the broth even if you're not canning other meat: some cans/jars of it are often a godsend. Freezing also is satisfactory for putting by soup stock, but it does take up valuable freezer space.

Pressure Canning only. Use Hot pack only. Use jars or plain cans.

Prepare and Pack Hot

Cover meat with lightly salted water (or omit salt; at any rate you may always season it to taste when you come to use it). Simmer until the meat falls away from the bones. Skim off all fat—lay absorbent paper toweling on the surface to pick up fat; or cool and refrigerate until the congealed fat may be lifted off in a hard sheet. Remove bones, but leave bits of meat and sediment. Reheat to boiling; continue boiling to concentrate the broth if it isn't strong enough.

IN JARS

Pour boiling hot stock into jars, leaving 1 inch of head-room. Wipe jar rims carefully. Adjust lids.

Pressure-process at 10 pounds (240 F.)—pints for 20 minutes, quarts for 25 minutes. Remove jars; complete seals if necessary.

IN PLAIN CANS

Fill cans with boiling hot broth, leaving *no* headroom. Wipe can rims carefully. Seal.

Pressure-process at 10 pounds (240 F.)—No. 2 cans for 20 minutes, No. 2½ cans for 25 minutes. Remove cans; cool quickly.

Before tasting home-canned meat and poultry—and their broths—boil the food hard for 20 minutes to destroy hidden toxins (stirring to distribute the heat, and adding water if necessary). If it foams un-duly or has a bad odor during boiling, destroy it completely so it can't be eaten by people or animals.

Canning Poultry and Small Game

The following instructions—which use chicken as the example for simplicity's sake—may be applied to canning domestic rabbits, wild birds and other small game, as well as canning other domestic poultry such as ducks, guinea hens, geese and turkeys, etc. *All these animals may be canned the same way:* general preparation (with specific exceptions as they come along), packing and processing are the same for all.

Note: Read the introduction under Canning Meat and Poultry for "Warning about Game," necessary equipment, and the special care required in handling all meat for processing.

If you refrigerate adequately, prevent contamination dur-ing handling, work quickly, and don't try short-cuts in packing and processing, poultry and small game may be canned satisfactorily.

Freezing of course is easier.

Use Pressure Canning only. Use Raw or Hot pack. Use straight-sided jars or plain cans.

WHERE CANNING OF POULTRY, ETC., IS DIFFERENT

Unlike the procedures given in the preceding section for packing meat, the methods that follow include canning with the bone left in.

Also, the skin on large pieces of birds—breast, thighs, drumsticks—is left on: processing at 240 F. compacts the surface of meat next to the sides of the container (making a pressure mark), so the skin you leave on acts as a cushion. Breast meat is skinned if packed in the center of jars/cans (surrounded by skin-on pieces that touch the containers' sides); so skin as you pack.

TO DRESS POULTRY, ETC.

Dressing involves two steps: (1) removing feathers by plucking or removing the fur pelt by skinning, and (2) drawing, which is removing the internal organs in one intact mass. Domestic birds are plucked before being drawn because they are handled for food immediately after they're killed.

However, a hunter *field-dresses* his kill by removing the innards on the spot, since they spoil a great deal more quickly than muscle tissue does; and he waits to skin it after he's home. Immediate drawing therefore reduces the chance of spoiling the rest of the meat en route home, and is especially necessary with mammals. Game birds may be held for several hours before being drawn; but if they cannot be got home for handling within half a day, they too should be drawn in the field, and plucked later.

170 F. = 77 C. / 240 F. = 116 C. / 1 U.S. cup (½ U.S. pint) = .24 L. / 1 U.S. pint (2 cups) = .48 L. / 1 U.S. quart = .95 L. / 1 U.S. tsp. = 5 ml. (5 cc.) / 1 U.S. T. = 15 ml. (15 cc.) / ½ inch = 1.25 cm. / 1 inch = 2.5 cm.

Plucking a Chicken

Pluck feathers from the still warm, fresh-killed and bled chicken, being careful to get all the pinfeathers. Hold the bird by its feet and pull the feathers toward the head, in the opposite direction to the way they lie naturally. Scalding the whole bird is not necessary, but if the feathers are resisting enough so you're afraid of tearing the skin, you can spot-scald: lift the chicken by the feet with its head dangling, and pour nearly boiling water into the base of the feathers, where it will be trapped momentarily against the skin.

Dry the bird and singe off the hairs. Wipe it clean.

Drawing a Chicken

Cut off the head of the fresh-killed and plucked bird if it is still on; remove feet at the "ankle" joint just below the drumstick. Cut out the oil sac at the top of the tail (it would flavor the meat unpleasantly, so don't break it).

Lay the chicken on its back, feet toward you. Using a sharp knife, cut a circle around the vent (anus), so it can be removed intact with the internal organs still attached. Cut deeply enough to free it, and be careful not to cut into the intestine that leads to it.

Insert the tip of the knife, with cutting edge upward, at the top of the circle around the vent, and cut through the thin ventral wall toward the bottom of the breast bone, making the slit long enough so you can draw the innards out through it with vent attached.

Reach clear to the front of the body cavity and gently pull out the mass of organs. Separate and save the heart, liver and gizzard.

Next, turn the chicken over and slit the skin lengthwise at the back of the neck—if you slit down the *front* of the neck you may cut into the crop; push away the skin and remove the crop and windpipe.

Cut the neck bone off close to the body. Wash the whole bird. Look inside it, and remove any bits of lung, etc., that may remain in the cavity.

Return to the giblets. From the liver, cut away the green gall sac, roots and all; and be mighty careful not to break it, because gall ruins the flavor of any meat it touches. Split one side of the gizzard, cutting until you see the tough inner lining. Press the gizzard open and peel the lining away and discard it and its contents. Trim the heart.

Refrigerate each dressed chicken, either whole or cut up, until you are ready to can it.

Cutting up Poultry

Lay the dressed, clean bird on its back and, using a sharp boning knife, disjoint the legs and wings from the body. Separate thighs from drumsticks at the "knee" joint. If the bird is very large—like a turkey or goose—separate the wing at its two joints, saving the two upper meaty sections for canning with bone in. (Very small birds, such as grouse, etc., may do best merely quartered with poultry shears.)

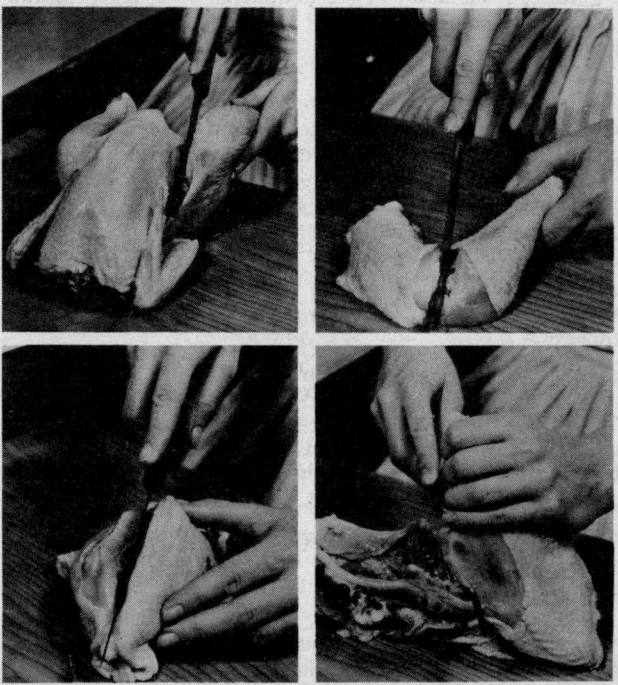

Turn the chicken crossways to you, hold the bottom of the breast section, and cut under it, through the ribs, until you reach the backbone; separate it from the backbone by cutting through the ribs. Poultry shears or heavy kitchen shears will be handy for use on the stronger bones at the shoulder joint.

Bone the breast meat by cutting down one side of the breast bone and easing the white meat off in a large piece; repeat on the other side.

Remove lumps of fat and any bits of broken bone from each piece of chicken and wipe it with a clean damp cloth.

Drawing, Skinning, Cutting up Rabbits

Lay the fresh-killed rabbit on its back, and proceed to draw it as if it were a chicken: cut around the vent carefully; make a slit in the abdominal wall, reach in and pull out the innards with vent attached; save the liver and heart.

Cut off the feet. Working from the hind legs upward, work the rabbit out of its skin, easing the job with your knife where you need to. The head may be skinned, but chances are you'll prefer to remove it with the skin when you reach it. Cut away the gall sac from the liver; trim the heart. Wash the dressed rabbit and pat it dry. *Refrigerate each dressed rabbit until you are ready to can it.*

If you raise rabbits for the table you'll find it simpler to skin them as soon as they are killed and before drawing them.

Because of its anatomy, think of a rabbit as making two forequarters, two hind quarters, and a saddle. Split the saddle down the backbone as you would split the breast of a chicken, boning it if you like. The size of the rabbit has a lot to do with whether you joint the quarters.

CANNING POULTRY, ETC., BONE IN

Prepare and Pack Raw

Pack meaty pieces loosely and more or less upright in the containers, putting breasts in the center (therefore skinned), surrounded by thighs and drumsticks (skin on) next to the sides of the jars/cans.

IN STRAIGHT-SIDED JARS

Pack, leaving 1 inch of headroom. Exhaust to 170 F. in a slow-boil bath (*c.* 75 minutes). Add ½ teaspoon salt to pints, 1 teaspoon salt to quarts. *Add no liquid:* there will be enough juice. Wipe jar rims carefully to remove any fat. Adjust lids.

Pressure-process at 10 pounds (240 F.)—pints for 65 minutes, quarts for 75 minutes. Remove jars; complete seals if necessary.

(In *this instance alone*—Raw pack in jars with bone in—some cooks omit the exhaust-bath-to-170 F. step, and Pressure-process at 10 pounds/240 F. for longer time: pints for 70 minutes, quarts for 80 minutes.)

IN PLAIN CANS

Pack, leaving *no* headroom. Exhaust to 170 F. in a slow-boil bath (*c.* 50 minutes). Add ½ teaspoon salt to No. 2 cans, ¾ teaspoon salt to No. 2½ cans. *Add no liquid.* Wipe can rims carefully to remove any fat. Seal.

Pressure-process at 10 pounds (240 F.)—No. 2 cans for 55 minutes, No. 2½ cans for 75 minutes. Remove cans; cool quickly.

Prepare and Pack Hot (Precooked), Bone In

Put raw meaty pieces in a large pan, cover with boiling water or boiling unseasoned (chicken) broth. Cover the pan and cook the meat slowly over moderate heat on top of the stove or in a 350 F. oven until Medium done. Pack hot meat with breasts preferably in the center (so skin them), surrounded by legs and thighs (unskinned, because they touch the sides of the containers).

IN STRAIGHT-SIDED JARS

Pack hot meat in loosely, leaving 1 inch of headroom. Add ½ teaspoon salt to pints, 1 teaspoon salt to quarts.

Cover with boiling unseasoned cooking liquid, leaving 1 inch of headroom. Wipe jar rims carefully. Adjust lids.

Pressure-process at 10 pounds (240 F.)—pints for 65 minutes, quarts for 75 minutes. Remove jars; complete seals if necessary.

IN PLAIN CANS

Pack hot meat in loosely, unskinned if they touch the side of the can; leave ½ inch of headroom. Add ½ teaspoon salt to No. 2 cans, ¾ teaspoon salt to No. 2½ cans. Add boiling unseasoned cooking water to the top of the cans, leaving *no* headroom. Wipe can rims carefully to remove any fat. Seal.

Pressure-process at 10 pounds (240 F.)—No. 2 cans for 55 minutes, No. 2½ cans for 75 minutes. Remove cans; cool quickly.

CANNING POULTRY, ETC., WITHOUT BONES

Whether you pack Raw or Hot, remove bones from good meaty pieces before you start. But leave skin on pieces of birds until you're actually filling the containers; then skin the pieces in the center of the pack (usually breast meat), and leave skin on the pieces that touch the side of the jar/can (usually legs).

Prepare and Pack Raw

IN STRAIGHT-SIDED JARS

Fill with boned raw pieces, skin taken off the ones in the center, skin left on the ones touching the side of the jar. Leave 1 inch of headroom. Exhaust to 170 F. in a slow-boil bath (*c.* 75 minutes). Add ½ teaspoon salt to pints, 1 teaspoon to quarts. *Add no liquid.* Wipe jar rims carefully. Adjust lids.

Pressure-process at 10 pounds (240 F.)—pints for 75 minutes, quarts for 90 minutes. Remove jars; complete seals if necessary.

IN PLAIN CANS

Pack, with skin on pieces that touch the side of the can, leaving *no* headroom. Exhaust to 170 F. in a slow-boil bath (*c.* 50 minutes). Add ½ teaspoon salt to No. 2 cans, ¾ teaspoon to No. 2½ cans. *Add no liquid.* Wipe can rims carefully. Seal.

Pressure-process at 10 pounds (240 F.)—No. 2 cans for 65 minutes, No. 2½ cans for 90 minutes. Remove cans; cool quickly.

Prepare and Pack Hot (Precooked), Without Bones

Remove bones from good meaty pieces, but leave skin on all pieces of poultry until you're filling the containers: then skin the ones in the center of the pack (usually breasts), leaving skin on the ones that touch the sides of jars/cans (usually legs).

Cover boned meat with boiling water or unseasoned broth and cook slowly on stove or in oven until Medium done as for Hot pack, Bone In, above.

IN STRAIGHT-SIDED JARS

Pack hot boned meat loosely (with outside pieces unskinned). Leave 1 inch of headroom. Add ½ teaspoon salt to pints, 1 teaspoon to quarts. Add boiling unseasoned broth, leaving 1 inch of headroom. Wipe jar rims carefully. Adjust lids.

Pressure-process at 10 pounds (240 F.)—pints for 75 minutes, quarts for 90 minutes. Remove jars; complete seals if necessary.

IN PLAIN CANS

Pack hot boned meat loosely, with outside pieces unskinned. Leave ½ inch of headroom. Add ½ teaspoon salt to No. 2 cans, ¾ teaspoon salt to No. 2½ cans. Add boiling unseasoned broth to the top of the cans, leaving *no* headroom. Wipe can rims carefully. Seal.

Pressure-process at 10 pounds (240 F.)—No. 2 cans for 65 minutes, No. 2½ cans for 90 minutes. Remove cans; cool quickly.

CANNING POULTRY GIBLETS

Giblets are more useful if they are canned together, rather than spread out among the cans of meat (use them chopped in gravies, meat sauces or spreads, as fillings for main-dish pies, on rice as a supper dish, etc.) Furthermore, the livers are better and handier if they are canned separately; and being so tender, they need much shorter precooking before Hot packing and processing.

Pressure Canning only. Use Hot pack only. Use straight-sided pint jars or No. 2 plain cans.

Prepare and Pack Hot (Precooked)

Cut clean gizzards in half, trimming off the gristle; cut smaller if necessary. Remove tops of hearts where the blood vessels come in; halve hearts if they're very large. Cover gizzards and hearts with hot water or hot unseasoned broth and cook until Medium done.

Remove all fat from the livers; cut away connecting tissue between the lobes. Cover livers with hot water or hot unseasoned broth and cook gently until firm and Medium done; stir occasionally to prevent sticking.

Pack gizzards and hearts together, pack livers separately.

IN STRAIGHT-SIDED PINT JARS

Fill with hot gizzards and hearts, or hot livers, leaving 1 inch of headroom. Add ½ teaspoon salt. Add boiling cooking liquid, leaving 1 inch of headroom. Wipe jar rims carefully. Adjust lids.

Pressure-process at 10 pounds (240 F.)—pint jars of either livers, or gizzards and hearts for 75 minutes. Remove jars; complete seals if necessary.

IN NO. 2 PLAIN CANS

Fill with hot gizzards and hearts, or hot livers, leaving ½ inch of headroom. Add ½ teaspoon salt. Add boiling liquid to the top of the cans, leaving *no* headroom. Wipe can rims carefully. Seal.

Pressure-process at 10 pounds (240 F.)—No. 2 cans of either gizzards and hearts, or livers for 65 minutes. Remove cans; cool quickly.

CANNING FROZEN POULTRY, ETC.

In case your freezer conks out, or you have a windfall of frozen poultry, domestic rabbits or small game, you may can it—*if:*

   (1) it is good quality and was properly frozen (see Freezing, and the table "Freezer Storage Life of Various Foods"), and
   (2) it is thawed slowly in the refrigerator below 40 F.

Then treat it as if it were fresh, using Pressure Canning only, Hot pack only, straight-sided jars or plain cans. Follow preparation and processing under Canning with Bone In/ Boned, above.

CANNING CHICKEN STOCK (BROTH)

Make broth from bony pieces of chicken (or other poultry, rabbits, wild birds, etc.) as you made it from meat, and pack and pressure-process it the same way (see Soup Stock under Canning Meat).

Because meat and poultry—and their broths—are very low-acid, boiling is the best way to determine if such home-canned food is safe to eat. So if there's the slightest possibility of spoilage, boil the food hard for 20 minutes in a covered pan to destroy possible hidden toxins. If it foams unduly or has an off-odor during boiling, destroy it completely so it can't be eaten by people or animals.

Canning Fish and Shellfish

Home-canned seafoods—that is, finned or shell-bearing creatures taken from salt or fresh water—are right up among the front-runners in the botulism sweepstakes. Add to this that in general they are the most perishable of all fresh foods and have great density of texture, and you see why fish and shellfish require faultless handling and longer Pressure-processing than do other foods that are canned at home.

So why can them? Why indeed, when proper freezing (q.v.) is an all-round better, and safer, means of putting them by?—or when even salt-curing followed by drying (q.v. again) and cold storage is, in the regions that practice this twofold method, less of a hazard?

But maybe you're faced with a surfeit of fresh seafood, and either freezing or curing-and-drying the excess is impossible. If such a bonanza is a repeated occurrence, you could plan ahead and organize a community kitchen, complete with good equipment and a skilled director in charge, as described early in this Chapter 2. If it's a once-in-a-lifetime event, though, and comes without warning (like the beach full of lobsters cast up by a hurricane that hit the Maine coast)—well, go ahead and can what you and your neighbors aren't able to eat, or swap for staples, or give to a public-service group near by.

The following procedures are for canning, without frills, some representative varieties of fresh fish and shellfish. We do not include canning fish that's been smoked (described at the end of Chapter 7, on Curing), or seafood in sauces, or chowders. For further information we recommend the following publications: U.S. Department of the Interior Fish and Wildlife Service *Conservation Bulletin No. 28, Home Canning of Fishery Products;* University of Wisconsin Sea Grant College Program *Report No. 120, Getting the Most from Your Great Lakes Salmon;* the Fisheries Experimental Commission of Alaska *Technical Report No. 4, Home Canning Alaska Fish and Shellfish;* and, because it is so well organized and well written, a book titled *Fish Handling and Processing*, published in Edinburgh by Her Majesty's Stationery Office (the British counterpart of our Government Printing Office).

170 F. = 77 C. / 240 F. = 116 C. / 1 U.S. cup (½ U.S. pint = .24 L. / 1 U.S. pint (2 cups) = .48 L. / 1 U.S. quart = .95 L. / 1 U.S. tsp. = 5 ml. (5 cc.) / 1 U.S. T. = 15 ml. (15 cc.) / ½ inch = 1.25 cm. / 1 inch = 2.5 cm.

EQUIPMENT FOR CANNING SEAFOOD

You cannot can fish or shellfish at home without an honest-to-goodness Pressure Canner. Not a pressure *saucepan*, regardless of what the makers of such saucepans give you leave to do: established safe timetables are kicked into a cocked hat by the quick-heating, quick-cooling little pans —which are fine for loose food.

And, just as for canning meats and poultry, you'll need a pencil-shaped glass thermometer because you'll be exhausting your jars here too.

In addition to the standard kitchen furnishings and the sharp good knives and cutting-boards you used for preparing meat, and the wherewithal to keep everything properly sanitary, you need:

Modern straight-sided ½-pint and pint canning jars in perfect condition, their 2-piece screwband lids ditto

Inexpensive styrofoam chest(s) in which to hold fish on ice

Hose or sprayer connected to your sink's drinking-water tap, for washing fish or shellfish

Fish scaler

Small wire brush for scrubbing shells, etc.

Big crocks or enameled vessels for soaking fish in salted water to remove blood from the tissues (dishpans will do)

Large enameled kettle for boiling shellfish, treating crabmeat, etc., to a mild salt/acid "blanch," or steaming open clams (your B-W Bath kettle is fine)

Special blunt knife for opening clams (if you shuck them raw)

Wire basket or rack for steaming

Shallow pans with perforated bottoms that will fit inside your Pressure Canner (for the so-called "tuna pack")

GENERAL HANDLING—PLUS REASONS WHY

From the sources mentioned above, and others, we have compiled the following stipulations, which must be followed by everyone who undertakes to can fish or shellfish at home.

All seafoods must be processed in a regular Pressure Canner for the *full long time* required, and *at the pressure given* (which of course is corrected for altitudes higher than 1,000 feet; see in Chapter 1). If the pressure drops below the recommended level at any time during the processing period, for safety's sake you must raise the pressure to the correct number of pounds, and start retiming as if you were starting the entire processing period from scratch.

Reason: The average natural acidity of seafood is so low that it flirts with Neutral on the *pH* scale (q.v. in Chapter 1). Therefore constant pressure for the full time is needed if enough heat is to penetrate the dense pack and sterilize the tissues, thus destroying dangerous bacteria.

Use only modern jars, manufactured for home-canning under pressure, that have 2-piece screwband lids. And use only ½-pint or pint jars (preferably straight-sided ones so the contents can slip out easily).

Reasons: For seafoods—or all home-canned products—it doesn't make sense to use makeshifts, or old-style jars and closures, or any other containers that have not been tested under the conditions required for safety by independent food scientists, and okayed. As for the 2-piece screwband lids, the flat metal discs indicate, by having snapped down to be concave, that you have obtained a proper seal. And finally, adequate processing cannot be assured for jars larger than 1 pint—or larger than ½-pint for certain fish or shellfish.

(We recommend *against cans* for home-processing of seafoods: (1) the correct sizes—like the commercial ones—are different from the cans used for other foods in this book; and (2) especially with the meat of lobsters and crabs, parchment-paper liners are usually needed to make an attractive product.)

All home-canned fish must be exhausted to a minimum of 170 F. at the center of the packed jar before it is Pressure-processed.

Reason: Before actual processing begins, we must drive air from the tissues of raw fish as well as from the pack to help ensure the seal and to prevent unwanted shrinkage of the food during processing. Fish in the so-called "tuna pack"—i.e., fully precooked —are cooled completely before packing; these packs also must be exhausted. (The completely precooked, and picked, meat of lobsters, crab, shrimp and clams is not exhausted when packed in small jars.)

Exhausting jars of fish is done best in the Pressure Canner at Zero pounds. Place filled jars on the rack in the bottom of the canner and pour hot water around them until it comes halfway up their sides. *Lay* the cover on and *leave the vent open.* Turn the heat up high, and when you hear the water boiling hard inside the canner and steam flows strongly in a steady stream from the vent—indicating that the temperature has reached 212 F. inside (see "Processing Adjustments for Altitude" in Chapter 1 if you're canning above 1,000 feet)—start counting the exhaust time. It will take 10 to 20 minutes for the center of the filled jars to reach the desired minimum of 170 F., depending on the size of the jar and the size of and solidness of the fish pieces; always insert your pencil thermometer deep in a test jar to make sure.

Water used in cleaning seafoods and preparing them for packing *must be of drinking quality*—whether it's the running water for washing them (which is always done under a tap, or with a spray or hose), or the water in a brine or anti-discoloration solution, or the canning liquid that goes in the jar.

Reason: It's easy to introduce dangerous bacteria, including *Cl. botulinum* itself, into the flesh by using polluted or contaminated water at any stage. Do not rinse fish in stream or lake water. Do not precook shellfish in seawater. If your household drinking water contains a lot of minerals, use bottled water at least for the canning liquid (iron, especially, reacts with the sulfur in the meat of shellfish and causes the product to darken).

All seafoods to be canned must be as fresh as is humanly possible to have them. Fish must be gutted as they're caught, and refrigerated or packed in ice immediately, to be kept cold until they are recooked or packed. Head shrimp immediately as they come from the water: if the head section is removed within 30 minutes after the shrimp is caught, the black "sand vein" (colon) will come out easily, attached to it; refrigerate shrimp or hold on ice. Keep lobsters, crabs and clams alive and cool until you prepare them for packing.

Reason: It takes only a couple of hours at room temperature to make dead seafood unfit to can; and spoilage is hastened if intestines and body wastes are not removed.

The flesh of all dressed and cleaned fish and shelled shrimp is given a preliminary brining; lobsters and crabs are precooked in brine and well rinsed before shelling. (The picked meat of lobsters, crabs and clams is given a further anti-darkening treatment in a mild acid "blanch" before packing.)

Reason: Brining draws diffused blood from the tissues, and reduces the chance that white curds of coagulated protein will occur in the processed jars. (Brines must be made up just before use, and should be *used only once.*)

The day after the seafood has been canned, store the jars in a cool, dry, dark place.

Reason: Storage that lets the jars freeze can also break the seals; storage over 50 F. courts spoilage; damp storage rusts metal closures and endangers the seals. During the 24 hours between processing and storing, check all seals, clean and label the jars.

CANNED SEAFOOD TROUBLES
AND WHAT TO DO

Do *NOT* reprocess jars of seafood found to have poor seals during the 24 hours of grace between canning and storage. And even if the contents are decanted into fresh containers and done over from scratch, the result is likely to be unsatisfactory (all the more reason for taking care in the first place).

After jars are stored you must be super-critical in examining them for *external* signs of spoilage: broken seal (flat lid no longer concave)—seepage around the closure —gassiness in the contents—cloudy, yeasty liquid or sediment at the bottom of the jar—contents an unnatural color or texture. *If any of these signs are present, destroy the food so it cannot be eaten by people or animals, sterilize the container and closure by boiling, and discard the sterilized closure.*

Even when the seal seems good and none of the trouble symptoms just listed is apparent, these are the signs of spoilage when you open a *jar of seafood:* pressure inside the container (instead of a vacuum), or spurting contents —fermentation—sour, cheesy odor—soft, mushy contents. If any of these signs is present, *destroy the food and sterilize the container and closure, as above.*

Before tasting any home-canned seafood, boil it for 20 minutes, stirring to distribute the heat and adding water to keep it from sticking to the pan. If the food foams unduly or smells bad during boiling, destroy it so it cannot be eaten by people or animals.

SPECIFIC SEAFOOD PRODUCTS

Salmon and Shad

Pressure Canning only. Use Raw pack and exhaust. Use pint jars (preferably straight-sided) with 2-piece screwband lids only.

Twenty-five pounds of round fish (i.e., whole and not dressed) will fill about 12 1-pint jars.

Dress, scale, scrub perfectly fresh fish; cut away the thin belly flap. Using a jar laid on its side as a measure, cut the fish across the grain in jar-length pieces—and *not one whit longer* lest they interfere with the seal (the fish will shrink in the jar to leave headroom). Prepare a cold brine of ¾ cup pure pickling salt dissolved in 1 gallon of water, an amount of brine that will do 25 pounds of prepared fish. Use enameled or non-metal tubs; use brine only once. Weight the fish pieces down in the brine for 60 minutes to draw out diffused blood and firm the flesh. Drain the pieces for 10 minutes; do not rinse.

Fill the jars solidly and in effect just to the top—this means no more that 1/16 inch below the sealing rim—

packing the pieces upright, skin side next to the glass, and carefully inserting slimmer pieces to fill vertical gaps. Do not crush down on the pieces or they'll spring back up later and endanger the seal.

Next, *half*-close the filled jars. To do this, place the flat lid on the sealing rim of the jar, and screw the band down *just until the band cannot be pulled up off the threads.* (Practice on an empty jar to get the feel of this half-closure.) Exhaust the jars in the Pressure Canner at Zero pounds (as described in "General Handling" above) until the center of the contents reaches a minimum of 170 F.—about 15 minutes.

When jars are exhausted, lift the canner off heat and finish screwing the bands firmly tight as for any processing. Return the canner to heat, put on the lid and let steam vent in a strong, steady flow for 10 minutes before closing the petcock/vent and starting to time the processing period. The amount of very hot water remaining in the canner after exhausting the jars should be ample for Pressure-processing.

Pressure-process at 10 pounds (240 F.) for 1 hour and 50 minutes. Remove jars; air-cool naturally.

Tunas, Large and King Mackerel

Pressure Canning only. Precook completely, cool, then exhaust. Use only ½-pint jars with 2-piece screwband lids.

Estimate 12 ½-pint jars for every 25 pounds of round fish. Dress, gut, scrub the fish; cut away the thin belly flap. Cut fish crosways in good-sized chunks (it will be cut for the jars after it's precooked).

For the precooking stage you will need several large round pans with perforated bottoms that can be stacked inside your Pressure Canner. Put 2½–3 inches of hot water in the canner; put a perforated support on the bottom of the canner—a wire cake rack, laid on some retired screwbands to help it take the weight of the fish; or an inverted metal pie pan with holes punched in its bottom; stack the pan-loads of fish in the canner. Put the lid on the canner; vent it (a strong, steady flow for 10 minutes); close the vent and Pressure-cook the fish at 10 pounds (240 F.) for 2 hours.

Remove fish, cool on large beds of cracked ice for several hours to ensure firm texture and good flavor. Then scrape away the skin, lift out the bones, remove dark streaks of flesh along the sides. Cut the cold chunks of fish ¾ inch shorter than the height of the containers (½-pints are about 4 inches tall, so the steak rounds of fish would be about 3¼ inches thick). Put ½ teaspoon salt and 3 tablespoons fresh water (or salad oil) in the bottom of each jar. Pack solidly with fish, using small flakes to fill gaps, and leave ½ inch of headroom.

To exhaust, *half*-close the lids as for Salmon, and boil hard in the Pressure Canner at Zero pounds (q.v. Salmon again) for 10 minutes: check center of contents of a test jar to make sure it has reached 170 F. When jars are exhausted, finish screwing the bands down firmly tight.

Pressure-process at *15 pounds (250 F.)* for 75 minutes.
Remove jars; air-cool naturally.

Lake Trout, Whitefish, Small Mackerel, Florida Mullet

Pressure Canning only. Raw pack and exhaust. Use pint jars (preferably straight-sided) with 2-piece screwband lids only.

About 35 pounds of fish, round weight, will fill 12 pint jars.

Dress, clean, scale, scrub perfectly fresh fish. Split the fish, leaving in the backbone; cut away the thin belly strip. Cut in jar-length pieces (as for Salmon, above), and brine the pieces for 60 minutes in a cold solution of ¾ cup pure pickling salt dissolved in 1 gallon of water. Remove the fish pieces, drain, and pack solidly just to the top of the jars—not more than 1/16 inch below the sealing rim—alternating head and tail sections upright in the jars for a firm pack, with skin sides next to the glass.

Use the kettle of your big enameled Boiling-Water Bath for exhausting the jars, because they will be boiled in a weak brine that would mar the metal surface of your Pressure Canner. Do not cap the filled jars; put them, open, on the rack in the B-W Bath kettle; pour in a fresh hot brine of ⅓ cup pure pickling salt dissolved in 1 gallon of water, until the brine comes 1 inch above the tops of the jars. Bring to boiling, and boil the jars briskly for 15 minutes, which should be enough to raise the temperature deep inside the jars to a minimum of 170 F. (check with your thermometer to make sure).

Remove the jars and invert them to drain on a wire cake-cooling rack for about 3 minutes. (Clap the slotted blade of a metal spatula over the mouth of the jar before you up-end it, and you won't have to worry about bits of fish sliding out.) Right the jars, wipe their rims carefully to remove any speck of material that would interfere with the seal; put on the lids and screw the bands down firmly tight.

Pressure-process at 10 pounds (240 F.) for 1 hour and 40 minutes. Remove jars; air-cool naturally.

Crab—Dungeness (Pacific) and Blue (Atlantic)

Pressure Canning only. Precook and exhaust. Use only ½-pint jars with 2-piece screwband lids.

To fill 12 ½-pint jars it will take about 25 pounds live weight of average-size Atlantic crabs, or 13 to 15 average Pacific crabs.

Use enameled or stainless-steel ware for boiling or acid-blanching shellfish—*never use* copper or iron. And *never use* seawater: use fresh drinking water to which you've added salt, etc.

For canning use only fresh-caught, frisky crabs in prime condition (not recently molted, not feeble or sickly). To avoid needless contamination of the meat by visceral matter you should butcher and clean them before precooking. Stun the live crab by submerging it in ice water for several minutes, then quickly twist off the legs, take off the back, remove gills and "butter" and the rest of the innards; clean out the body cavity under a strong flow of fresh, cool drinking water. Save claws and bodies of Atlantic crabs, discarding their legs as too small to bother with; save legs as well as claws and bodies of Pacific crabs, which are larger.

Meanwhile prepare and have heating in your biggest enameled B-W Bath kettle enough brine to cover the broken crabs you're dealing with, made in the proportion of 1 cup pure pickling salt and ¼ cup lemon juice to each gallon of fresh water. Dump crab pieces in the brine, bring back to boiling, and boil hard for 15 minutes by your timer. Quickly dip out the crab pieces and cool them quickly under cold running water *just until* they're cool enough to handle (the meat comes more easily from the shells if it's still warm). Pick out the meat, keeping body meat separate from leg and claw meat. Wash the meat piecemeal under a gentle spray to get rid of any curds of coagulated protein, etc., and press excess moisture out with your hands.

The following acid-blanch is designed to prevent natural—and harmless but unsightly—sulfur compounds present in shellfish from darkening the meat during processing. Therefore prepare beforehand and have ready a cold mixture in the proportions of 1 cup lemon juice (or of the citric-acid solution described in Chapter 1, or of distilled white vinegar in a pinch) to 1 cup salt dissolved in 1 gallon of water. This amount will treat 15 pounds of picked meat; make up fresh brine for each batch. Dealing with a colander-ful at a time, immerse leg meat for 2 minutes, body meat for 1 minute. Drain well, pressing out excess moisture with your hand.

Fill ½-pint jars firmly with meat, making a solid pack with attractive pieces next to the glass; leave ½ inch of headroom. Add boiling water to cover the meat—it won't take much—leaving ½ inch of headroom. Half-close jars and exhaust at Zero pounds (see Salmon, above) until the inside of the pack reaches a minimum of 170 F. on your thermometer—about 10 minutes. Finish screwing down bands firmly tight.

Pressure-process ½-pint jars at 10 pounds (240 F.) for 65 minutes. Remove jars; air-cool naturally.

Lobsters

Pressure Canning only. Precook and exhaust. Use only ½-pint jars with 2-piece screwband lids.

To fill 12 ½-pint jars, figure on 7 to 10 lobsters—depending on size and whether they're the huge-clawed Atlantic lobster of cold North American waters or the bigger-tailed spiny lobster without claws.

Can only fresh-caught, healthy, lively lobsters. Cook

and then cool them in separate containers of brine made of 2 tablespoons pure pickling salt to each 1 gallon of fresh drinking water: *never cook or cool lobsters in seawater;* make up fresh cooking/cooling brines for each batch.

In your biggest B-W Bath kettle, bring to boiling 3 to 4 gallons of the brine just described. Plunge live lobsters head first into the boiling salted water and, when it returns to boiling, boil them until their entire shells are bright red —about 20 minutes on the average. Lift them from the kettle and immerse them immediately in a tub of very cold brine (also made as above) to cool as fast as possible. When well cooled, each lobster is split, cleaned under running water, and the meat picked from the shell. Quickly and gently spray the picked meat as necessary to remove curds of coagulated protein. Press out excess liquid. Dip the picked meat, a small amount at a time, in a fresh acid-blanch as for Crab (in the proportions of 1 cup lemon juice and 1 cup salt dissolved in 1 gallon of water). Press out extra liquid and pack attractively in ½-pint jars, fitting claw and tail meat carefully to get a firm, solid pack; leave ½ inch of headroom. Just cover the meat with boiling fresh brine made of 1¼ teaspoons salt to each 1 quart of water; leave ½ inch of headroom. Exhaust as for Crab, above. Finish screwing down bands firmly tight.

Pressure-process ½-pint jars at 10 pounds (240 F.) for 70 minutes. Remove jars; air-cool naturally.

Shrimp

Pressure Canning only. Precook and exhaust. Use only ½-pint jars with 2-piece screwband lids.

About 10 pounds of fresh-caught headless shrimp will fill 12 ½-pint jars.

If shrimp are headed within 30 minutes after catching, the "sand vein" will come out with the head section. At any rate, pack raw shrimp immediately in crushed ice and hold them on ice to retard spoilage—and also to make peeling them easier. Remove heads, peel off shells, take out the sand vein; wash the meat quickly in fresh water and drain thoroughly.

In a large enameled kettle make enough cold brine to cover the meat, in the proportion of 2 cups of pure pickling salt dissolved in each 1 gallon of water. Hold the meat in the brine 20 to 30 minutes depending on the size of the shrimp, stirring from time to time to make this first brining uniform. Remove the meats and drain thoroughly.

Meanwhile prepare an acid-blanch of 1 cup lemon juice and 1 cup pure pickling salt for each 1 gallon of fresh water (the same as for Crab and Lobster, above); make enough to deal with all the shrimp you're working with, because you must use a fresh lot of the solution for each blanching-basket's worth of shrimp meat; otherwise the liquid will become ropy from diffused blood, etc. Bring to boiling enough of the acid-blanch to cover a household deep-frying basket *half*-filled with shrimp. Boil the meat 6 to 8 minutes (again depending on size of the shrimp) after the liquid returns to the boil. Lift out the shrimp, spread them on wire racks to air-dry and cool. An electric fan blowing across the shrimp will hurry the process: shrimp must be cool and surface-dry when packed.

Fill ½-pint jars with shrimp, fitting them in carefully to get a solid pack—but *don't crush them down;* leave ½ inch of headroom. Add boiling water just to cover the shrimp, leaving ½ inch of headroom. Half-close and exhaust as for Salmon, Crab and Lobster. Finish screwing down bands firmly tight.

Pressure-process ½-pint jars at 10 pounds (240 F.) for 35 minutes. Remove jars; air-cool naturally.

Hard-shell Clams (Littleneck, Butter, Razor, Quahaugs) Whole or Minced

Pressure Canning only. Blanch, pack and exhaust. Use only ½-pint jars with 2-piece screwband lids.

To fill 12 ½-pint jars with *whole* clam meats (including their juice), you'll need about 6 quarts of raw shucked meats; about 12 quarts of raw shucked meats will fill 12 ½-pint jars with *minced* clams.

The early steps for preparing clams for canning are the same for either whole or minced meats. We'll describe the complete procedure for canning whole clams, and give the variations for minced clams separately later.

Have ready some large vessels filled with clean salt water in which to hold your clams from 12 to 24 hours, so they'll have time to get rid of any sand in their stomachs. *Do not use seawater:* instead, approximate the necessary salinity by making a mild brine in the proportion of ¼ cup pure pickling salt for every 1 gallon of drinking water; and make enough to cover them by several inches.

Pick over the clams, choosing only those with tightly closed shells or ones that quickly retract their siphons (necks) when touched; discard any with broken or open shells. Scrub them with a stiff brush, rinse quickly, and put them in your mild holding brine. Sprinkle a few handfuls of cornmeal in the brine and swish it around; during the night the critters will eat the cornmeal and spit out sand.

Take the clams from their holding brine, throwing away any that don't have shells closed fiercely tight. Open the clams by steaming, or by shucking the live clams with a blunt knife as described in "Freezing Oysters, Clams, etc." in Chapter 3. If you open them live, work over a bowl to catch the juice (which you save, strain through cheesecloth, boil down to ⅔ its original volume, and use for canning liquid in the jars). *To steam open:* Take clams from their holding brine, spray-rinse, and pile them wet on a rack in the bottom of a big steel or enameled kettle with a tight lid; work with about ½ peck at a time, because their volume increases as they open. Cover the kettle, put it on high heat; reduce the heat to medium when the liquid from the clams starts to boil, and let them steam until their shells are part way open—up to 20 minutes.

From each opened clam, remove the dark gasket-like membrane that runs around the inside edge of the shell and encloses the siphon/neck; snip off the dark tip of the neck. Keep the dark stomach mass if you like: it's nutritious, but it also could give the canned product an unappetizing color or odor. Wash the dressed meats thoroughly in a fresh brine made in the proportion of 3 tablespoons pickling salt to each 1 gallon of water; make enough so you can change the washing-brine often.

Acid-blanch the meats in a boiling solution of 2 teaspoons pure citric acid powder dissolved in 1 gallon of water. making enough so you can change the blanch often. Half-fill a deep-frying basket with clam meats and hold them submerged in the acid-blanch for 2 minutes after the liquid returns to boiling. Lift out the basket of meats; drain.

Pack in ½-pint jars, leaving ¾ inch of headroom; do not add salt. Add boiling-hot, reduced clam juice to cover the meats, leaving ½ inch of headroom. Half-close the lids and exhaust as for Crab, Lobster, Shrimp, above. Finish screwing bands down firmly tight.

Pressure-process ½-pint jars at 10 pounds (240 F.) for 1 hour and 10 minutes. Remove jars; air-cool naturally.

MINCED CLAMS

Remove the dark stomach mass as you dress the meats after the clams are opened. Proceed with washing the meats in brine and acid-blanching, as above. Drain. Put the meats through a food grinder, using a plate with ⅜-inch holes. Strain the clam broth through cheesecloth, bring to boiling. Pack minced clams in ½-pint jars, leaving ¾ inch of headroom; do not add salt. Add hot clam broth, leaving ½ inch of headroom.

Exhaust and Pressure-process as for whole meats.

Freezing   3

Any food which is able to be canned is able to be frozen. In addition, some things which are chancy to can, such as seafood, freeze well. Further, a freezer holds safely for later use many table-ready dishes, meal leftovers and home-prepared convenience foods which may not be canned at all.

And usually—provided that preparation for freezing was adequate and that the freezer is working right—frozen food retains more of its original fresh flavor and texture, and generally keeps more of its nutrients.

However, freezing does NOT kill the organisms that cause spoilage, as canning does: it merely stops their growth temporarily. When they become suitably warm again, they multiply as quickly as ever.

Freezing as a way of preserving food means subjecting each sealed-from-the-air parcel/container of food to the sharpest cold we can—ideally even colder than 20 degrees below Zero Fahrenheit—for 24 hours, and then storing it at a sustained Zero F. Even so, frozen food has a limited storage life; and the warmer the freezer, the shorter the time you can hold it (see "Temperature *vs.* Spoilage" in Chapter 1, and "How Long in the Freezer?" coming in a minute).

Other points to bear in mind about freezing are that preparing food for a freezer is quicker than any other method of putting it by except for root-cellaring certain vegetables—and that home freezers represent a hefty capital investment.

CONSIDERATIONS IN GETTING A FREEZER

A proper sharp-freeze at −20 F. followed by Zero F. storage require a mechanical freezer, either one of your own, at home, or as space rented in a freezer-locker plant (see under Warehouses—Cold Storage, in the Yellow Pages). It would be a good idea for young families on a tight budget to rent a drawer in a nearby locker for a

THOSE FROZEN-FOOD/FREEZER PLANS

There are a number of companies that sell, on time payments, a package deal of freezer and food supply replenished over a period of several years. Some of these plans involve a membership fee of several hundred dollars to join, payable either as a lump sum or by installments; in addition, the buyer contracts in advance to purchase food at stated intervals. Still other plans offer only a food-purchase contract without the freezer.

All such plans undoubtedly are convenient: the householder orders ahead of time, and therefore has always on hand a certain quantity and type of food.

We recommend that everyone who is considering joining any such food/freezer plan check the company's offer with the local Better Business Bureau or other consumer protection group. Any contract which the householder is tempted to sign deserves careful investigation as to quality of product, be it freezer or food; terms of purchase, including carrying charges; and reliability of the firm, including what recourse either the company or the buyer may have if the contract appears to have been violated.

This recommendation is not to be construed as a blanket condemnation, even implied, of group-purchasing agreements. Prior investigation simply makes sense if the family is to get a suitable plan sponsored by a reliable organization.

For tips on selecting freezers and on some ins and outs of frozen-food plans, see the USDA's *Home and Garden Bulletin No. 48, Home Freezers, Their Selection and Use,* and *Consumer and Marketing Service Home and Garden Bulletin No. 166, How to Buy Meat for Your Freezer;* also, ask your County Agent for the latest information available from your state Consumer Education Specialist.

while: with such a trial run, they'll (1) decide on a feasible food-storage pattern, (2) learn what size freezer to buy come the time they can afford one, and (3) get the benefit of such special services as professional meat-cutting and commercial wrapping, if they want them. And it takes only a little extra bother to trot down to their locker drawer with a carton of home-packaged specialties or a case of good frozen product bought at an authentic saving during the frequent supermarket sales.

How big a freezer?

Some advisers recommend 6 cubic feet of freezer space for each person in the family, but farm families who raise most of their food and store it in their freezer might need 10 cubic feet per person. A cubic foot holds about 35 pounds of food.

> *Note:* Manufacturers' sales brochures describing cubic-foot sizes in freezers are sometimes approximate, so check the specifications and measurements to determine the actual usable interior space. See the USDA bulletin on seelcting a home freezer, referred to above. There is also an interesting discussion of freezers' cubic-foot measurements in the *Consumer Reports Buying Guide for 1974*, which summarizes material in the January 1974 issue of *Consumer Reports*.

Of all the ways to put food by, freezing limits storage room most severely, so what you freeze should be given careful thinking-through beforehand. If your freezer is to be more than a place to stash random things you're not going to use right away, figure out a system of priorities. A good rule-of-thumb is to assign freezer space first to the more expensive and heavier foods, and to ones that can't be preserved so well any other way. Therefore, plan to freeze meats and seafood; and plan to freeze certain vegetables (prime example, broccoli) and certain fruits (prime example, strawberries—assuming you don't want all of them in jam), and mentally allot room for some favorite main-dish combinations or desserts.

You needn't be rigid about it, because if you make the best use of your freezer you'll be continually rotating its contents.

What type of freezer?

To do its job, the freezer must have adequate controls, no warm spots ("warm" being a constant temperature higher than the rest of the storage area), and the ability to provide the initial sharp-freeze for 24 hours at −20 F.

The small enclosed space for ice cubes or below 32 F. storage in some refrigerators is not adequate unless it has its own controls for sharp-freeze and storage, and has its own outside door.

CHEST

The chest type with one or more top openings offers best use of the space within it, and it holds its cold better than the others (cold air sinks to the bottom, hence the open frozen-food bins at the supermarket).

To think about: More care is needed when filling it, so that all foods may be arranged in easy-to-reach fashion. Especially if the little woman is a little woman.

UPRIGHT

The upright ones with a refrigerator-type door of course take up less space in the room, they are easier to load and unload than the chests, and the little shelves on the door are handy for temporary storage of dabs and snippets.

To think about: More cold spills out when the door is opened than is lost with opening a chest; and irregular-shaped packages may tumble out at the same time.

TWO-DOOR COMBINATIONS

Side-by-side or stacked (one above the other) freezer-refrigerator combinations save floor space, and the freezer section is an adequate, though usually smaller, variation of the upright type with its own controls.

To think about: The freezer space is usually too limited to accommodate more than a quite modest freezing program and storage for bargains from markets' sales. With

care, though, it could hold a balanced supply of food types
and varieties for a very small family.

*And don't ignore the possible advantages in renting a
freezer-locker drawer.*

Where to put it?

If possible, locate the freezer in or near the kitchen:
you'll use it more (which means using it better) than if it's
in the cellar or other remote spot.

If you have a choice of convenient locations, choose
the cooler one—so long as the place isn't actually freezing.

Put it in a relatively dry room, because moisture rusts
the mechanism and can build up frost inside non-defrost-
ing models, particularly upright ones.

And *please* place it away from a back wall, so there's
adequate air circulation and you can get under and around
it to get out the fluff. Lack of air and a build-up of dust
can overheat the motor and even cause fire. Mounting it
on small rollers is a help in cleaning.

Complete metric conversion tables are in Chapter 1, but the
following apply particularly to this section: −20 Fahrenheit =
−28.9 Celsius / 0 F. = −17.8 C. / 212 F. = 100 C. / 1 U.S. cup
(½ U.S. pint) = .24 Liter / 1 U.S. pint (2 cups) = .48 L. / 1 U.S.
quart (4 cups) = .95 L. / 1 U.S. teaspoon = 5 milliliters (5
cubic centimeters / 1 U.S. tablespoon = 15 ml. (15 cc.) / ½
inch = 1.25 centimeters / 1 inch = 2.5 cm.

USING A FREEZER

Operating costs per pound of food are less if the
freezer is kept at least ¾ full at all times.

Operating temperatures

For the initial sharp-freeze, set the control at the lowest
possible point: at or below −20 F., the temperature that

makes smaller ice crystals in the food and gives a better finished product.

Place packages in single layers in contact with the parts of the box which cover the freezing coils—these would be certain shelves or walls or parts of the floor—and leave the food spread there for 24 hours before stacking the packages compactly for storage. For best results, don't try to sharp-freeze, at one time, more than 2 to 3 pounds of food for each 1 cubic foot of available freezer space. Later, when you're not sharp-freezing but simply storing, turn the controls back to no higher than Zero F.

For storing food after it's sharp-frozen, stack packages close together, and keep the storage section temperature at Zero F. or lower.

HOW LONG IN THE FREEZER?

Frozen foods lose quality when subjected to freezer temperatures above Zero F. While the storage life of different products varies, it can be stated generally that each rise of 10 degrees Fahrenheit cuts the storage life in half. (Thus if a food has a storage life of 8 months maintained at Zero F., its safe storage life will be only 4 months maintained at 10 F., and only 2 months if maintained at 20 F.) Foods maintained at −10 F. (10 degrees below Zero Fahrenheit) *in general* will keep their quality longer than shown in the following table—although keeping an item more than 12 months is uneconomical use of freezer space.

Times shown in the table are approximate, and are for foods of good quality that are properly processed/packaged when put in freezer storage.

Keeping an inventory

Running a freezer is like running a small store—you have to know what you have on hand and how long you've had it.

First, label and date each package of frozen food so you know how much of what is in each parcel; and when it was put by, so you'll use it while it's still top quality.

Freezer Storage Life of Various Foods Maintained at Zero Fahrenheit (−18 Celsius)

FOOD	MONTHS at 0 F.	FOOD	MONTHS at 0 F.
Fruits		*Fish and Shellfish*	
Apricots	12	Fatty Fish	3
Peaches	12	(Mackerel, Salmon,	
Raspberries	12	Swordfish, etc.)	
Strawberries	12	Lean fish	6
		(Haddock, Cod,	
Vegetables		Flounder, Trout, etc.)	
Asparagus	8 to 12	Lobster, Crabs	2
Beans, green	8 to 12	Shrimp	6
Beans, Lima	12	Oysters	3 to 4
Broccoli	12	Scallops	3 to 4
Carrots	12	Clams	3 to 4
Cauliflower	12		
Corn on the cob	8 to 10	*Bakery Goods (Precooked)*	
Corn, cut	12	Bread:	
Mushrooms	8 to 10	Quick	2 to 4
Peas	12	Yeast	6 to 12
Spinach	12	Rolls	2 to 4
Squash	12	Cakes:	
		Angel	4 to 6
Meats		Gingerbread	4 to 6
		Sponge	4 to 6
Beef:		Chiffon	4 to 6
Roasts, Steaks	12	Cheese	4 to 6
Ground	8	Fruit	12
Cubed, Pieces	10 to 12	Cookies	4 to 6
Veal:		Pies:	
Roasts, Chops	10 to 12	Fruit	12
Cutlets, Cubes	8 to 10	Mince	4 to 8
Lamb:		Chiffon	1
Roasts, Chops	12	Pumpkin	1
Pork:			
Roasts, Chops	6 to 8	*Other Precooked Foods*	
Ground, Sausage	4	Combination	4 to 8
Pork or Ham, smoked	5 to 7	dishes (Stews, Casseroles, etc.)	
Bacon	3	Potatoes:	
Variety meats	Up to 4	French-fried	4 to 8
(Liver, Kidneys, Brains, etc.)		Scalloped	1
		Soups	4 to 6
Poultry	6 to 12	Sandwiches	2

Store similar foods together, and you won't end up with a hodge-podge you have to paw through to find what you're looking for.

Devise some sort of inventory sheet or board that lets you keep track of food going in and coming out of your freezer. Some cooks make a sort of pegboard arrangement by driving partway into a board one nail for each type of food, labeling the nail, and slipping on it some sort of marker (a plastic ring, a washer, even a paper clip) for each package of that food when it is put in the freezer. As a package is taken out, a marker from the corresponding nail is removed.

Check the contents of your freezer every so often, and put maverick or to-be-used-soon items in places where you can't overlook them.

CARING FOR A FREEZER

Freezers need little care—just respect.

Treat the outside the same way that you do your refrigerator. Keep the surface and the door gaskets wiped clean. Clean behind and under it, and keep the protective grid over the motor free of dust.

Many are self-defrosting, but it is easy to do the defrosting yourself. Do it once or twice a year at times when the food supply is low. Disconnect the freezer, remove the food and wrap it in newspapers and blankets to keep it frozen. Use a wooden paddle—or similar tool that won't scratch the finish—to scrape the condensation (frost) from the walls of the cabinet onto papers or towels. Wipe the box with a clean cloth wrung out in water and baking soda. Dry the box before restarting the motor. About once a year, really wash the inside of the freezer.

By the way, have you read the manufacturer's instruction book lately?

WHEN THE FREEZER STOPS FREEZING

Now and again everybody's electricity fails for a time. Or, Heaven forbid, someone accidentally disconnects the freezer. Or the motor isn't working properly. Resist the impulse to open the door to check everything: make a plan of action first.

Find out, if you can, how long your freezer has been, or is likely to be, stopped. If it can be running in a few hours, don't worry: food in a fully loaded, closed freezer will keep for two days; if it's less than half loaded, the food won't keep longer than one day.

A freezer full of meat does not warm up as fast as a freezer full of baked foods.

The colder the food, the longer it will keep.

The larger the freezer, the longer the food will stay frozen.

But you must not open the door.

Emergency measures for a long stoppage

Call your local freezer-locker plant (in the Yellow Pages, try under Warehouses—Cold Storage, or under Butchering) to see if they can take care of your food while the emergency lasts. Then take out all the food, wrap it in newspapers and blankets, or pack it closely in insulated cartons, and hurry it down to the locker.

If your good friends have extra space, ask their help, and divvy your food among their freezers—writing your name on each package beforehand, and insulating it well for the trip to your neighbors'.

If you can't parcel it out for temporary storage, and your freezer will be stopped more than one day, try to buy 25 pounds of dry ice. (In the Yellow Pages, look under Dry Ice, or call an ice and fuel company). This amount will hold the inside of a partly full 10-cubic-foot freezer below freezing for 2 or 3 days, and hold longer a 10-foot one that's full.

Consolidate the food packages into a compact pile. Put heavy cardboard directly on the food packages and lay the dry ice on the cardboard. Careful! Handle dry ice with gloves: it burns—or, more accurately, it freezes.

Then cover the entire freezer with blankets, but leave the air-vent openings free so the motor won't overheat in case the current comes on unexpectedly.

But if all that food thaws . . .

If, despite your emergency measures, the food in your freezer thawed, there are several things you can do. You can refreeze some of it, you can cook up some of it and freeze the cooked dishes from scratch; some of it you may can. And some you may have to destroy.

WHEN TO REFREEZE

This calls for good judgment, and a definition of terms.

When food has *thawed,* it still contains many ice crystals; individual pieces may be able to be separated, but they still contain ice in their tissues; and dense foods, or ones that pack solidly, might have a firm-to-hard core of ice in the middle of the package, in addition to the crystals in the tissues. Many thawed foods may be refrozen. Be sure to re-label them for limited storage.

When food has *defrosted,* all the ice crystals in its tissues have warmed to liquid. *No defrosted foods except very strong-acid fruits should be refrozen:* defrosted low-acid foods, if refrozen, are possible sources of food poisoning. Certain defrosted foods, if they have defrosted only within an hour or so, and hence are still quite cold, may be canned; but the resulting quality might not warrant the effort. Remember, however, that defrosted low-acid foods, vegetables, shellfish and precooked dishes may be spoiled although they have no telltale odor; if spoiled, they'd be nasty if pressure-canned, and could be downright dangerous to boot if merely cooked up and served.

WHAT CAN BE SAFELY REFROZEN

The first check of thawed food in a package or a non-rigid container is to squeeze it. *Don't open it.* Squeeze it: if you can feel good, firm crystals inside, the package is O.K. to refreeze—*provided the food is not highly perishable in the first place.*

Of course food in rigid containers must be opened to be inspected for adequate ice crystals.

Even though they're defrosted, strong-acid fruits may be refrozen if they're still cold; there will be definite loss in quality, however.

Refreeze thawed vegetables only if they contain plenty of ice crystals.

Give wrapped meat packages the squeeze test. Beef, veal and lamb that are firm with ice crystals may be refrozen, but you can always cook them up in convenience dishes and freeze them from scratch. The salt in merely thawed short-storage cured pork helps the ice crystals, and it can be refrozen, with a noticeable loss in quality. Unless fresh pork has quite a lot of crystals, it's better to cook it and serve it, or cook it fully in a combination dish, and freeze the dish for later reheating.

If packages of meat seem at all soft in the squeeze test, open them and examine the contents for unnatural color or an off-odor: on the slightest suspicion that they're not perfectly fresh, *destroy them.*

Thawed poultry, being more perishable than beef, veal or lamb to start with, would do better if fully cooked and then refrozen, either as is or in a made dish. Thawed variety meats—liver, kidneys, heart, sweetbreads, brains— are highly perishable and should be cooked and served, rather than cooked and frozen again as a convenience dish; you could make a fully cooked liver loaf (see Recipes), though, and freeze that. Seafood, being extremely perishable, should be cooked and served (cooked clams and oysters get tough in refreezing anyway).

Never refreeze melted ice cream. Never refreeze cream pies, eclairs, or similar foods. But you can refreeze unfrosted cakes, uncooked fruit pies, bread, rolls, etc.

So what do you do with all the good thawed food that you don't want to refreeze, and that wouldn't can well, and that your family can't manage to eat right away? Give it to friends who will enjoy it—of course explaining that it's thawed. Or how about throwing the biggest impromptu dinner party you can imagine?

FREEZER PACKAGING MATERIALS

Freezer packaging materials usually are found at freezer-locker plants, hardware stores, feed and grain stores, and, often, in supermarkets and discount stores.

First, read the fine print

There is no simple definition we can give the homemaker that will enable her to recognize whether the wrappings and bags and rigid containers she should use in

reezing food for her family are truly moisture-vapor-proof
or moisture-vapor-resistant. But manufacturers are constantly researching improvements in their products, so we
say: Read the makers' labels to learn if the materials are
designed for freezer use.

The prime purpose of freezer packaging is to keep
frozen food from drying out ("freezer-burning"), and to
preserve nutritive value, flavor, texture and color. To do
this, packaging should be moisture-vapor-*proof*—or at least
resistant—and be easy to seal. And the seal should do its
job too.

And don't be sad if . . .

Don't expect perfect results from all your work if you
package the food for your freezer in household waxed
paper or regular aluminum foil or wrappings that are
intended for short-term storage in the refrigerator.

And don't expect perfect results if you make-do with
those coated-paper cartons that cottage cheese or milk or
ice cream came from the store in.

And don't expect perfect results if you seal your good
food with the sticky tape you use on Christmas parcels.
The adhesive used on made-for-the-freezer tape remains
effective at temperatures way below Zero Fahrenheit—
and the stuff on regular household tapes does not.

Rigid containers

As the name implies, these hold their shape and may
be stood upright, and are suitable for all foods except those
with irregular shape (a whole chicken, say); and they are
the best packaging for liquids.

Made of aluminum, glass, plastic, or plastic-coated
cardboard, these boxes, tubs, jars and pans come fitted
with tight-sealing covers. If the rims and lids remain
smooth they often may be re-used; however, the aluminum
ones have a tendency to bend as the packages are opened.

A leading maker of home-canning jars has come out
with a very sturdy plastic freezer box which, with proper
care, stands re-use remarkably well.

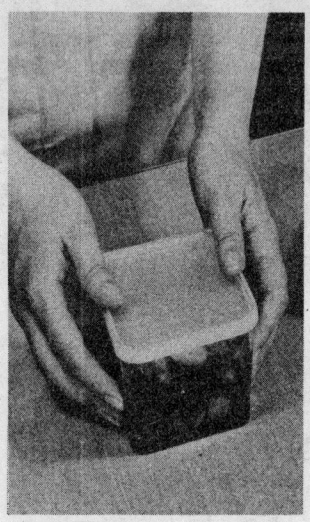

Three rigid containers and the inner bag for a fourth. The strawberries and broccoli are in plastic boxes, above; the peaches, below, are in a can/freeze jar (be sure that it's made for the freezer too). The special funnel and rack aren't required, but they do help make the best use of the cardboard freezer box that will protect the bag.

Some modern glass canning jars may also be used for freezing most fruits and vegetables. The wide-top jars with tapered sides are advised for liquid packs: the contents will slide out easily without having to be fully defrosted.

Non-rigid containers

Non-rigid containers are the moisture-vapor-resistant bags and sheet materials used for Dry pack fruits and vegetables, meats and poultry, fish, and sometimes liquids. They are made of cellophane, heavy aluminum foil, plio-film, polyethylene or laminates of paper, metal foil, glassine, cellophane and rubber latex.

The best ways of using sheet wrapping are the *butcher wrap* and the *drugstore fold*—both shown in Freezing Roasts.

Usually food in bags, and sometimes sheet-wrapped foods, are stored in a cardboard carton (re-usable) for protection and easier stacking in the freezer.

And then there are the so-called "cook-in" or "boil-in" pouches or bags, which are not as readily available as the regular non-boilable ones. Made of a tougher plastic to withstand hard boiling for up to 30 minutes, they are a good deal more expensive than conventional freezer bags; in addition, they come only in relatively small sizes. Also, because they're too stiff to twist and tie tight like the regular bags, they must be heat-sealed with a special appliance. Cook-in-pouch foods are described at the end of this chapter in "Freezing Convenience Foods."

Sealing and labeling

The packaging is no better than the sealing that closes it.

SEALING RIGID CONTAINERS

Some rigid freezing containers are automatically sealed by their lids, or by screw-type bands, or by flanged snap-on plastic covers.

SOME FREEZING TERMS

Anti-oxidant—Chemical agent, such as ascorbic acid (Vitamin C), added to sugar or sirup to control discoloration of fruits.

Blanch (scald)—To heat vegetables in boiling water or steam in order to slow or stop the action of enzymes.

Dry pack—To package without added liquid or sugar.

Enzymes—Naturally occurring substances that help promote organic change (ripening, decomposition, etc.) in vegetable or animal tissues. Their action must be stopped before certain food is frozen in order to prevent loss of quality, flavor and color.

Freezer burn—Dehydration of improperly wrapped food, leading to loss of color, flavor and texture.

Freezer tape—A pressure-sensitive adhesive tape designed to stick tightly at freezer temperatures (when ordinary tape comes off).

Glaze—To dip frozen food quickly in water iced to just above freezing, so that a thin coat of ice forms over the food.

Headroom—Space left at top of container to allow for expansion of food as it freezes.

Heat seal—To seal closure with pressure from a warm flatiron or special heating device.

Moisture-vapor-proof—Term applied to freezer packaging materials specially treated to prevent evaporation from food or transfer of moisture from the freezer to the food.

Overwrap—The second covering, usually sheet material (and water-vapor-resistant), wrapped around the first sealed container or sealed package, and then completely sealed with freezer tape.

Wet pack—To package fruits in sugar sirup, or with plain sugar (which draws juice to form liquid).

Then there's the waxed-cardboard freezer box with a tuck-in top that is sealed tight shut with freezer tape. If the contents are already sealed in an inner bag, though, you don't have to seal the box top.

There is also the re-usable container like a coffee can with an extra plastic lid: for your seal, tape the lid to the can all around.

The lids for glass jars must have an attached rubber-composition ring or a separate rubber ring to make a seal.

The lids for cans are put on with their own special sealing machine, as effectively for the freezer as for the Pressure Canner.

SEALING NON-RIGID CONTAINERS

Freezing bags are best sealed by twisting and folding the top, and fastening them with string, a rubber band, or a strip of coated wire.

Heat-sealing is possible—but it's tricky unless you have the special equipment for doing it. (See "Cook-in-Pouch Freezing" later.)

Non-rigid sheet wrappings sometimes can be heat-sealed, but people more often seal all the edges with freezer tape.

LABELING

Use a wide, indelible marking pen to label each package with the name of the contents, the amount, and the date it is packaged.

Freezing Fruits

The quality that comes out of the freezer is not one whit better than what goes in it. Use only prime fruit and treat it with respect.

Before freezing specific fruits: look at the chart "Headroom for Fruits and Vegetables," and "Sugar Sirups for Freezing Fruits."

Press air out of bags, then twist and tie the tops. Below crumpled wrap keeps peaches from floating up to the head room and darkening.

HEADROOM FOR FRUITS
AND VEGETABLES

It is necessary to allow ample headroom for foods packed in containers because the contents expand during freezing, and if extra space hasn't been provided, the expansion forces the closures off.

Loosely packed vegetables (like asparagus and broccoli) require no extra headroom, since space between the pieces is adequate for expansion.

The wide-top containers referred to are tall, with sides either straight or slightly flared.

The narrow-top containers referred to include canning jars, which may be used for freezing most fruits and vegetables that are *not packed in water*.

	Wide-top Container		Narrow-top Container	
Type of Pack	*Pint*	*Quart*	*Pint*	*Quart*
Liquid pack: (Fruit in juice, sugar, sirup or water; or crushed or puréed)	½ inch	1 inch	¾ inch	1½ inches
Fruit juice	½ inch	1 inch	1½ inches	1½ inches
Dry pack: (Fruit/vegetable packed without added sugar or liquid)	½ inch	½ inch	½ inch	½ inch

GENERAL PREPARATION

Handle only a small quantity at a time—2 or 3 quarts. A good way: put them in a wire basket and dunk it up and down several times in deep, cold water. After peeling, trimming, pitting and such, fix fruits much as you would for serving. Cut large fruits to convenient size, or crush. Small ones, such as berries, usually are left whole, or just crushed.

Crush soft fruits with a wire potato masher or pastry blender, firm ones in a food chopper. To make purée, press fruits through a colander, food mill or strainer (blenders can liquify too much).

TO PREVENT DARKENING

Unlike the choice of anti-discoloration procedures used most often in canning apples, apricots, peaches, nectarines and pears (Chapter 2), these and other oxidizing foods are kept from darkening by the addition of ascorbic acid (Vitamin C) either crystalline or tablets; commercial mixtures with an ascorbic acid base; crystalline citric acid, or plain lemon juice. Or steam-blanching for 3 to 5 minutes, depending on size of the pieces, may be used (see how, under Freezing Vegetables).

Fruit has a tendency to float to the top, where it changes color when exposed to the trapped air; so crumple some moisture-resistant sheet wrapping and put it on the top of the packaged fruit to hold it below the sirup. Seal and freeze.

ASCORBIC ACID, CRYSTALLINE

Usually available from drugstores, or from some freezer-locker plants. If bought in ounces, figure that 1 ounce will give roughly 40 ¼-teaspoons or 20 ½-teaspoons (these being the common amounts called for in freezing).

There is no record of known undesirable side effects from using ascorbic acid to hold the color of processed foods. It is Vitamin C. (There's more about it in Drying, and in Curing Meats.)

And not only is it the *safest* anti-oxidant to use, it is also the most effective of the agents employed in freezing to prevent darkening, because it will not change the flavor of the food, as the larger amounts needed of citric or lemon juice will do.

It dissolves easily in cold water or juices. Figure how much you'll need for one session at a time (see individual instructions), and prepare enough.

For wet pack with sirup. Add dissolved ascorbic acid to cold sirup and stir gently.

For wet pack with sugar. Just before packing, sprinkle the needed amount of ascorbic acid—dissolved in 2 to 3 tablespoons of cold water—over the fruit before you add the sugar.

For wet pack in crushed fruits and purées. Add dissolved ascorbic acid to the prepared fruit; stir well.

In fruit juices. Add dry ascorbic acid to the juice and stir to dissolve.

In dry pack (no sugar). Just before packing, sprinkle dissolved ascorbic acid over the fruit; mix gently but thoroughly to coat each piece.

ASCORBIC ACID TABLETS

It takes 3,000 milligrams (mg.) worth of ascorbic acid to equal 1 teaspoon of the crystalline form. Crush the tablets and dissolve them in the small amount of water.

COMMERCIAL ASCORBIC ACID MIXTURES

These mixtures of crystalline ascorbic acid and sugar—or ascorbic acid, sugar and citric acid—are sold under trade names. They are not quite as effective, volume for volume, as plain ascorbic acid, but are readily available and easy to use.

Follow the manufacturer's instructions.

CITRIC ACID

Drugstores carry this in pure crystalline form (see more about it in Chapter 1, where it's also discussed under "Acids"). You need three times as much of this than ascorbic acid to help prevent discoloration.

Dissolve the required amount in 2 or 3 tablespoons of cold water. Following directions for the individual fruits, add it as for dissolved ascorbic acid, above.

LEMON JUICE

A long-time favorite, it contains both citric and ascorbic acids. An equal amount of crystalline ascorbic acid is six times as effective than, lemon juice—which also imparts its own flavor to the food.

STEAMING

Steaming in a single layer over boiling water is enough to retard darkening in some fruits (for example, apples). The treatments described above are easier, though.

THE VARIOUS PACKS

A few fruits freeze well without sweetening, but most have a better texture and flavor when packed in sugar or a sugar sirup.

Unlike canning fruits, the size and texture influence the form in which you pack them for freezing. The intended future use is your final deciding factor.

Fruit to freeze whole, in pieces, juiced, crushed or puréed, is packed Dry or Wet.

DRY PACK (ALWAYS SUGARLESS)

The simplest way is just to put whole or cut-up firm fruits in containers (do not add a thing), seal and freeze. This is especially good for blueberries, cranberries, currants, figs, gooseberries and rhubarb.

If you have the space, spread raspberries, blueberries, currants or other similar berries one layer deep on a tray or cookie sheet and set in the freezer. When berries are frozen hard, pour them into polyethylene bags and seal. They won't stick together. Later the bag may be opened; the needed amount taken out, the bag reclosed and returned to the freezer.

Versatile Dry pack lets you use the fruits as if they were fresh.

WET PACKS

This means adding some liquid—such as its own natural juice, sugar sirup, crushed fruit, or water.

Wet pack with sugar. Plain sugar is sprinkled over and gently mixed with the prepared fruit until juice is drawn out and sugar is dissolved. Then you pack and freeze. Fruit fixed this way is especially good for cooked dishes and fruit cocktails. This has less liquid than the Wet pack with Sirup.

Wet pack with sirup. Fruit, whole or in pieces, is packed in containers and covered with a cold sugar sirup to improve their flavor and make a delightful sauce around the fruit. Generally best for dessert dishes.

Plan on using ⅓ to ½ cup of sirup for each pint package of fruit.

A 40 Percent Sirup (see "Sugar Sirups for Freezing Fruit") is used for most fruits, but to keep the delicate flavor of the milder ones, use a thinner sirup. A 50 to 60 Percent Sirup is best for sour fruits such as pie cherries.

Wet pack with fruit juice. The fruit—whole, crushed, or in pieces—is packed in the container and covered with juice extracted from good parts of less perfect fruit, and treated with ascorbic acid to prevent darkening. Place with a piece of crumbled moisture-resistant sheet wrapping on top to hold fruit below the liquid. Seal and freeze. People on sugar-restricted diets can enjoy this unsweetened fruit; or artificial sweeteners *approved by their doctor* may be dissolved in the juice to enhance the flavor.

For general use, sugar may be dissolved in the juice in the same proportion used in making a sugar sirup suitable for the particular fruit.

A greater degree of natural flavor is kept in the Juice pack, either sweetened or unsweetened, than in the Sirup pack.

Wet pack, purée. Fruit is puréed by forcing it through a food mill, strainer or colander. Dissolved ascorbic acid or lemon juice is mixed with the purée before packing to prevent darkening. Sweetening may or may not be added.

Wet pack with water. This is similar to the Juice and
Sirup packs, except that the added liquid is cold
water in which ascorbic acid has been dissolved. The
flavor is not as satisfactory as it is in Juice or Sirup
packs.

−20 F. = −28.9 C. / 0 F. = −17.8 C. / 212 F. = 100 C. / 1 U.S. cup (½
U.S. pint) = .24 L. / 1 U.S. pint (2 cups) = .48 L. / 1 U.S. quart =
.95 L. / 1 U.S. tsp. = 5 ml (5 cc.) / 1 U.S. T. = 15 ml. (15 cc.) / ½
inch = 1.25 cm. / 1 inch = 2.5 cm.

SUGAR SIRUPS FOR FREEZING FRUITS

Designation	Sugar	Water	Yield
30 percent (thin)	2 cups	4 cups	5 cups
35 percent	2½ cups	4 cups	5⅓ cups
40 percent (medium)	3 cups	4 cups	5½ cups
50 percent	4¾ cups	4 cups	6½ cups
60 percent (very heavy)	7 cups	4 cups	7¾ cups
65 percent	8¾ cups	4 cups	8⅔ cups

Dissolve the sugar thoroughly in cold or hot
water (if hot, chill it thoroughly before packing).
Sirup can be made the day before and stored in the
refrigerator: it must be kept quite cold.

Roughly, estimate ½ to ⅔ cup of sirup for each
pint container of fruit.

Substitutions: Generally, ¼ of the sugar may be
replaced by light corn sirup without affecting the
flavor of the fruit; indeed, the additional blandness
is often desirable for delicately flavored fruits, and
some cooks prefer substituting even more corn sirup.

Honey or maple sirup may also replace ¼ of the
sugar—if the family likes the different flavor either
imparts. Raw sugar of course affects the color and,
to some degree, the flavor.

Apples

Apples, more so than most produce, store well by several methods: fresh in a root cellar, or dried, or as canned applesauce or dessert slices (all q.v.). But you may want to freeze a few for late-season cooked dishes—especially in a package shaped for a pie.

SLICES

Prepare. Peel, core and slice. As you go, treat against darkening by coating the slices with ½ teaspoon pure ascorbic acid dissolved in each 3 tablespoons of cold water. Or steam-blanch. Less satisfactory but easier: drop slices in a solution of 2 tablespoons salt to each 1 gallon of water (no vinegar) for no longer than 20 minutes; rinse well and drain before packing.

Dry pack, no sugar (for pies). Arrange in a pie plate as for a pie, slip the filled plate into a plastic bag and freeze. Remove the solid chunk of slices from the plate as soon as frozen and overwrap it tightly in moisture-vapor-proof material—as if it were a piece of meat (q.v.)—and return to the freezer. (Handy at pie-making time because you lay the pie-shaped chunk of slices right in your pastry, put on the sugar and seasonings, top with a crust and bake.)

Wet pack, sugar. Sprinkle ¼ cup sugar over each 1 quart of slices for pie-making. Leave appropriate headroom.

Seal; freeze.

Wet pack, sirup. Cover with 40 Percent Sirup for use in fruit cocktail or serving uncooked. Leave appropriate headroom.

Seal; freeze.

SAUCE

Prepare. Make applesauce as you like it—strained, chunky, sweetened or unsweetened.

Wet pack, puréed. Fill containers. Leave ½ inch of head-room.

Seal; freeze.

Apricots

HALVES AND SLICES

Prepare. If apricots do not need peeling, heat them for 30 seconds in boiling water to keep their skins from toughening. Cool immediately in cold water. Cut up as you like.

Wet pack, sirup. Pack in a container, cover with 40 Per-cent Sirup to which has been added ¾ teaspoon ascorbic acid to each 1 quart of sirup. Leave appropriate headroom.

Seal; freeze.

Wet pack, sugar. Sprinkle ¼ teaspoon ascorbic acid dissolved in ¼ cup water over each 1 quart of apricots. Mix ½ cup of sugar with each 1 quart of fruit, stir until sugar dissolves. Pack in containers with appropriate headroom.

Seal; freeze.

CRUSHED OR PURÉED

Prepare. Wash and treat skins as for whole fruit. Crush or put through a sieve or food mill.

Wet pack, sugar. Mix 1 cup sugar with each 1 quart of crushed or sieved fruit. Add anti-darkening agent. Pack, leaving ½ inch of headroom.

Seal; freeze.

Avocados

Most versatile if frozen plain. Sweetening for milk shakes or ice cream, etc., may be added when you make them up. And for Guacamole, the further seasonings—

minced onion, tomatoes, green peppers, etc.—are also best added shortly before serving this delicious dip/spread.

Prepare. Peel and mash. If intended for future sweet dishes, add 1/8 teaspoon crystalline ascorbic acid to each 1 quart of purée to prevent darkening. If for Guacamole, add 1 tablespoon lemon juice (anti-darkening plus flavoring) and a dash of salt for each 2 avocados as you mash them.

Wet pack, puréed. Leave 1/2 inch of headroom.

Seal; freeze.

Most Soft Berries

WHOLE

Prepare. Sort, wash gently and drain: blackberries, boysenberries, dewberries, loganberries, youngberries.

Dry pack, no sugar. Pack in containers, leaving 1/2 inch of headroom. Or see the alternate Dry method in "The Various Packs."

Seal; freeze.

Wet pack, sirup. (For berries to be served uncooked.) Pack and cover with 40 to 50 Percent Sirup. Leave 1/2 inch of headroom.

Seal; freeze.

Wet pack, sugar. (For berries to be used in cooked dishes.) In a bowl mix 3/4 cup sugar with each 1 quart of berries. Mix until sugar dissolves. Pack; leave 1/2 inch of headroom.

Seal; freeze.

CRUSHED OR PURÉED

Wet pack, puréed. Add 1 cup sugar to each 1 quart crushed or puréed berries. Mix well. Pack; leave 1/2 inch of headroom.

Seal; freeze.

Most Firm Berries

Prepare. Sort and wash blueberries, elderberries, huckle-berries. Optional: steam berries for 1 minute to tenderize skins.

WHOLE

Dry pack, no sugar. (For berries to be used in cooked dishes.) Pack; leave ½ inch of headroom.

Seal; freeze.

Wet pack, sirup. (For berries to be served uncooked.) Pack; cover with 40 Percent Sirup, leaving ½ inch of headroom.

Seal; freeze.

CRUSHED OR PURÉED

Wet pack, puréed. Add 1 to 1½ cups sugar to each 1 quart of crushed or puréed berries; stir to dissolve. Leave appropriate headroom.

Seal; freeze.

Cranberries

WHOLE

Prepare. Wash and drain.

Dry pack, no sugar. Fill containers with clean berries. Leave ½ inch of headroom.

Seal; freeze.

Wet pack, sirup. Cover with 50 Percent Sirup. Leave appropriate headroom.

Seal; freeze.

PURÉED

Prepare. Wash and drain berries. Add 2 cups water to each 1 quart (1 pound) berries and boil until skins burst.

Press through a sieve and add 2 cups sugar to each 1
quart purée. Mix.

Wet pack, puréed. Pack; leave appropriate headroom.

Seal; freeze.

Currants

Prepare. Wash; remove stems.

WHOLE

Dry pack, no sugar. Treat like Cranberries.

Seal; freeze.

Wet pack, sugar. Add ¾ cup sugar to each 1 quart of
fruit; stir gently to dissolve. Pack; leave appropriate
headroom.

Seal; freeze.

Wet pack, sirup. Treat like Cranberries.

Seal; freeze.

CRUSHED

Wet pack, crushed. Add 1⅛ cups sugar to each 1 quart
crushed currants; stir to dissolve sugar. Pack; leave
appropriate headroom.

Seal; freeze.

JUICE

For beverages, use ripe currants. For future jellies,
mix in some slightly underripe currants for added pectin.

Prepare. Crush currants and warm to 165 F. over low heat.
Drain through a jelly bag. Cool.

Wet pack, juice. Sweeten with ¾ to 1 cup sugar to each 1
quart of juice, or pack unsweetened. Leave appropriate
headroom.

Seal; freeze.

Gooseberries

Prepare. Wash; remove stems and tails.

Dry pack, no sugar. (Best for future pies and preserves.) Pack whole berries; leave ½ inch of headroom.

Seal; freeze.

Wet pack, sirup. Cover whole berries with 50 Percent Sirup. Leave appropriate headroom.

Seal; freeze.

Raspberries

The versatile and very tender raspberries freeze even better than strawberries do. The wild ones, though small, have fine flavor. Real seedy berries are best used in purée or as juice.

WHOLE

Prepare. Sort; wash very carefully in cold water and drain thoroughly.

Dry pack, no sugar. Fill containers gently, leaving ½ inch of headroom.

Seal; freeze.

Wet pack, sugar. In a shallow pan, carefully mix ¾ cup sugar with each 1 quart of berries so as to avoid crushing. Pack; leave ½ inch of headroom.

Seal; freeze.

Wet pack, sirup. Cover with 40 Percent Sirup. Leave appropriate headroom.

Seal; freeze.

CRUSHED OR PURÉED

Prepare. Crush or sieve washed berries.

Wet pack, juice. Add ¾ to 1 cup sugar to each 1 quart of berry pulp; mix to dissolve sugar. Pack; leave appropriate headroom.

Seal; freeze.

JUICE

Prepare. Select fully ripe raspberries. Crush and slightly heat berries to start juice flowing. Strain through a jelly bag.

Wet pack, juice. For beverage, sweeten with ½ to 1 cup sugar to each 1 quart of juice. (For future jelly, do not sweeten.) Pour into containers; leave appropriate headroom.

Seal; freeze.

Strawberries

Choose slightly tart firm berries with solid red centers. Plan to slice or crush the very large ones. Sweetened strawberries hold better than unsweetened.

WHOLE

Prepare. Sort; wash in cold water; drain. Remove hulls.

Wet pack, sugar. In a shallow pan, add ¾ cup sugar to each 1 quart of berries and mix thoroughly. Pack; leave ½ inch of headroom.

Seal; freeze.

Wet pack, sirup. Cover berries with cold 50 Percent Sirup. Leave appropriate headroom.

Seal; freeze.

Wet pack, water (unsweetened). To protect the color of the berries, cover them with water in which 1 teaspoon crystalline ascorbic acid to each 1 quart of water has been dissolved. Leave appropriate headroom.

Seal; freeze.

SLICED OR CRUSHED

Prepare. Wash and hull as for whole berries, then slice or crush partially or completely. ·

Wet pack, sugar. Add ¾ cup sugar to each 1 quart of berries in a shallow pan. Mix thoroughly. Pack; leave appropriate headroom.

Seal; freeze.

JUICE

Prepare. Crush berries; drain juice through a jelly bag.

Wet pack, juice. Add ⅔ to 1 cup sugar to each 1 quart of juice—or omit sugar if you wish. Pour into containers; leave appropriate headroom.

Seal; freeze.

Cherries, Sour (for Pie)

As pie timber these are better canned; but if you want to freeze them, here's how.

WHOLE

Prepare. Use only tree-ripened cherries. Stem, wash, drain and pit. (The rounded end of a clean paper clip makes a good cherry-pitter.)

Wet pack, sugar. Add ¾ cup sugar to 1 quart of pitted cherries; stir until dissolved. Pack, leave appropriate headroom.

Seal; freeze.

Wet pack, sirup. Cover pitted cherries with cold 60 to 65 Percent Sirup. Leave appropriate headroom.

Seal; freeze.

CRUSHED

Wet pack, juice. Add 1 to 1½ cups sugar to each 1 quart of crushed cherries. Mix well. Pack; leave appropriate headroom.

Seal; freeze.

PURÉED

Wet pack, juice. Crush cherries; heat just to boiling and press through a sieve or food mill. Add ¾ cup sugar to each 1 quart of purée. Pack; leave appropriate headroom.

Seal; freeze.

JUICE

Home-made cherry juice in a party punch makes it exceptional!

Wet pack, juice. Crush cherries, heat slightly (*do not boil*) to start juice flowing. Strain through a jelly bag. Add 1½ to 2 cups sugar to each 1 quart of juice, or pack unsweetened. Pour into containers; leave appropriate headroom.

Seal; freeze.

Cherries, Sweet

The dark red and "black" varieties are best for freezing but do handle them quickly to prevent color and flavor changes. Use only tree-ripened fruit and remove the pits: they give an almond flavor to cherries when frozen.

WHOLE

Wet pack, sirup. Cover pitted cherries with 40 Percent Sirup in which you've dissolved ½ teaspoon crystalline ascorbic acid to each 1 quart of sirup. Leave appropriate headroom.

Seal; freeze.

CRUSHED

Wet pack, juice. To each 1 quart of crushed cherries add 1½ cups sugar and ¼ teaspoon crystalline ascorbic acid; mix well. Pack; leave appropriate headroom.

Seal; freeze.

JUICE

Sweet red cherries and sweet white cherries are handled differently for juice.

Prepare. Heat sweet *red* cherries slightly (to 165 F.) to start the juice. Strain through a jelly bag.

Crush sweet *white* cherries *without heating*. Strain through a jelly bag. Then warm this juice in a double boiler or over low heat to 165 F. Cool the red or the white juice and let it stand covered overnight.

Wet pack, juice. Pour off the clear juice into containers, being careful not to include any sediment from the bottom of the kettle. Add 1 cup sugar to each 1 quart of juice; or leave unsweetened if you prefer. Leave appropriate headroom.

Seal; freeze. (Sweet cherry juice by itself is pretty blah. So mix some *sour* cherry juice with the sweet to make a better beverage.)

Coconut

If you have a windfall, freeze some simply for fun—you may want to have a Mainland luau!

Prepare. Puncture the "eye" of the coconut; drain out and save the milk. Remove the meat from the broken-open shell. Shred it, or put it through a food chopper.

Wet pack, juice. Cover shredded meat with coconut milk. Leave appropriate headroom.

Seal; freeze.

Dates

Prepare. Wash, if necessary, and dry on paper toweling; remove pits.

Dry pack, no sugar. Pack in containers with no headroom.

Seal; freeze.

Figs

WHOLE OR SLICED

Prepare. Only tree-ripened, soft-ripe fruit, please; and check a sample for good flavor clear through the flesh. Sort, wash and cut off stems. Peeling is optional.

Dry pack, no sugar. Fill containers with the prepared figs; leave appropriate headroom.

Seal; freeze.

Wet pack, sirup. Cover with 35 Percent Sirup to which you have added ¾ teaspoon crystalline ascorbic acid—or ½ cup lemon juice—to each 1 quart of sirup.

Seal; freeze.

Wet pack, water. Pack figs; cover with water to which you have added ¾ teaspoon crystalline ascorbic acid to each 1 quart of water. Leave appropriate headroom.

Seal; freeze.

CRUSHED

Wet pack, juice. Crush prepared figs. Mix ⅔ cup sugar and ¼ teaspoon crystalline ascorbic acid with each 1 quart of crushed fruit. Leave appropriate headroom.

Seal; freeze.

Fruit Cocktail (for Compôte)

Freezing is excellent for your favorite combinations of fruit to serve either as an appetizer or dessert. A few added blueberries or dark sweet cherries make a nice color contrast.

Prepare. Use any combination of fruits peeled, cored, etc., and cut to suitable size.

Wet pack, sirup. Pack. Cover with cold 30 to 40 Percent Sirup in which ¾ teaspoon crystalline ascorbic acid to each 1 quart of sirup has been dissolved. If cut-up oranges are in the mixture, the ascorbic acid may be omitted. Leave appropriate headroom.

Seal; freeze.

Grapefruit (and Oranges)

Commercial processors do a fine job with citrus fruits. It's hardly worthwhile to compete unless you've a surplus of grapefruit and/or oranges.

Use heavy, blemish-free, tree-ripened fruits.

SECTIONS OR SLICES

Prepare. Wash; peel, cutting off the outside membranes. Cut a thin slice from each end. With a sharp, thin-bladed knife, cut down each side of the membranes and lift out the whole sections. Work over a large bowl to catch the juice. Remove seeds. Oranges may be sliced.

Wet pack, sirup. Cover fruit with 40 Percent Sirup made with excess fruit juice, and water if needed. (For better quality, add ½ teaspoon crystalline ascorbic acid to each 1 quart of sirup before packing.) Leave appropriate headroom.

Seal; freeze.

JUICE

Prepare. Use good tree-ripened fruits. Squeeze, using a squeezer that does not press oil from the rind.

Wet pack, juice. Either sweeten with 2 tablespoons sugar to each 1 quart of juice, or pack unsweetened. (For best quality, add ¾ teaspoon crystalline ascorbic acid to each 1 gallon of juice before packing.) Pour into glass freezing jars. Leave appropriate headroom.

Seal; freeze.

Grapes

Canning is probably smarter for these—but you may want to freeze some for gelatine salads and desserts. Juice is likely to be the best frozen use of grapes.

WHOLE OR HALVES

Prepare. Use firm-ripe grapes with tender skins and nice color and flavor. Wash and stem. Leave seedless grapes whole; cut other varieties in half and remove their seeds.

Dry pack, no sugar. Leave appropriate headroom.

Seal; freeze.

Wet pack, sirup. Cover grapes with cold 40 Percent Sirup. Leave appropriate headroom.

Seal; freeze.

JUICE

For a beverage or future jelly-making, use firm-ripe grapes.

Prepare. Wash, stem, crush. Do *not* heat. Strain through a jelly bag. Allow juice to stand overnight in the refrigerator while sediment settles to the bottom. Carefully pour off the clear juice.

Wet pack, juice. Pour into containers; leave appropriate headroom.

Seal; freeze.

(If tartrate crystals—the basis for cream of tartar—form in frozen juice, strain them out after the juice thaws.)

Melons

SLICES, CUBES OR BALLS

Prepare. Cut firm-ripe melons in half; remove seeds and soft tissues holding them. If for slices or cubes, cut off all rind; cut to shape. If for balls, do not cut off rind, but scoop out with a baller, taking care not to include any rind.

Wet pack, sirup. Cover with 30 Percent Sirup. Leave appropriate headroom.

Seal; freeze.

CRUSHED (NOT FOR WATERMELON)

Prepare. Halve, cut off rind; remove seeds and their soft tissue. Crush or put through the food chopper, using a coarse knife.

Wet pack, juice. Add 1 tablespoon sugar to each 1 quart of crushed melon, if you wish (and an added 1 teaspoon lemon juice points up the flavor). Stir to dissolve. Pack; leave appropriate headroom.

Seal; freeze.

Nectarines

These are not as satisfactory frozen as most other fruits are.

HALVES, QUARTERS OR SLICES

Choose only firm, fully ripe nectarines—avoiding overripe ones, which often develop a disagreeable flavor in the freezer.

Prepare. Wash and pit. Peeling is optional.

Wet pack, sirup. Put ½ cup of 40 Percent Sirup in each container and cut fruit directly into it. (For a better product add ½ teaspoon crystalline ascorbic acid to each 1 quart of sirup before packing.) Gently press fruit down and add extra sirup to cover. Top with crumpled moisture-resistant wrap to hold fruit in place. Leave appropriate headroom.

Seal; freeze.

PURÉED

Treat like Peach Purée.

Peaches

Peaches are excellent either canned or frozen.

HALVES AND SLICES

Prepare. Use firm, ripe peaches without any green color on their skins. Wash, pit, and peel. (They are less ragged if peeled without the boiling-water dip.)

Wet pack, sugar. Coat cut peaches with a solution of ¼ teaspoon crystalline ascorbic acid dissolved in each ¼ cup of water to prevent darkening. Add ⅔ cup of sugar to each 1 quart of fruit, and mix gently. Pack, leaving appropriate headroom.

Wet pack, sirup. Put ½ cup 40 Percent Sirup in the bottom of each container. Cut peaches directly into it. (For a better product add ½ teaspoon crystalline ascorbic acid to each 1 quart of the sirup before packing.) Gently press fruit down and add extra sirup to cover. Top with crumpled moisture-resistant wrap to hold fruit in place. Leave appropriate headroom.

Seal; freeze.

Wet pack, water. Cover cut peaches with water in which 1 teaspoon crystalline ascorbic acid has been dissolved in each 1 quart of water. Leave appropriate headroom.

Seal; freeze.

CRUSHED OR PURÉED

Prepare. Loosen skins by dipping peaches in boiling water for 30 to 60 seconds. Cool immediately in cold water; peel and pit.

Crush coarsely. For purée, press through a sieve or food mill; it's easier to make the purée if you heat the peaches in a very little water for 4 minutes before you sieve them.

Wet pack, juice. Mix 1 cup sugar and ⅛ teaspoon crystalline ascorbic acid with each 1 quart of peaches. Pack; leave appropriate headroom.

Seal; freeze.

Pears

Use Bartlett or a similar variety—not any of the so-called winter pears, which keep in cold storage (see Root-Cellaring).

HALVES OR QUARTERS

Prepare. Choose well-ripened pears, firm but not hard. Wash, cut in halves and quarters; core. Cover them with cold water to prevent their oxidizing during preparation (leaching is negligible because immersion time is so short).

Wet pack, sirup. Handling no more than 3 pints at a time in a deep-fry basket, low cut-up pears into boiling 40 Percent Sirup for 1 to 2 minutes. Drain; cool. (Save the hot sirup for another load of fruit.) To pack, cover cooled pears with cold 40 Percent Sirup to which has been added ¾ teaspoon crystalline ascorbic acid to each 1 quart of sirup. Leave appropriate headroom.

Seal; freeze.

PURÉED

Prepare. Wash well-ripened pears that are not hard or gritty. Peeling is optional. Proceed as for Peach Purée.

Persimmons

PURÉED

Purée made from late-ripening native ones needs no sweetening, but nursery varieties may be packed with or without sugar.

Prepare. Choose orange-colored, soft-ripe persimmons. Sort, wash, peel and cut in sections. Press through a sieve or food mill. Mix $\frac{1}{8}$ teaspoon crystalline ascorbic acid—or $1\frac{1}{2}$ teaspoons crystalline citric acid—with each 1 quart of purée.

Wet pack, juice (unsweetened). Pack unsweetened purée. Leave appropriate headroom.

Seal; freeze.

Wet pack, juice (sweetened). Mix 1 cup sugar with each 1 quart of purée; leave appropriate headroom.

Seal; freeze.

Pineapple

Prepare. Use firm, ripe pineapple with full flavor and aroma. Pare, removing eyes, and core. Slice, dice, crush or cut in wedges or sticks.

Wet pack, sirup. Pack fruit tightly. Cover with 30 Percent Sirup made with pineapple juice, if available, or water. Leave appropriate headroom.

Seal; freeze.

Wet pack, juice (unsweetened). Pack fruit tightly without sugar; enough juice will squeeze out to fill the crevices. Leave appropriate headroom.

Seal; freeze.

Plums (and Prunes)

Frozen plums and prunes are good in pies and jams, salads and desserts. Use the unsweetened pack for future jams. To serve unsweetened whole plums raw, see below.

WHOLE, HALVES OR QUARTERS

Prepare. Choose tree-ripened fruit with deep color. Wash. Cut as desired. Leave pits in fruits you freeze whole.

Wet pack, sirup. Cover with cold 40 to 50 Percent Sirup in which is dissolved $\frac{1}{2}$ teaspoon crystalline ascorbic acid to each 1 quart of sirup. Leave appropriate head-room.

Seal; freeze.

Wet pack, juice (unsweetened). Pack plums tightly. Leave appropriate headroom.

Seal; freeze. (To serve whole plums uncooked, dip them in cold water for 5 to 10 seconds; remove skins, and cover with 40 Percent Sirup to thaw. Serve in the sirup.)

PURÉED

Purée may be made from heated or unheated fruit, de-pending on its softness.

Prepare. Wash plums, cut in half and pit. *Unheated fruit:* press raw through a sieve or food mill. Add ¼ tea-spoon crystalline ascorbic acid—or ½ teaspoon crystal-line citric acid—to each 1 quart of purée. *Heated fruit* (the firm ones): add 1 cup water to each 4 quarts of plums; boil for 2 minutes; cool, and press through a sieve or food mill.

Wet pack, juice. Mix ½ to 1 cup sugar with each 1 quart of purée. Pack; leave appropriate headroom.

Seal; freeze.

JUICE

Prepare. Wash plums, simmer until soft in enough water barely to cover. Strain through a jelly bag and cool the juice.

Wet pack, juice. Add 1 to 2 cups sugar to each 1 quart of juice. Pour into containers; leave appropriate head-room.

Seal; freeze.

Rhubarb

Freeze only firm, young, well-colored stalks with good flavor and few fibers. (See also Canning.)

PIECES

Prepare. Wash, trim and cut in 1- to 2-inch pieces, or longer to fit the package. Heating rhubarb in boiling water for 1 minute and cooling immediately in cold water helps to set the color and flavor.

Dry pack, no sugar. Pack either raw or preheated (and now cold) rhubarb tightly in containers. Leave appropriate headroom.

Seal; freeze.

Wet pack, sirup. Pack either raw or preheated (and now cold) rhubarb tightly. Cover with cold 45 Percent Sirup. Leave appropriate headroom.

Seal; freeze.

PURÉED

Prepare. Prepare as pieces. Add 1 cup water to each 6 cups of rhubarb and boil 2 minutes. Cool immediately; press through a sieve or food mill.

Wet pack, juice. Add ⅔ cup sugar to each 1 quart of purée. Pack; leave appropriate headroom.

Seal; freeze.

JUICE

Prepare. Select as for pieces. Wash, trim, and cut in 4- or 5-inch lengths. Add 4 cups water to each 4 quarts of rhubarb, and bring just to a boil. Strain through a jelly bag.

Wet pack, juice. Pour into containers. Leave appropriate headroom.

Seal; freeze.

Tomatoes, see under Vegetables.

Freezing Vegetables

With a very few exceptions, any vegetable that cans well freezes equally well at home, if not better. The exceptions at this writing (all raw) are whole tomatoes, greens for salads, white (Irish) potatoes and cabbage. Because they have a high water content, formation of ice crystals ruptures their flesh, and the result is loss of texture or shape when defrosted. The extremely low temperature now being used by some frozen-food companies bypasses the crystal stage in freezing, so commercially frozen white potatoes or whole tomatoes are infinitely superior than those done at home.

Certain vegetable varieties are better for freezing than others, so read your seed catalogs carefully to see which ones you'll have the most luck with. Or ask your County Agent for a listing in your area. Or a truck gardener can tell you (but sometimes the person tending his roadside stand can't).

Because of the investment in nutrition and money that freezing entails, you'll want to freeze only prime vegetables that are garden-fresh and tender-young (younger, usually, than for canning). If you can't freeze them the day they're picked, refrigerate them overnight.

Note: Procedures for cooking frozen vegetables come at the end of this section.

GENERAL PREPARATION

The first step, after you've gathered your packaging, etc., is to wash the vegetables (only peas, lima beans and others that are protected by pods may not need to be washed). Use cold water and lift the vegetables out of it to leave any grit in the bottom of the pan.

You may need to take a further step to draw out possible insects in broccoli, Brussels sprouts and cauliflower: simply soak them for ½ hour in a solution of 4 tablespoons salt to 1 gallon of cold water.

Sort the vegetables according to size, or cut them to uniform pieces. Peel, trim and cut as needed.

Blanching

Even after vegetables are picked, the enzymes in them make them lose flavor and color and sometimes make them tough—*even at freezer temperatures.* Therefore the enzymes must be stopped in their tracks by being heated for a few minutes (how many minutes depends on the size and texture of the vegetable) before the vegetables are cooled quickly and packed. This preheating is necessary for virtually all vegetables: green (sweet) peppers are the notable exception.

IN BOILING WATER

Practically all vegetables are safely blanched in boiling water. Use a large kettle which has a wire basket that fits down in it. Put in at least 4 quarts of water and bring it to a boil. Put prepared vegetables in the basket—not more than a pound or two at a time—and lower it into the boiling water. Start counting the time at once. Shake the basket to let heat reach all parts of its load. When time is up, lift out the basket and immediately dunk the vegetables in ice-cold water to cool them fast.

Blanching in hard water tends to toughen vegetables.

If you live 5,000 feet or more above sea level, preheat vegetables 1 minute longer than the time called for.

Cool food as fast as possible after it's blanched.

BLANCHING IN STEAM

A few vegetables are better if heated in steam, and some may be done in either steam or boiling water.

For steaming, use a large kettle with a tight lid and a rack that holds a steaming basket at least 3 inches above the bottom of the kettle. Put in 1 or 2 inches of water and bring it to a boil.

Put your prepared vegetables in the basket in only a single layer, so the steam can reach all parts quickly. Cover the kettle and keep heat high. Start counting the time as soon as the cover is on.

For altitudes of 5,000 feet or more above sea level, add 1 minute to steaming time.

OTHER WAYS TO PREHEAT

Pumpkins, squash and sweet potatoes may be heated in a pressure cooker or in the oven before freezing. Mushrooms may be heated in fat in the skillet. And tomatoes for juice may be simmered.

Cool after blanching

Cool all vegetables as quickly as possible after they've been preheated. Use plenty of cold water—ice-water is ideal—and change it often to keep it cold. It takes about the same time to cool vegetables as it did to blanch them.

When they are completely cooled, drain them well on clean, absorbent toweling: you want as few crystals as possible in the pack.

THE PACKS

Vegetables for freezing may be packed either dry or in brine. The Dry pack is easier and lets you use the vegetables as you would if they were fresh, so Dry pack is the method we'll use the most.

Incidentally, a trick borrowed from commercial processors will make your packaging of dry-packed vegetables (and some fruits) easier. Just place a single layer of any freezer-ready small vegetable on a tray and sharp-freeze it fast (below —20 F.). Then pour the frozen vegetable into a freezer-type container and seal. Because the pieces are not stuck to each other, you can pour out the amount needed, reclose and seal the container, and return it and its partial contents to the freezer.

COOKING FROZEN VEGETABLES

The secret of cooking frozen vegetables (if it is a secret), is to cook them in a small amount of liquid, and only until they are tender. When you blanched them for the freezer, you already did a small part of the cooking.

So treat your frozen vegetables like fresh ones—except for a shorter cooking time. This way you'll keep more of the nutrients, as well as more of the natural color, flavor and texture.

TO THAW OR NOT TO THAW

Most are best cooked *without* thawing.

Defrost the leafy ones just enough to separate the leaves.

Fully defrost corn-on-the-cob, else the cob will not be heated through and the cooked kernels will cool too soon at the table. Open only the amount needed of any style corn and cook and serve it at once: holding it either before or after cooking makes it soggy.

BOILING

Generally you bring to the boil ½ cup water for each 1 pint of frozen vegetables. Add the vegetables, cover, and begin to count cooking time when water returns to the boil.

Exceptions: 1 cup water to each 1 pint of lima beans; water to cover for corn-on-the-cob.

Cooking times. Spinach—3 minutes; turnip greens—15 to 20 minutes; all other greens—8 to 12 minutes.

Depending on size of pieces: large lima beans, cut green/snap/wax beans, broccoli, carrots, cauliflower, corn in all forms, green peas—all from 3 to 10 minutes.

Kohlrabi (and similar-textured vegetables)—8 to 10 minutes.

Summer squash—up to 12 minutes.

At high altitudes, boil a bit longer—water boils at 2 degrees *lower* than 212 F. for each 1,000 feet above sea level.

PRESSURE-COOKING

The best guide is the manufacturer's instructions which come with the pressure saucepan.

BAKING

Most frozen vegetables can be cooked well by baking in a covered casserole. (It takes longer than boiling, but if your oven is running anyway, why not try it?)

Partially defrost the vegetable to separate the pieces; put it in a buttered casserole. Add the seasonings you like. Cover and bake at about 350 F. Most thawed vegetables cook in about 45 minutes.

For corn-on-the-cob, brush the completely defrosted ears with butter or margarine and salt, and then roast at 400 F. about 20 minutes.

PAN-FRYING

Use a heavy skillet with a cover. Put in about 1 tablespoon of table fat for each 1 pint of the frozen vegetable (which has thawed enough to separate in pieces). Cook tightly covered with moderate heat, stirring occasionally, until tender. Season to taste, and serve right away.

Asparagus, broccoli and peas will cook tender in about 10 minutes. Mushrooms will be done in 15 minutes. Green/snap/wax beans pan-fry to tenderness in 15 to 20 minutes.

REHEATING

Those vegetables fully cooked before freezing—usually leftovers—just need gentle reheating to serving temperature.

Frozen vegetables used in made dishes are treated like fresh ones. They're good creamed or scalloped, served au gratin or added to soufflés, cream soups and salads. See Recipes.

-20 F. = -28.9 C. / 0 F. = -17.8 C. / 212 F. = 100 C. / 1 U.S. cup (½ U.S. pint) = .24 L. / 1 U.S. pint (2 cups) = .48 L. / 1 U.S. quart = .95 L. / 1 U.S. tsp. = 5 ml. (5 cc.) / 1 U.S. T. = 15 ml. (15 cc.) / ½ inch = 1.25 cm. / 1 inch = 2.5 cm.

Asparagus

Prepare. Sort for size, wash well. Peel slightly tough ends c. 2 inches back from the bottom, cut off the really tough ends. Leave spears in lengths to fit the package, or cut in 2-inch pieces.

Blanch. In boiling water—small-diameter stalks for 2 minutes, medium stalks for 3 minutes, thick ones for 4 minutes. Cool immediately; drain.

Pack. Leave no headroom. With spears, alternate tips and stem ends; if it's a wide-top container, pack tips down.

Seal; freeze.

Beans, Lima

These are handier canned: but freeze the tenderest ones, if you can afford the space.

Prepare. Shell and sort for size.

Blanch. In boiling water—small beans for 2 minutes, medium for 3 minutes, large for 4 minutes. Cool immediately and drain.

Pack. Leave ½ inch of headroom.

Seal; freeze.

Beans, fresh Shell

Prepare. Shell and wash.

Blanch. In boiling water—1 minute. Cool immediately and drain.

Pack. Leave ½ inch of headroom.
Seal; freeze.

Beans—Snap/String/Green/Italian

These also can well. Fancy young tender ones are better frozen.

Prepare. Cut in 1- or 2-inch pieces, or in lengthwise strips (frenching), or leave whole if they're very young and tender.

Blanch. In boiling water—for 3 minutes. Cool immediately, drain well.

Pack. Leave ½ inch of headroom.

Seal; freeze.

Beets

Baby ones are worth freezer space. (Why not can larger ones plain or pickled?)

Prepare. Wash and sort for size—maximum 3 inches, small are best. Leave on tails and ½ inch of stem so their juice won't bleed out while boiling.

Boil. Until tender—25 to 30 minutes for small beets, 45 to 50 for medium. Cool quickly. Slip off skins; trim and cut in slices or cubes.

Pack. Leave ½ inch of headroom for cubes; no headroom for whole or sliced.

Seal; freeze.

Broccoli

Prepare. Peel coarse stalks, trimming off leaves and blemishes; split if necessary. Salt-soak for ½ hour (1 tablespoon salt for each 1 quart cold water) to drive out bugs; wash well. Sort for uniform spears, or cut up.

Blanch. In steam—5 minutes for stalks; in boiling water—3 minutes for stalks. (Reduce blanching time for cutup or chopped.) Cool immediately; drain.

Pack. Leave no headroom for spears or large chunks; arrange stalks so blossom ends are divided between either end of the container. Leave ½ inch of headroom for cut-up or chopped (they have less air space).

Seal; freeze.

Brussels Sprouts

Give freezer space only to the best heads.

Prepare. Salt-soak as for Broccoli if necessary. Wash well. Trim off outer leaves. Sort for size.

Blanch. In boiling water—small heads for 3 minutes, medium heads for 4 minutes, large heads for 5 minutes. Cool immediately, drain well.

Pack. Leave no headroom.

Seal; freeze.

Cabbage (and Chinese Cabbage)

Plan to use these only in cooked dishes: after being frozen they aren't crisp enough for salads.

Prepare. Trim off coarse outer leaves; cut heads in medium or coarse shreds or thin wedges, or separate into leaves.

Blanch. In boiling water—1½ minutes. Cool immediately and drain.

Pack. Leave 1½ inch of headroom.

Seal; freeze.

Carrots

These cold-store and can well, so freeze only the fancy young ones (preferably whole).

Prepare. Remove tops, wash and peel. Leave baby ones whole; cut others into ¼-inch cubes, thin slices or lengthwise strips.

Blanch. In boiling water—tiny whole ones for 5 minutes; dice, slices, or lengthwise strips for 2 minutes. Cool immediately; drain.

Pack. Leave ½ inch of headroom.

Seal; freeze.

Cauliflower

Infinitely better frozen than canned.

Prepare. Break or cut flowerets apart in pieces *c.* 1 inch across. Salt-soak as for Broccoli for ½ hour to get rid of bugs, etc. Wash thoroughly; drain.

Blanch. In boiling water (1 teaspoon salt to each 1 quart of water)—3 minutes. Cool immediately; drain.

Pack. Leave no headroom.

Seal; freeze.

Celery

Usable only in cooked dishes, so assign it freezer space accordingly. See also "Dabs and Snippets" at the end of Combination Dishes, in the Freezing section.

Prepare. Strip any coarse strings from any young stalks; wash well, trim, and cut in 1-inch pieces.

Blanch. In boiling water—3 minutes. Cool immediately; drain.

Pack. Leave ½ inch of headroom.

Seal; freeze.

Corn

Feasibility for freezing sweet corn; whole-kernel, Yes (it's better than canning); cream-style, Maybe (it's certainly handier canned, and there's not much difference in the product); on-the-cob, No—unless you've got loads of freezer space and don't mind thawing it before cooking it for the table (it shouldn't be popped frozen into the pot because the kernels will be cooked to death by the time the core of the cob is hot through). See Recipes.

WHOLE-KERNEL

Prepare. Choose ears with thin, sweet milk; husk, de-silk and wash. (Cut from cob *after* blanching.)

Blanch. In boiling water—4 minutes. Cool ears immediately; drain.

Pack. Cut from cob about ⅔ the depth of the kernels, and don't scrape in any milk. Leave ½ inch of headroom.

Seal; freeze.

CREAM-STYLE

Prepare. Choose ears with thick and starchy milk. Husk, de-silk and wash. (Cut from cob *after* blanching.)

Blanch. In boiling water—4 minutes. Cool immediately; drain.

Pack. Cut from the cob at about the center of the kernels, then scrape the cobs with the back of the knife to force out the hearts of the kernels and the juice (milk); mix with cut corn. Pack, leaving ½ inch of headroom.

Seal; freeze.

ON-THE-COB

Prepare. Choose ears with thin, sweet milk (as for whole-kernel). Husk, de-silk, wash; sort for size.

Blanch. In boiling water—small ears (1¼ inches or less in diameter) for 7 minutes, medium ears (to 1½ inches) for 9 minutes, large ears (over 1½ inches) for 11 minutes. Drain on terry cloth, and refrigerate immediately on dry toweling, in a single layer.

Pack. In containers, or wrap in moisture-vapor-resistant material.

Seal; freeze.

MAVERICK FREEZING IN THE HUSK

People who know the *Why's* and the *How's* of freezing say: "Never freeze corn without blanching it first to stop enzymatic action." But one hears of corn frozen successfully in its husk (though de-silked), without blanching.

Note: "Successfully" doesn't mean much unless you make the comparison by using identical ears from the same crop, picked and treated and frozen at the same time and for the same period, and cooked the same way for the same meal. Compare it right yourself—prepare it as described below (but eat it before Thanksgiving, because its storage life is bound to be short).

Prepare and pack. Without husking, pull out the silk; and, to save freezer space, remove a little of the outer husk. Do not blanch. Put ears in plastic bags simply to keep the freezer clean.

Freeze.

Greens, Garden

Prepare. Remove imperfect leaves, trim away tough midribs and tough stems; cut large leaves (like chard) in pieces. Wash carefully, lifting from the water to let silt settle.

Blanch. In boiling water—spinach, New Zealand spinach, kale, chard, mustard and beet and turnip greens: all for 2 minutes; collards for 3 minutes. Cool immediately; drain.

Pack. Leave ½ inch of headroom.

Seal; freeze.

Greens, Wild

Prepare. Collect and clean fiddleheads (ostrich fern) according to directions given in Recipes, blanch for 2 minutes. Cool and drain.

Collect and clean dandelions according to directions given in Recipes. If you like the slightly bitter taste, merely blanch the very tenderest leaves for 1½ minutes; otherwise boil in two or more waters. Cool and drain.

Collect milkweed and boil in several waters according to directions given in Recipes. Cool and drain.

Collect American cowslips (the marsh-marigold) and bring to boiling in several waters, cooking thoroughly to get rid of toxin. Cool and drain.

Pack. Leave ½ inch of headroom.

Seal; freeze.

Jerusalem Artichokes

Treat like Kohlrabi or small Turnips (q.v.).

Kohlrabi

Prepare. Cut off the tops and roots of small to medium kohlrabi. Wash, peel; leave whole or dice in ½-inch cubes.

Blanch. In boiling water—whole for 3 minutes, cubes 1 minute. Cool immediately and drain.

Pack. Whole in containers or wrap in moisture-vapor-resistant material. Cubes in containers, leaving ½ inch of headroom.

Seal; freeze.

Mushrooms

Prepare. Wash carefully in cold water. Cut off ends of stems. Leave stems on fancy small buttons if you like; if mushrooms are larger than 1 inch across the caps, slice or quarter them. If serving cold (in salads, etc.), blanch in steam; if serving hot (as garnish for meats, or in combination dishes), precook.

Blanch. In one layer, over steam—whole for 5 minutes, quarters or small caps for 3½ minutes, slices for 3 minutes. (This also prevents darkening; see "To Prevent Darkening" at the start of Freezing Fruits.) Cool immediately; drain.

Precooking. In table fat—sauté in a skillet until nearly done. Air-cool, or set the skillet in cold water (you'll freeze them in the good buttery juice from the pan).

Pack. Leave ½ inch of headroom.

Seal; freeze.

Okra (Gumbo)

Use in soups and stews.

Prepare. Wash. Cut off stems, being careful not to open the seed cells.

Blanch. In boiling water—small pods 3 minutes, large pods 4 minutes. Cool immediately; drain. Leave whole, or cut in crosswise slices.

Pack. Leave ½ inch of headroom.

Seal; freeze.

Parsnips

Really best left in the ground over winter for the first treat of spring—freezing is only a second choice. Treat like Carrots.

Peas, Black-eyed (Cowpeas, Black-eyed Beans)

Prepare. Shell; save only the tender peas.

Blanch. In boiling water—for 2 minutes. Cool immediately, drain well.

Pack. Leave ½ inch of headroom.

Seal; freeze.

Peas, Green

Prepare. Shell; use only sweet, tender peas.

Blanch. In boiling water—for 1½ minutes. Cool immediately; drain.

Pack. Leave ½ inch of headroom.

Seal; freeze.

Peppers, Green (Bell, Sweet)

Here is a vegetable that *does not require* blanching: the brief precooking described below is designed to make them more limp, so you can pack more peppers in the container—and it's for large-ish pieces you plan to use in cooked dishes, at that.

If you plan to serve them raw (for instance in thin rings as a garnish, or diced in a salad), don't bother to blanch.

Prepare. Wash; cut out stems, cut in half, remove seeds. Leave in halves, or cut in slices, strips, rings or dice (depending on future use).

If blanched. In boiling water—halves for 3 minutes, slices for 2 minutes. Cool immediately; drain.

Pack. Blanched, leave ½ inch of headroom. Raw, leave no headroom.

Seal; freeze.

Peppers, Hot

Prepare. Wash and stem.

Blanch. No.

Pack. Leave no headroom.

Seal; freeze.

Pimientos

Prepare. Wash and dry crisp, thick-walled pimientos.

Roast. In a 400 F. oven—for 3 to 4 minutes. Rinse and rub off charred skins in cold water. Drain.

Pack. Leave ½ inch of headroom.

Seal; freeze.

Pumpkin

Pumpkin makes fine pies and breads (see Recipes), but is seldom used as a table vegetable. Why not can it instead?

Prepare. Wash whole pumpkin; cut or break in pieces. Remove seeds. Do not peel.

Precook. Until soft—in boiling water, steam, a pressure cooker or in the oven. Scrape pulp from rind; mash through a sieve. Cool immediately.

Pack. Leave ½ inch of headroom.

Seal; freeze.

Rutabagas

Prepare. Cut off tops of young, medium-sized, rutabagas; wash and peel. Cut in cubes to freeze merely blanched, or in large chunks to cook and mash before freezing.

Blanch (for cubes). In boiling water—for 2 minutes. Cool immediately; drain.

Cook (chunks to mash). In boiling water until tender. Drain; mash or sieve. Cool immediately.

Pack. Leave ½ inch of headroom for either cubed or mashed.

Seal; freeze.

Soybeans

Prepare. To serve as a vegetable, wash firm, well-filled, bright-green pods (shell *after* blanching).

Blanch. In boiling water—5 minutes. Cool quickly. Squeeze beans out of pods.

Pack. Leave ½ inch of headroom.

Seal; freeze.

Sprouts, see Chapter 8.

Squash, Summer (and Zucchini)

Only young squash with small seeds and tender rinds are suitable for freezing.

Prepare. Cut off blossom and stem ends; wash and cut in slices.

Blanch. In boiling water—for 3 minutes. Cool immediately in ice water; drain well.

Pack. Leave ½ inch of headroom.

Seal; freeze.

Squash, Winter

Root-cellar mature squash with hard rinds. Treat it like Pumpkin if you do freeze it, though.

Sweet Potatoes (and Yams)

Use medium to large sweet potatoes that have air-dried (to cure) after being dug. Pack whole, sliced or mashed.

Prepare. Sort for size; wash. Leave skins on.

Precook. Cook, until almost tender, in water, steam, a pressure cooker or an oven. Cool at room temperature. Peel; cut in halves, slices or mash.

Prevent darkening. Dip whole peeled sweet potatoes or slices for 5 seconds in a solution of 1 tablespoon citric acid or ½ cup lemon juice and 1 quart of water. For mashed sweet potatoes mix 2 tablespoons orange or lemon juice with each quart.

Pack. Leave ½ inch of headroom.

Pack variations. Roll slices in sugar. Or cover whole or sliced with a cold 50 Percent Sirup. In either case, leave appropriate headroom.

Seal; freeze.

Tomatoes

Tomatoes are so easy to can—and are so handy in several table-ready forms—that we question the feasibility of freezing them. And aside from taking up a good deal of freezer space, a frozen whole tomato has limited appeal: its tender flesh is ruptured by ice crystals, and you have a deflated mush when you defrost it.

STEWED TOMATOES

Prepare. Remove stem ends and cores of ripe tomatoes; peel and quarter.

Cook. In a covered enameled or stainless-steel kettle, cook gently in their own juice until tender—10 to 20 minutes. Set the kettle bodily in cold water to cool the contents.

Pack. Leave appropriate headroom.

Seal; freeze.

TOMATO JUICE

Prepare. Cut vine-ripened tomatoes in quarters or smaller. In an enamel or stainless-steel kettle simmer them in their own juice for 5 to 10 minutes—or until tender with a good deal of liquid. Put through a sieve or food mill. Season with ½ teaspoon salt to each pint of juice, or 1 teaspoon to each quart.

Pack. Leave appropriate headroom.

Seal; freeze.

Turnips, White

Turnips are similar to rutabagas, but they mature more quickly. Freeze them in cubes or fully cooked and mashed. They also keep well in the root cellar.

Cubes: treat like Rutabagas.

Mashed: treat like Winter Squash or Pumpkin.

Freezing Meat

For handling meat from large animals, see the introduction to Canning Meat and Poultry: the same safeguards and methods for cutting meat apply to meat that is to be frozen.

To avoid disappointment, before you buy a whole, half or quarter of an animal, find out what cuts it will yield—and how many. There is 20 to 25 percent waste, regardless; and usually there will be more pounds of stewing and/or ground meat than pounds of steaks and roasts and chops.

COOKING FROZEN MEAT

Generally, any cut of meat may be cooked either frozen or thawed—which leaves the decision up to you. How do you plan to serve it?

THAW THESE:

—Meat to be coated with crumbs before cooking.
—Meat to be browned as the first step in cooking.
—Ground meat that must be shaped for cooking.
—Large roasts: they can overcook on the outside before the inner part is done, if they're not defrosted first.

APPROXIMATE THAWING TIMES FOR MEAT

In the refrigerator: large roasts—4 to 5 hours per pound; small roasts—3 hours per pound.

At room temperature: 1 hour per pound for roasts and packages of stewing-size pieces or ground meat.

THESE MAY BE COOKED FROZEN:

—Preshaped ground meat patties.
—Meat loaves.
—Thin steaks or chops.
—Meatballs in their own gravy or broth.

APPROXIMATE TIMES FOR COOKING
FROZEN MEAT

Juices rich in B vitamins seep out of all frozen meat and poultry as they defrost. Therefore, if possible thaw meat completely before cooking it, and save the dripped-out juice for the pan gravy. However, juice can be kept in chops or ground-meat patties if they are cooked as soon as ice crystals have disappeared from their *surfaces*. Also, large pieces (roasts) may be put in a preheated oven when the surface yields to the pressure of your hand.

Roasting. If you're caught short of time and must roast a big piece of frozen meat, do it in a preheated oven *about 25 degrees lower* than generally used for roasting unfrozen meat (that is, do it in an oven not more than 300 F.), and *increase the roasting time by one-half.* Neither adjustment need be made for day-long or overnight roasting of frozen meat at below 200 F. temperatures.

Broiling. Boil frozen meat of any thickness *at least 5 to 6 inches below* the heat source, and *increase broiling time by one-half.*

Pan-broiling. Cook frozen *thin* hamburgers, chops and steaks in a *very hot skillet* with a small amount of fat swished around to keep meat from sticking.

Start to cook frozen *thicker* patties, chops and steaks in a *warm skillet* with 1 tablespoon of fat. Heat the meat slowly and turn it until thawed. Then *increase the heat* and pan-broil the meat as for unfrozen thin cuts.

−20 F. = −28.9 C. / 0 F. = −17.8 C. / 212 F. = 100 C. / 1 U.S. cup (½ U.S. pint) = .24 L. / 1 U.S. pint (2 cups) = .48 L. / 1 U.S. quart = .95 L. / 1 U.S. tsp. = 5 ml. (5 cc.) / 1 U.S. T. = 15 ml. (15 cc.) / ½ inch = 1.25 cm. / 1 inch = 2.5 cm.

Freezing Roasts

Prepare. Trim away excess fat. Wipe with a clean damp cloth. Pad protruding sharp bones with fat or with extra wrapping, so they can't pierce the package.

THE DRUGSTORE FOLD

THE BUTCHER WRAP

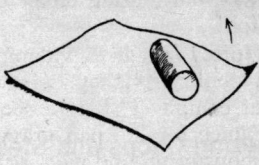

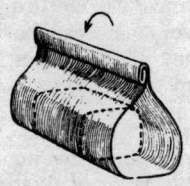

Roll folded edge down, turn over

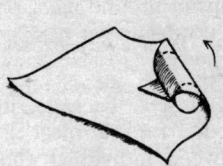

Fold ends of roll down

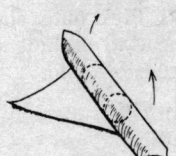

Ends up and over, tuck tight

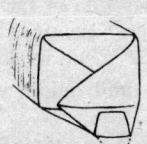

Sides over end

Tuck sides in

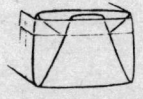

Fold tip of point over

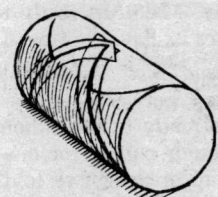

*Roll to end of paper—
seal open edges with tape*

Fold up and tape

Pack and seal. Package individual roasts tightly in sheet wrapping, using either the butcher wrap or drugstore fold.

Label; freeze.

Freezing Chops and Steaks

Prepare. Trim away excess fat. Wipe with a clean damp cloth.

Pack and seal. Package, in sheet wrapping, the number needed for one meal. Put a double layer of wrapping between individual chops/steaks or layers of chops/steaks. Press outer sheet wrapping closely to the bundle of meat to exclude air. Use either the butcher wrap or drugstore fold.

Label; freeze.

Freezing Ground Meat

Use only freshly ground meat to freeze as patties, loaves or in bulk. Freshly made pork sausage (see recipe in the Roundup section) also may be frozen; but its freezer life is short because of its high fat content.

PATTIES

Prepare. Make up ready to cook—but omit onion, which loses its flavor in frozen foods.

Pack and seal. Put double layers of lightweight freezer wrap between patties for easy separation when you are ready to cook them. In each bundle, tightly wrap enough patties for one meal, using either the butcher wrap or drugstore fold.

Label; freeze.

LOAVES, COOKED

Prepare. Cool cooked loaves, remove from baking pan.

Pack and seal. Wrap tightly, using either the butcher wrap or drugstore fold.

Label; freeze.

LOAVES, UNCOOKED

Prepare. Mix loaves as for baking. Line loaf pans with foil; fill with meat-loaf mixture and fold ends of foil over meat. *Freeze.*

Pack and seal. Remove loaves from pans when frozen, and overwrap tightly, using either the butcher wrap or drugstore fold.

Label; store in freezer.

BULK

Pack and seal. Put meal-size quantities in freezer boxes or bags, excluding air. Seal tightly.

Label; freeze.

Freezing Stew Meat

Prepare. Cut in cubes. For easier use later, pan-brown them; but they may be packed without browning.

Pack and seal. Fill rigid containers with meal-size portions of browned cubes. Cover with pan liquid or broth, leaving ½ inch of headroom. Seal.

Pack unbrowned cubes in rigid containers, freezer bags or sheet wrapping, excluding air. Seal tightly.

Label; freeze.

Freezing Cooked Meat

Prepare. It's better to freeze cooked meat or poultry in large pieces (so less surface may be exposed to air). Slices of meat or poultry keep best if covered with broth or gravy.

Pack and seal. Large pieces are wrapped tightly, using either the butcher wrap or drugstore fold. Slices are

stored in rigid containers of suitable size and covered
with broth or gravy, then closely covered and sealed.
Label; freeze.

Freezing Store-bought Cuts

(Meaning those prepackaged fresh meats from the
market's display case.)

Pack and seal. Remove the store wrapping—even though
it is well sealed; discard the tray, and rewrap and
seal the meat closely in your own freezing materials.
This will close out air and give the meat a more durable cover. (There's too much air held in store packages—and this causes freezer burn; also, the clear
film that's O.K. to sell it in is not strong enough for
freezer storage.)

Label; freeze.

Freezing Poultry and Small Game

*For handling domestic poultry, wild birds, domestic
rabbits and small game, see the introduction to Canning
Meat: the same safeguards apply to freezing these foods.
For specifics on dressing—plucking/skinning and drawing
—and cutting up the meat, see Canning Poultry.*

All freshly killed and dressed birds are better if stored
in the refrigerator for 12 hours to develop their greatest
tenderness before freezing.

COOKING FROZEN POULTRY

Here again, "poultry" applies to domestic and wild
birds, domestic rabbits and small game.

For best results, thaw before cooking (unless you're
boiling it to use in a fricassee or such): roasting and broiling is more uniform if the poultry is thawed first, and the
meat is less likely to be dry or rubbery. Pieces to be coated
before frying, or browned before stewing, should always

be thawed beforehand. It's easier to stuff a thawed bird
than one that's still frozen.
Cook all poultry immediately after thawing.

APPROXIMATE THAWING TIMES
FOR FROZEN POULTRY

Thaw it in its freezer wrappings.

In the refrigerator: 2 hours per pound.

At room temperature: 1 hour per pound.

Before a cold-air fan: 20 minutes per pound. Don't try to
hasten thawing by using a hot-air fan.

Freezing Birds Whole

Any bird may be frozen whole for future stuffing and
roasting.

Prepare. Tie legs of dressed, washed birds together with
thighs close to body; press wings snugly against breast.

Pack and seal. Put bird in a freeze bag, press out air and
tightly close the bag top. Or wrap the bird in moisture-
vapor-resistant material (see wrapping illustrations);
seal tightly. Pack and freeze giblets separately.

Label; freeze.

STUFF IT LATER

Even if the cook is careful at every step, dangerous bacteria causing food spoilage can develop in
poultry stuffed at home and then frozen: *normal
roasting will not kill heat-resistant bacteria in the
center of the densely stuffed cavity.* Prestuffed frozen
birds sold by big commercial processors are prepared
under controlled conditions of temperature and humidity, etc., that cannot be duplicated in the home.

Freezing Birds in Halves

Prepare. Split dressed, washed birds lengthwise and cut off the backbone (use it in soup stock).

Pack and seal. Put a double layer of freezer paper between the halves. Pack and seal in a freezer bag or wrap as for Whole.

Label; freeze.

Freezing Birds in Smaller Pieces

Prepare. Cut in pieces suitable for intended use (see Canning Poultry).

Pack and seal. Put a double layer of lightweight wrap between meaty pieces and pack them snugly together in freezer bag or carton; wrap tightly in sheet material. Seal.

Label; freeze.

Freezing Birds in Cooked Dishes, see Convenience Dishes.

Freezing Fish and Shellfish

Freezing is the best method for preserving the fresh qualities of fish and shellfish (but see Canning, and also Drying and Curing).

For ideal results, freeze all seafood right after it's caught. Or at least within 24 hours, and keep it packed in ice or in the refrigerator while it waits. Detailed instructions for immediate handling of fish and shellfish are given in the Canning Seafood section: do read them.

Any fish or shellfish are safely and easily frozen, although certain species' characteristics dictate slight differences in the way you prepare them for the freezer.

All fish and shellfish must be stored at ZERO FAHRENHEIT (−18 Celsius) *after initial sharp freezing at* c. −20 F.

PRELIMINARIES TO FREEZING FISH

For handling, fish may be divided into two categories: Lean and Fat.

The Fat—mackerel, pink and chum salmon, ocean perch, smelt, herring, lake trout, flounder, shad and tuna— are more perishable than the leaner varieties; plan to freezer-store these not more than 3 months.

The Lean fish—cod, haddock, halibut, yellow pike, yellow perch, fresh-water herring, Coho and King and red salmon—all keep well in frozen storage up to 6 months.

DRESSING (CLEANING)

Scale the fish (or skin it, depending on the variety); remove fins and tail. Slit the belly with a thin-bladed sharp knife and remove the entrails, saving any roe. Cut off the head if you wish. Wash the fish in cold water.

FLAVOR-PROTECTING DIPS

The Fat fish (and roe) are given a 20-second dip in an *ascorbic-acid* solution—2 teaspoons crystalline ascorbic acid dissolved in 1 quart of cold water—to lessen the chance of rancidity and flavor change during storage.

The Lean fish are dipped for 20 seconds in a *brine* of 1 cup salt to 1 gallon of cold water; this firms the flesh and reduces leakage when the fish thaws.

GLAZING WITH ICE

Sometimes whole fish or pieces of fish are ice-glazed before wrapping. This helps keep the air away, thus saving the flavor. The fish is frozen until solid, then dipped quickly in and out of ice-cold water, whereupon a thin coat of ice will form on the fish. Repeat several times to thicken the ice, then wrap the fish for storage.

CUTTING TO SIZE

Fish are frozen whole if they are small enough (under 2 pounds); or are cut in steaks—crosswise slices about 1

inch thick, or are filleted. Exception: large-ish fish you
expect to bake whole, you freeze whole.

Fillets are made usually from fish weighing 2 to 4
pounds. Lay the cleaned fish on its side on a clean cutting-
board. Run a thin-bladed sharp knife the length of the
backbone and slightly above it, and continue cutting to
separate the side of the fish from the backbone and ribs;
repeat on the opposite side. (This works on most fish; but
not on shad—whose build is so complicated that it takes
special skill to fillet them.)

COOKING FROZEN FISH AND SHELLFISH

With two exceptions, frozen seafood may be cooked
when still frozen—the exceptions being a large whole fish
you're baking and pieces that are to be crumbed or coated
with batter before cooking.

Small whole fish—under ½ pound—may be defrosted
just enough to separate them before they're fried (without
crumbs or batter coating) or broiled on a greased broiler.

Fish fillets and steaks are baked or poached from the
frozen state; they're partially defrosted before broiling or
frying (without crumbs or batter coating).

Shellfish and fish for stews, chowders and Newburgs
are cooked still frozen.

−20 F. = −28.9 C. / 0 F. = −17.8 C. / 212 F. = 100 C. / 1 U.S. cup (½
U.S. pint) = .24 L. / 1 U.S. pint (2 cups) = .48 L. / 1 U.S. quart = .95
L. / 1 U.S. tsp. = 5 ml. (5 cc.) / 1 U.S. T. = 15 ml. (15 cc.) / ½ inch =
1.25 cm. / 1 inch = 2.5 cm.

Freezing Large Whole Fish

Prepare. Dress (clean) as above, removing the head if you
wish.

Pack and seal. Freeze-glaze with ice (q.v. above). Wrap
snugly with moisture-vapor-proof covering, using the
butcher wrap or drugstore fold—then overwrap for
security. Seal.

Label; store in freezer.

Freezing Small Whole Fish

Prepare. Dress as for large whole fish, leaving on heads if you like.

Pack and seal. Small whole fish are most often packed in rigid containers with added cold water to fill crevices between the fish. Hold the lid tightly on the container with freezer tape wrapped around the rim, and over-wrap with moisture-vapor-proof freezer paper, using the butcher wrap or drugstore fold. Seal.

For easy separation in thawing, individual small fish may be enclosed in a household plastic bag or other clear wrapping before going into the rigid freezer containers. Proceed with overwrap, and seal.

Sport fishermen often freeze their catch covered with water in large bread pans or the like, the whole thing sealed in a freezer bag, and frozen. When solid, the block of fish-in-ice is removed from the pan and tightly wrapped in moisture-vapor-proof material, closed with the butcher wrap or drugstore fold, then sealed.

Label; store in freezer.

Freezing Fish Fillets and Steaks

Prepare. Dress and cut up strictly fresh fish. Treat Fat fish pieces with the ascorbic-acid dip, or Lean fish pieces with the brine dip (q.v.). Fillets may also be glazed with ice before wrapping.

Pack and seal. Fill rigid containers with layers of fillets or steaks, dividing layers with double sheets of freezer wrap for easy separation when frozen. Cover and seal.

For even greater odor prevention, overwrap the container with sheet freezer material, using the butcher wrap or drugstore fold.

Layers of fillets and steaks may also be wrapped in bundles and sealed instead of going into rigid containers; the bundles are then overwrapped and sealed.

Label; freeze.

Freezing Fish Roe

Roe is more perishable than the rest of the fish, so it should be frozen and stored separately from the fish.

Prepare. Carefully wash each set of roe from strictly fresh fish and prick the covering membrane in several places with a sterilized fine needle. Treat the roe with an ascorbic-acid dip (q.v.), even though it may come from a Lean fish.

Pack and seal. Wrap each set of roe closely in lightweight plastic for easy separation when frozen, smoothing out all air. Pack in flat layers in rigid containers and seal; then overwrap the containers in moisture-vapor-proof material, using the butcher wrap or drugstore fold. Seal.

Label; freeze. (Sharp-freeze at −20 F.; store at Zero F. or below for not more than 3 months before using.)

Freezing Eels

Prepare. Skin the eel. Tie a stout cord tightly around the fish below the head and secure the end of the cord to a strong, fixed support (a post or whatever). About 3 inches behind the head, cut completely through the skin around the body of the eel, necklace fashion. Grip the cut edge of the skin and pull it downward, removing the entire skin inside out.

Remove the entrails; wash the eel.

Cut in fillets or in the more usual steak-type rounds.

Because eel is a Fat fish, treat the pieces with an ascorbic-acid dip. Pack as for fillets or steaks of other fish, above. Seal.

Label; freeze.

Freezing Crab and Lobster Meat

Prepare. Scrub frisky live crabs and lobsters, butcher; cook the meat as described fully in preparation for Canning Shellfish (q.v.). Rinse and cool under running water;

pick the meat carefully, removing all bits of shell and tendon.

Pack and seal. Fill rigid containers solidly with meal-size amounts; add no liquid, but leave ½ inch of headroom. Seal.

Label; freeze.

Freezing Shrimp

As with other shellfish, shrimp you freeze must be absolutely fresh. They are best frozen raw, though they may be precooked as for the table before you freeze them.

RAW SHRIMP

Prepare. Wash, cut off the heads and take out the sand vein. Shelling is optional. Wash again in a mild salt solution of 1 teaspoon salt to each 1 quart of water. Drain well.

Pack and seal. Pack snugly in rigid freezer containers without any headroom. Seal tightly.

Label; freeze.

COOKED SHRIMP

Prepare. Wash in a mild salt solution of 1 teaspoon salt to each 1 quart of water; remove heads. Boil gently in lightly salted water for 10 minutes. Cool. Slit the shell and remove the sand vein (for table-ready use remove shells and vein). Rinse quickly. Drain.

Pack and seal. Pack snugly in rigid freezer containers, without any headroom. Seal tightly.

Label; freeze.

Freezing Oysters, Clams, Mussels and Scallops

Probably the most perishable of the shellfish, these should be frozen within hours of the time they leave the sea and *held at refrigerator temperature* (c. *36 F.*) during any waiting period.

Cooked oysters, clams and mussels toughen in the freezer: freeze them raw.

Prepare. Wash in cold water while still in their shells to rid them of sand. Shuck them over a bowl to catch the natural liquid. Wash them quickly again in a brine of 4 tablespoons salt to 1 gallon of water.

> Shucking is removing the shells. Since shucking bivalves (oysters, clams, etc.) involves severing the two strong muscles that close the two halves of the shell, you can cut yourself badly if you go about it wrong. *DO NOT USE a sharp or pointed knife.* Instead, use a dull blade with a rounded tip; insert it between the lips of the shell just beyond one end of the hinge, twist to cut the muscle at that point, and repeat at the other end of the hinge. A good shucker does it in one continuous *safe* motion: get someone who knows how, to show you.

Pack and seal. Put in rigid containers and cover with their own juice, extended with a weak brine of 1 teaspoon salt to 1 cup of water. Scallops are packed tightly, then covered with the brine (they have little juice). Leave ½ inch of headroom—unless the container is larger than 3-cup size (more than 3-cup is not advised: it takes too long to defrost; see the table "Headroom for Freezing Fruits and Vegetables"). Seal tightly.

Label; freeze.

Freezing Dairy Products and Eggs

Freezing Milk

Homogenized milk freezes smoothly and may be stored for up to 3 months. But the fat in milk that is not homogenized separates out as flakes, and these won't blend again when the milk thaws.

Prepare and seal. Homogenized milk in its tightly sealed carton—just as it comes from the dairy—is ready for

the freezer. Or it may be transferred to rigid freezer
containers and sealed with appropriate headroom.

Label; freeze.

Freezing Cream

Frozen heavy cream whips nicely, but it separates and
sends an oily film to the top of hot coffee. Probably it is
frozen best as dollops of whipped cream to be used on
individual cold desserts.

Prepare and seal. Whip; sweeten if you like. Drop in small
mounds on a cookie sheet. Freeze uncovered. Remove;
pack with a double layer of plastic wrap between the
dollops in rigid freezer containers. Seal.

Label; freeze.

Freezing Butter

Salted butter. Freeze in its store carton for a month or two.
For longer storage, overwrap and seal.

Unsalted (fresh) butter loses flavor more quickly than salted
does. Overwrap, seal, and freeze it immediately after
you make it or get it.

Freezing Ice Cream

Either purchased or home-made ice cream keeps its
quality up to 2 months in the freezer.

Prepare and seal. Purchase or make the ice cream. Pack in
rigid, tight-sealing containers. Seal.

Label; freeze. (When using up only part of a container,
cover the remainder with a sheet of plastic to keep air
away, so crystals won't form.)

Freezing Cheese

Cheeses which freeze well are Camembert, Port du
Salut, Mozzarella, Liederkrantz, and their cousins, and
Parmesan.

Cheese with a high fat content (such as Cheddar, Swiss, and American brick, etc.), are best kept at refrigerator temperatures (32 to 40 F.). If you have more than you can use soon, though, cut it in ½-pound (or less) pieces, wrap each piece tightly, label and freeze.

Plain cream cheese (fatty) mixed with cream for dips, etc., will freeze satisfactorily.

If the curds of cottage cheese are *not washed*, it keeps quite well. This means you can freeze home-made cottage cheese, but not the commercial kind.

Freezing Eggs

Freezing is a fine way to put by seasonably abundant eggs (see also Waterglassing), but frozen eggs have limited use: because they may be bearers of organisms that cause acute intestinal infections, for safety's sake they should be used *only* in long-cooked or long-baked foods.

Bacteria enter nest-cracked eggs or cling to the shells of unbroken ones. So avoid cracked eggs, and wash sound eggs carefully and wipe them dry before breaking the shells. Then, for extra insurance, reserve defrosted eggs for use in cakes, fancy breads and similar things (*not* in mayonnaise, eggnogs, custard, scrambled eggs or omelets, etc.).

Use only strictly fresh eggs—those kept refrigerated from nest to freezer.

Open the eggs (freezing expansion would break the shells) and pack them in meal- or recipe-size quantities.

If you're freezing in bulk, though, about 10 whole eggs *or* 16 large egg whites *or* 24 egg yolks fill a 1-pint container.

Note: Examine each egg for quality before adding it to others, whether you're freezing them whole or separated. For whole eggs, break each into a saucer, and look for desirable firm whites and plump, high-standing yolks; for separated eggs, put each white and each yolk in a saucer before adding either to the batch you are freezing.

WHOLE EGGS

Prepare. Examine each egg as above. Break all yolks with a fork, then slowly and carefully mix them with the whites so as to get *no foam*: beaten-in air bubbles will dry out the eggs.

To keep the yolks from thickening add 1 teaspoon salt to each 1 pint for eventual main dishes, etc., or 1 tablespoon sugar to each 1 pint of eggs for future desserts.

Pack and seal. Fill rigid containers, leaving ½ inch of headroom. Seal.

Label; freeze. (Be sure the label tells how many eggs, and whether salt or sugar was added.)

EGG YOLKS ALONE

Prepare. Mix yolks well but without foaming; add ½ teaspoon salt to each ½-pint of yolks for main dishes or 1½ teaspoons sugar to each ½-pint for future dessert uses.

Pack and seal. Fill rigid containers, leaving ½ inch of headroom. Seal.

Label; freeze.

EGG WHITES ONLY

Prepare. Separate whites from yolks of good-quality eggs.

Pack and seal. Just fill containers *without* added salt or sugar, leaving ½ inch of headroom. Seal.

Label; freeze.

Thawing Frozen Eggs

It's best to thaw eggs in the refrigerator, where it takes 8 to 10 hours to thaw a pint. They may be thawed in their containers under cold *running* water in 2 to 3 hours.

After defrosting, use them like fresh eggs, but in long-cooked dishes only.

Never refreeze thawed eggs!

Freezing Convenience Dishes

Keep your freezer busy—at least ¾ full. Restock it with cooked or ready-to-cook dishes as the traditional meats and vegetables, etc., are used up. You'll be ready for unexpected guests, parties, emergencies or a "goof-off" day.

The pros and cons and procedure for doing foods frozen in the so-called cook-in pouches or bags are discussed at the end of this section.

MAKE YOUR OWN BONUS

Fix a larger amount than normal of a food your family likes; eat some and freeze the surplus in meal-size packages. Certain slow-cooked foods, or ones calling for special equipment that's a nuisance to use, are freezer-fodder. Best feature: you can work at your convenience.

PACKAGING

Cool food quickly. Pack in family-size amounts, using appropriately sealed freezer wraps and/or containers; and, for surest results, overwrap and seal most convenience foods with a second layer of freezer wrapping. Store at Zero F. or lower.

A *good trick for casseroles:* Line the dish you'll later reheat and serve it in with heavy-duty foil, fill with food, and freeze until solid; remove the shaped block of food and its foil liner and overwrap it and seal. Label; store in the freezer.

Some favorite casseroles that freeze especially well are included in Recipes (Chapter 10).

Freezing Main Dishes, in General

EASY TO FREEZE

Creamed foods, stews, most casseroles, meat pies, croquettes and spaghetti sauce freeze well, as do cooked meats if covered with broth or gravy.

Undercook any vegetables or pastas you're including in such a main dish: they finish cooking when the whole thing is reheated.

BEST THICKENERS

If you're making a flour-thickened sauce that you plan to freeze, use *waxy* corn flour or *waxy* rice flour instead of the usual wheat flour: they make a clearer sauce, and it's less likely to separate when it's defrosted and reheated. These waxy flours are sometimes hard to find (corn*starch* is not the answer, because it separates too); so try natural-food stores, or the gourmet sections of very large supermarkets.

If you do find the waxy flours, use them measure for measure like wheat flour, but double the amount if you're substituting them for cornstarch.

PITFALLS TO SUCCESS:

—Fried foods almost always tend to become rancid, tough and dry; store them for only a short time.

—Sauces heavy in fat are likely to separate when reheated; stirring well usually recombines them.

—Sauces with much milk or cheese often curdle from freezing (but they're still good to eat).

—*Un*cooked potatoes change texture unpleasantly when frozen.

—Hard-cooked egg *whites* become tough when frozen.

—Flavors of garlic and clove grow stronger during freezing; onion, salt and herb flavors vanish.

—Crumb-and-cheese toppings on casseroles get soggy in the freezer, so add them when food is to be reheated.

TO REHEAT

Stews and creamed dishes are reheated in a double-boiler, or in a heat-proof dish set in a pan of hot water in a 350 F. oven; stir as little as possible. Precooked casseroles also may be reheated in a pan of water if the freezer container is oven-proof.

Thaw croquettes or other breaded foods uncovered in the refrigerator before heating or frying.

Freezing Baked Goods

BREADS

Fresh-cooked muffins, biscuits, yeast bread and rolls, thoroughly cooled and snugly wrapped, freeze nicely.

Loaves of garlic-buttered bread can be prepared, foil-wrapped, and frozen for later heating.

Uncooked *thin-rolled* biscuits can be frozen.

Uncooked loaves of bread dough—made with double the usual quantity of yeast—may be frozen. Let them rise after defrosting (which takes about 4 hours) at room temperature before baking them.

PIE PASTRY

Pie pastry you plan to freeze should have the maximum amount of shortening your recipe calls for, so the crust will have the desired tenderness (freezing could make the texture cardboardy if you scamped on the fat).

Both baked and unbaked pie crusts freeze well.

Baked crusts. These break less if wrapped and frozen in their pie pans.

Unbaked crusts. Roll these out to size and stack on heavy cardboard cut to the same size and covered with waxed paper. Lay on it a flat rolled crust, cover the crust with *two* sheets of waxed paper, then add another rolled crust. Up to 6 crusts may be stacked, ending with a waxed-paper cover. Wrap and seal the entire pile; label and freeze.

To use uncooked pastry, remove one or more crusts; let thaw for 15 minutes with the waxed paper still above and below each crust to keep moisture from condensing on crust as it thaws.

Label; freeze.

FILLED PIES, UNCOOKED

Fresh fruit or mincemeat are the best fillings for freezing.

Prepare and seal. Prepare pies as usual—except brush the bottom crust with shortening to prevent the filling from soaking into it; and brush the top crust of the finished pie with shortening (*not* milk, water or egg), to keep it from drying out.

Wrap tightly in freezer wrap, using the butcher wrap or drugstore fold. Seal. An outer cardboard box safeguards against breakage in storage.

Label; freeze.

FILLED PIES, COOKED

Cooled fresh fruit, mincemeat and chiffon pies freeze well.

Prepare and seal. Make and bake the pies as for immediate use. *Thoroughly cool them,* and then wrap securely with freezer wrap, using the butcher wrap or drugstore fold. Seal.

Label; freeze.

TO COOK FROZEN PIES

Bake *uncooked* pies in the frozen state in a 450 F. oven for 15 to 20 minutes, then reduce heat to 375 F. until done.

Precooked pies to be served cold are thawed at room temperature for 8 hours. To serve hot, heat the frozen pie in a 400 F. oven for 30 to 35 minutes.

FREEZING CAKES AND COOKIES

Butter, sponge, angel, pound, chiffon and fruit cakes and almost all cookies, are successfully frozen after baking.

Icing is better put on when you are ready to serve the cakes—except for the butter-confectioner's-sugar type (which does freeze well).

Prepare and seal. Make by your favorite recipe. Remove from pan; cool thoroughly. Use the butcher wrap or drugstore fold; use an outer rigid container for cookies. Seal.

Label; freeze. (To defrost: leave, wrapped, at room temperature.)

COOKIE DOUGH

Prepare and Seal. Shape the dough in rolls for easy slicing (or in bulk—which must be thawed before cookies can be shaped).

Also, cookies may be cut, frozen on a tray, then packaged with freezer paper between layers for easy handling later.

Package rolls of dough or cut cookies tightly with freezer wrap and seal.

Label; freeze.

FREEZING YOUR OWN TV DINNERS

Meal leftovers can make better-than-store-bought TV dinners if they are carefully frozen. Freeze the portions either assembled on divided aluminum-foil plates, or freeze them separately and assemble them later when you're reheating them to serve. If you're using divided foil plates or trays, be most careful to press your covering of freezer foil tightly to the food to expel as much air as possible, and to seal the edges completely. Overwrap and seal. Label carefully; freeze.

Ideas for combination plate dinners:

—Swiss steak, french-fried potatoes, peas.
—Beef stew, hot bread, asparagus.
—Meat loaf, candied sweet potatoes, spinach.
—Corned-beef-hash patties, pan-fried potatoes, string beans.
—Ham steak, candied sweet potatoes, peas.
—Breaded veal cutlet, mashed potatoes, carrots.
—Turkey and dressing, giblet gravy, mashed sweet potatoes, peas.
—Pot roast, gravy, brown potatoes, corn, peas.

—Fried chicken, mixed vegetables, mashed potatoes, gravy.

—Ham croquettes, stuffed baked potatoes, broccoli.

FREEZING DABS AND SNIPPETS

When you're chopping onions or celery or green peppers anyway, cut up extra amounts and package them individually or combined in packets the right size for future dishes.

Other foods good to have ready to use in a minute are snipped chives, chopped nuts and breadcrumbs. And freeze a bunch of tied-together parsley. Just snip off what you need while the bunch is still frozen, then rewrap the remainder and return it quickly to the freezer before it has a chance to defrost.

—But let your own originality take over! Try what tempts you.

Cook-in-Pouch Freezing

There is nothing basically different about freezing cook-in-pouch foods, but they do call for special packaging—and success depends on it: the food must be put into extra-impervious bags, which are then heat-sealed. These bags (usually polyester and always with super-strong seams) are made to withstand hard boiling for up to 30 minutes to defrost/heat or to finish cooking the contents. *Regular freezing bags won't work for this.*

Feasible?—Maybe

We'll start with the drawbacks, so you can weigh them against the benefits to your family food program.

Equipment is expensive. Because a hot flatiron cannot be guaranteed to make an adequate seal even when wielded with care, you should have a special electric heat-sealer. This costs about as much as a very good automatic toaster; the price is high for just one more fun gadget that takes up space, but it's fair for a sturdy appliance you plan to use often. The leading mail-order firms offer sealers under proprietary names; you should also be able to get them in well-stocked housewares departments of large stores.

The special pouches/bags are included with the sealer, and they can be ordered separately from the outfits that sell the sealers. Usually listed in a range of small sizes, they cost roughly four times as much as conventional freezer bags—which regularly come in much larger sizes to boot. To get unusual sizes of the special pouches, ask your local office-supplies store to order them for you from the "3M" Company regional distributor.

The preparation and packaging is more complicated than for conventional freezing. Cook-in-pouch foods are either fully cooked (to be reheated), or are at least half-cooked (to be finished), before they are sealed and frozen in relatively small portions. Therefore you should get organized for a fairly concentrated session, and this means more planning ahead.

You'll wash the pot anyway. We doubt that anyone truly interested in putting food by will go in for cook-in-pouch freezing just because the admen's brochure stress that it saves washing a saucepan. We wish they'd stop, and anyway it doesn't: boiling one of the convenient little bags for 15+ minutes will leave a "bathtub ring" even if your tap water is soft.

On the plus side, though, are these advantages to having cook-in-pouch foods in your freezer.

For special diets. It's not much of a trick to prepare without salt, say, a regular quantity of some good dish and seal it, in individual servings, in boilable bags. Or perhaps the texture of the food must be different from that which is served to the rest of the family. Or maybe it's for a special treat to tempt the appetite of a convalescent; or maybe the portions should be carefully pre-measured.

For special occasions. These handy pouches of food are notably husband- and teenage-proof, because anyone who can read the boil-in time required (included on the label before it went in the freezer) can hardly go wrong in preparing it. They can be a godsend for family members working odd shifts, or for a crowd of guests needing staggered meals; the alternative could be dining off sandwiches or bits and pieces scrounged from the refrigerator.

FILLING POUCHES AND SEALING

Three hands are better than two here. And the sealer can be mounted on a wall or the side of a cupboard; or, as shown in the picture, it can rest on a chopping block to allow the weight of the food to pull the pouch down and help to expel air.

Our sealer came with an elliptical plastic collar that fits inside the pouches (all of which are the same width— it's their height that varies for the different capacities) and holds the bag open wide for filling. The collar also prevents dribbles from reaching the inside sealing surfaces, which must be entirely free of food particles in order to seal right.

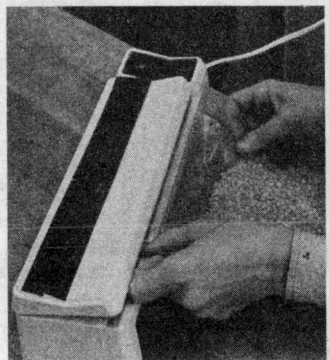
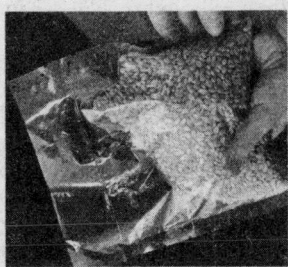

Heat-sealing the special cook-in pouch; note headroom flattened, above.

If you don't have the collar, make some by cutting the bottoms from small-size aluminum-foil bread pans, then pinching little tucks in the foil to reduce the circumference of the pan to the right size. These handy makeshifts let you or your helper keep several propped-open bags ready for filling.

Rest the bottom of the collared bag on a pie dish; put in the food, *leaving 2 inches of headroom.* Remove the collar and, holding the bag by the sides near the top, snap it shut, thus forcing air out of the headroom.

Many sealers have pegs to help hold bags steady for sealing, and many pouches have perforations high on their sides to catch on the pegs (see photo). So, holding the bag upright and stretched between your hands, place its closed top in the sealer—being careful to maintain the

2-inch headroom at the sealing point—and operate the
heat/pressure according to the manufacturer's directions.

Cool the sealed pouches briefly in the refrigerator.
Then label and lay them flat in the freezer, preferably only
one layer deep until they're frozen; pat each bag to distribute the contents to be *uniformly about 1 inch thick.*
This even thickness is important: boil-in times are based
on it.

WHAT TO FREEZE IN COOK-IN POUCHES

Unless you have particular needs in mind, your likeliest
foods for cook-in-pouch freezing would be programmed
extras, sealed in 1-pint bags (about 3 average servings) or
1½-pint bags (5 servings). The mail-order sealer we're
most familiar with offers extra 1-quart bags, but a casual
consensus from the householders who actually go in for
this type of freezing is that amounts larger than 4 to
5 servings are probably handled more easily over-all if
they're packaged conventionally for freezing, and are decanted for regular reheating on the stove-top or in the
oven.

These extras can be grouped roughly as main dishes
(fully precooked, just to be reheated); side-dish vegetables
(partly precooked, to be finished in the boil-in bag); and
dessert fruits (raw, to be defrosted by putting the sealed
pouch in a bowl of warm water). The manufacturers' pam-
phlets include many more dishes, but a number of them
seem like too much fuss for the benefit, frankly; you'll judge
for yourself.

Main Dishes

Thin slices of meat in its gravy, stews, chicken à la king,
fillets of fish in a favorite sauce, creamed things—the list
goes on. Prepare as for the table: *all are fully precooked.*
(Remember that any thickened sauces are best made with
waxy flour, as mentioned earlier in the Convenience Foods
introduction.)

Cool the food slightly—just enough so you can handle
the filled pouches—and pack it immediately, leaving 2 inches

of headroom for both 1-pint and 1½-pint bags (this air-space will flatten away as the bag is held to the sealer).

Boneless meats and fish take 18 minutes' boil-in time for 1-pint bags, 20 minutes for 1½-pint bags.

Casseroles and pastas take 13 minutes' boil-in time for pints, 15 minutes for 1½-pint pouches.

SIDE-DISH VEGETABLES

Of course all your vegetables will be perfectly fresh, and young and tender-crisp; carefully washed and cut/trimmed, etc., as for serving.

Precooking times given below are average for *half-* cooking the individual vegetables—which will then be finished during the boil-in time of 15 minutes for pints, 18 minutes for 1½-pint pouches.

Use only enough water to cover. Don't salt it now because salt, onion and herb flavors tend to disappear in freezing.

When it's half-cooked, the vegetable should be cooled only enough to allow you to handle the filled pouches comfortably. Leave 2 inches of headroom as for Main Dishes. Refrigerate the sealed bags for an hour before freezing them.

Unless you use stick butter or margarine that's easily sliced, slowly melt your butter or whatever, pour it ⅛ to ¼ inch deep in a bread tin; chill quickly until it's solid, and cut squares to insert in the pouches when you fill them.

Asparagus. Choose uniformly slender spears; trim to length to fit your pouches with 2 inches of headroom, or cut small. Cook gently in your usual manner for 5–8 minutes, depending on length of the pieces. Drain, cool slightly; pack and seal.

Beans—green/Italian/snap/string/wax. Cook gently for 10 minutes. Drain, cool slightly; pack and seal.

Broccoli. Cook split young spears gently for 5 minutes. Drain, cool slightly; pack and seal.

Carrots. Cook slices gently for 15 minutes. Drain, cool slightly; pack and seal.

Cauliflower. Cook prepared flowerets gently for 5 minutes. Drain, cool slightly; pack and seal.

Corn, whole-kernel. Husk, remove silk, wash. Cut from the cob over a bowl to catch the milk. In its milk—plus only enough water to keep from sticking—cook gently for 3 minutes. Drain, cool slightly; pack and seal.

Peas, green. Cook shelled peas gently for 10 minutes. Drain, cool slightly; pack and seal.

Spinach, etc. Remove stems and any tough midribs, cut large leaves in several pieces. Boil gently for 15 minutes. (Or, for very tender leaves, shake off extra water and steam-sauté in a little oil, covered, for half the full cooking time you use for this method—about 4 minutes.) Drain, cool slightly; pack and seal.

The Preserving Kettle   4

Beginning homemakers get their early canning experience via the preserving kettle; then, with confidence gained, they graduate to canning basic foods. And meanwhile the preserves add so much to any table—plus an extra thrill if they've won a prize at the fair!

There are many good sources for recipes. Most women's magazines offer them in season; many cookbooks have chapters on all types of preserves and pickles. Standbys are the USDA's *H&G Bulletin No. 56, How to Make Jellies, Jams, and Preserves at Home;* and *H&G Bulletin No. 92, Making Pickles and Relishes at Home;* Agriculture Canada's *Publication No. 992, Jams, Jellies and Pickles;* the *Ball Blue Book,* sold by the Ball Corporation, Muncie, Indiana 47302; the *Kerr Home Canning and Freezing*

THE FINISHING B-W BATH

Preserves (including jams and marmalades) and pickles can suffer from mold and other spoilage micro-organisms when storage is not the ideal 32 to 50 F. and dark and dry—conditions not always possible in warm, humid climates or in modern centrally heated homes. Therefore we recommend that all such foods from the preserving kettle *except jellies and the so-called Diet or Freezer jams* be packed in conventional ½-pint or pint canning jars with ½ inch of headroom. Then adjust the lids and process in a short Boiling-Water Bath (212 F.), complete seals if necessary, cool and store.

Book, sold by Kerr Glass Manufacturing Corporation, Sand Springs, Oklahoma 74063; and the *Bernardin Home Canning Guide,* sold by Bernardin, Inc., Evansville, Indiana 47701, and Bernardin of Canada, Ltd., Toronto 550, Ontario, Canada. Also, the containers of store-bought pectin include folders of recipes for jellies, jams and the like.

Some of the following "receipts," to use the pleasant old term, are from Beatrice Vaughan's heirloom collection, but are translated into today's methods. Ingredients for all are given for small, easy-to-work-with batches; especially with jellies, you get best results if you handle no more than 3 to 6 cups of juice at a time.

Complete metric conversion tables are in Chapter 1, but the following apply particularly to this section: 212 Fahrenheit = 100 Celsius / 220 F. = 104 C. / 1 U.S. cup (½ U.S. pint) = .24 Liter / 1 U.S. pint (2 cups) = .48 L. / 1 U.S. teaspoon = 5 milliliters (5 cubic centimeters) / 1 U.S. tablespoon = 15 ml. (15 cc.) / ½ inch = 1.25 centimeters

---

Jellies, Jams and Sweet Preserves

Jellies, jams, preserves, conserves, marmalades and butters are the six cousins of the fruit world. All have fruit and sugar in common, but differences in texture and fruit-form distinguish one from another.

Jelly. Made from fruit juice, it is clear and tenderly firm. Quiveringly, it holds its shape when turned out of the jar.

Jam. Made from crushed or ground fruit, it almost holds its shape, but is not jelly-firm.

Preserves. These are whole fruits or large pieces of fruit in a thick sirup that sometimes is slightly jellied.

Conserves. These glorified jams are made from a mixture of fruits, usually including citrus. Raisins and nuts also are frequent additions.

EQUIPMENT FOR JELLIES, JAMS, ETC.

If your jelling and jamming is modest in amount, your regular kitchen utensils will be adequate. The one essential addition is a jelly bag. To make it, fold ½ yard of 30- to 36-inch wide, sturdy unbleached muslin so selvage edges are together. Machine-stitch with durable thread down the side seam and across the bottom, leaving top open for filling. Wash before using, to remove any filler in the fabric. Have ready a strong yard-long white cord to tie the filled bag closed and suspend it so as to let the juice drip into a wide container.

Essential:

> Boiling-Water Bath container for processing jams, etc.
>
> 6- to 8-quart stainless steel or enameled kettle
>
> Jars/glasses in prime condition, with lids/sealers/gaskets ditto
>
> Minute-timer with warning bell to time processing periods
>
> Jelly thermometer
>
> Shallow pans (dishpans are fine)
>
> Ladle or dipper
>
> Long-handled wooden or metal spoon for stirring
>
> Wide-mouth funnel for filling containers
>
> Jar-lifter
>
> Siever or strainer for de-seeding blackberries or other large-seeded fruits
>
> Colander, for draining
>
> Large measuring cups, and measuring spoons
>
> Muslin bag for straining juices
>
> Plenty of clean dry potholders, dish cloths and towels
>
> Expendable small metal spouted pot in which to melt paraffin over water (or a small conventional double boiler)

Nice, but not absolutely necessary:

> Household scales
>
> Food mill, for puréeing
>
> Large trays

Marmalade. This is a tender jelly with small pieces of citrus fruit distributed evenly throughout.

Butters. These are fruit pulps cooked with sugar until thick.

THE FOUR ESSENTIAL INGREDIENTS

FRUIT

This gives each product its special flavor, and provides at least a part of the pectin and acid that combine with added sugar to make successful gels.

Full-flavored, just-ripe fruits are preferred, because their flavor is diluted by the large proportion of sugar that is added for good consistency and keeping quality. Never use overripe fruit.

Unsweetened frozen fruit makes good jelly and jam.

PECTIN

This substance, which combines with added sugar—or other sweeteners, *except* artificial ones—and natural or added acid to produce a gel, is found naturally in most fruits. (See the table "Pectin/Acid Content of Common Fruits," coming in a minute.) Pectin content is highest in lightly underripe fruit, and diminishes as the fruit becomes fully ripe; overripe fruit, lacking adequate pectin of its own, is responsible for a good deal of runny jam and jelly. Pectin is concentrated in the skins and cores of the various fruits: this is why many recipes say to use skins and cores in preparing fruit for juicing or pulping.

This natural pectin in the fruit can be activated only by cooking—but *cooking quickly,* both in heating the fruit to help start the juice, and later when juice or pulp is boiled together with the sugar. And *too-slow cooking,* or *boiling too long,* can reduce the gelling property of the pectin, whether natural or added.

In the old days apple juice was added ot less pectin-rich juices to make them gel, and this combination still works. Today, though, the readily available commercial powdered or liquid pectins take the guesswork out of jellies, jams, and the like.

Testing for pectin content. There are several tests, but the simplest one uses ready-to-hand materials. In a cup, stir together 1 teaspoon cooked fruit juice with 1 tablespoon rubbing alcohol (everyday 70 percent kind). No extra pectin is needed if the juice forms one big clot that can be picked up with a fork. If the fruit juice is too low in pectin, it will make several small dabs that do not clump together. DON'T EVER TASTE THE SAMPLES—rubbing alcohol must never be taken by mouth.

HOMEMADE PECTIN

Apple pectin can be made at home to add to the non-gelling juices of strawberries, cherries, rhubarb or pineapple to set a perfect jelly.

For about a pint of pectin you will need 10 pounds of apples, a 10- to 12-quart preserving kettle, and a good strong jelly bag.

Wash apples, cut in quarters *without peeling or coring,* but remove stems and blossom ends. Put the apples in the kettle, barely cover with cold water, and set over moderate heat; cover and cook slowly for about 30 miuntes, or until the fruit is quite soft. Turn the cooked apples into the dampened jelly bag, letting it hang overnight to extract as much juice as possible without squeezing. This will produce around 3 quarts of juice, which you boil down to about ⅙ or ⅛ its original volume to make 1½ to 2 cups of heavy, almost ropy, sirup. Strain before using.

To store for later use. Pour the hot, strained pectin into hot ½-pint canning jars; adjust the lids and process in a Boiling-Water Bath for 5 minutes. Remove jars; complete seals if necessary. Store in a cool, dry, dark place.

To use homemade pectin. Add ½ to ¾ cup apple pectin to 4 cups low-pectin fruit juice in a large kettle and bring to a boil; after 2 or 3 minutes of boiling add 2 to 3 cups of sugar for the amount of fruit juice used (4 cups, above) and boil rapidly until the jelly stage is reached (see "Testing for Doneness"). Skim, ladle into hot glasses; seal.

ACID

None of the fruits will gel or thicken without acid. The acid content of fruit varies, and is *higher in underripe* than in the fully ripe fruit.

Taste-test for acid content. This is a comparison. If the prepared fruit juice is not so tart as a mixture of 1 teaspoon lemon juice, 3 tablespoons water and ½ teaspoon sugar, your juice needs extra acid to form a successful gel. A rule-of-thumb addition would be 1 tablespoon lemon juice or homemade citric acid solution (for how to make it, see "Adding Acid to Canned Tomatoes") to each 1 cup prepared juice.

SWEETENERS

Sugar. This helps the gel to form, is a preserving aid and increases flavor in the final product. The sugar called for in the recipes for jellies, jams, and other preserves is, unless otherwise specified, refined white cane or beet sugar.

The semi-refined brown sugars (sometimes called "raw") differ in sweetening power from white sugar—a difference that could upset the balance needed for a successful gel in recipes calling for added pectin. The color and pronounced flavor of brown sugars will affect the looks and taste of jellies and jams, etc.

Corn sirup (light only). As a general rule, in recipes *without added pectin* you may substitute light corn sirup for ¼ the sugar called for in jellies, and up to ½ the sugar used in jams and preserves. Add it when you add the sugar. And be prepared to boil the mixture longer than usual to evaporate the extra moisture contained in the corn sirup.

In recipes using *powdered pectin,* light corn sirup may replace ½ the sugar needed in either jellies or jams. Where *liquid pectin* is used, light corn sirup may replace up to 2 cups of the sugar.

Honey (light-colored and mild-flavored seems to work best). Although some groups promoting the use of honey recommend substituting it measure-for-measure for sugar in making jellies and jams, we have not been satisfied with our results when we did so. Honey generally has *nearly twice the sweetening power* of the same amount of white sugar; this property, coupled with the distinctive flavor imparted by even mild honey in such quantity, was enough to eclipse the delicate fruit taste we like in jellies and jams. However, we were more successful when we replaced only part of the called-for sugar with honey, adding it when we added the sugar, and cooking the jelly/jam a bit longer to get rid of the extra moisture in the honey.

In recipes *without added pectin*, we suggest substituting no more than ½ the sugar with a mild-flavored honey. In recipes *with added pectin*, we replace no more than 2 cups of the required sugar with an equal measure of honey. *Caution:* In small batches (5- or 6-glass yield), no more than 1 cup of the sugar should be replaced by honey.

Sorghum and *molasses* are not recommended for making preserves because their flavors are so strong, and their relative sweetening powers are varied.

Artificial sweeteners require fairly detailed specific treatment when used in preserves, and are dealt with later under "Diet Jellies" and "Diet Jams." *Use only the artificial sweeteners recommended by your doctor:* Continuing research may have turned up undesirable side-effects even from sweetening agents formerly accepted as medically safe to use. And do read the labels to learn if their sweetening power is affected by processing or storage.

PECTIN/ACID CONTENT OF COMMON FRUITS

Group I. These fruits if not overripe usually contain enough natural pectin and acid to gel with only added sugar: apples (sour), blackberries (sour), crabapples, cranberries, currants, gooseberries, grapes (Eastern Concord), lemons, loganberries, plums (except Italian), quinces.

Group II. These fruits usually are low in natural acid or pectin, and *may* need added acid or pectin: apples (ripe), blackberries (ripe), cherries (sour), choke-

cherries, elderberries, grapefruit, bottled grape juice
(Eastern Concord), grapes (California), loquats,
oranges.

Group III. These fruits *always* need added acid or pectin, or
both: apricots, figs, grapes (Western Concord), guavas,
peaches, pears, prunes (Italian), raspberries, strawberries.

Jellies

STEPS IN MAKING COOKED JELLY

The recipes that follow are for cooked jellies—that is,
ones boiled with sugar and pectin as indicated. (Uncooked
jellies are discussed later.)

Always work with the recommended batch. The quantities given are tailored for success: the longer boiling
needed for larger amounts can prevent desired flavor and
texture in the finished product.

Preparing fruit and extracting juice

A rough, very rough, rule-of-thumb for estimating how
much fruit will be needed to make a particular batch of
jelly is: 1 pound of prepared fruit (i.e., washed, stemmed/
trimmed/cut as the recipe says to) will make 1 cup of
juice.

It's impossible to come up with one workable translation of weight to volume for all the fruits used in jelly-
making: the best help we can be is to give volume generally for fruits usually sold in small quantity by the
container (berries, sometimes currants, sometimes sweet
cherries, etc.), and give weight for fruits usually sold in
small quantity by the pound (apples, crabapples, quinces,
guavas, often grapes, sometimes sour cherries).

Plan to process the fruit as soon as possible after it's
picked or bought; refrigerate, for no more than 1 day, soft
fruits and berries if you can't handle them right away.
When you do start, keep at it and work right along.

Pick over the fruit carefully, discarding any that is overripe or has rotten spots. For a successful gel from recipes that have no pectin added, make up the amount called for with ¼ the total in underripe and ¾ in just-ripe fruit.

Wash the fruit quickly but thoroughly. Don't let it soak; lift it out of the basin of fresh water, don't pour it with the water into a strainer. The lighter and quicker you are in handling berries, the better. And always use good, clean drinking water for washing your fruit.

Remove the stems and blossom ends of apples and quinces and guavas, but retain their skins and cores. The skins of plums and grapes also contain a good deal of pectin, so keep them too. The stems and pits of cherries and berries need not be removed: the jelly bag will take care of them when the pulp is strained.

TO EXTRACT JUICE

Sparkling clear, firm jelly calls for carefully strained juice, which is not so diluted as to prevent a good gel. Therefore most modern recipes describe the way the juice is to be extracted—simply by crushing; or by short heating, with or without "enough water to keep from sticking"; or by longer cooking with more water added—and these instructions should be followed.

Sometimes, though, you will like the sound of an older recipe that's not explicit about method, so we offer the following rules-of-thumb as a help in figuring out what to do.

Always start heating the fruit at a fairly high temperature.

To heat ripe soft berries without any water, crush a layer in the bottom of the kettle to start the juice (mashing them with the bottom of a drinking glass, or with a pastry-blender); pile on the remainder and put the kettle on fairly high heat, stirring to mix the contents; reduce heat to moderate and boil gently and stir until all the fruit is soft—5 to 10 minutes.

To heat soft berries that are slightly underripe, Concord and wild grapes, currants, add no more than ¼ cup water to each 1 cup of prepared fruit. With currants, cook until they are translucent and faded. Add ½ cup of water to chokecherries and wild cherries; add a scant ¼ cup to juicy sour cherries.

To cut-up (but unpitted) plums, add water to *just below the top layer* in the kettle, and cook until soft—about 15 minutes.

To prepared apples, crabapples, quinces and guavas, add water *just to cover*, and cook until soft—20 to 25 minutes.

Strain all crushed raw or cooked fruit through a jelly bag. Dampen the bag to encourage the juice to start dripping through it; bunch the top together and tie it with strong string. Hang it high enough over a big mixing bowl so the tip of the bag cannot touch the strained juice (a broomstick laid across the tops of two kitchen chairs makes a good height).

Squeezing the jelly bag forces through bits of pulp that will cloud the jelly, but pressing the back of a wooden spoon against the bag will often quicken the flow without clouding the juice.

If there is traffic through the room, with attendant insects and dust, drape a clean sheet over the whole business.

Be fussy about washing the jelly bag after each use and rinsing it well; even a little diluted juice left in the fabric will spoil, and a musty, winey bag will hurt the next juice that's strained in it.

Refrigerate, in a tightly covered sterilized container, any juice left over from measuring for the batch of jelly.

Sugar and pectin

When you add the sugar depends on the type of commercial pectin you use. Each recipe stipulates the type—*and they are not interchangeable.* Always follow the recipe exactly, because time and quantity variations almost always bring failure.

Powdered pectin is added to the strained juice *before* heating. Heat rapidly, bringing to a full rolling boil—i.e., a boil which cannot be stirred down; *then add the sugar*, bring again to a full rolling boil, and boil for 1 minute.

Liquid pectin (except for homemade pectin, above) is added to the strained juice and sugar *after* the mixture is brought to a full boil. Stir constantly during heating. Add pectin, bring again to a full rolling boil, boil for 1 minute.

WITHOUT ADDED COMMERCIAL PECTIN

Jellies with enough natural pectin (like Basic Apple Jelly) require less sugar per cup of juice than jellies with added *store-bought* pectin do. The longer cooking needed to reach the jelly stage produces the right proportion of sweetness, acid and pectin.

Testing for doneness

Because barometric pressure as well as altitude affects the boiling point, make necessary adjustments for heights above 1,000 feet above sea level, and for whether the day is close and damp, or clear and dry.

Jelly with added pectin will be done if boiled as the individual instructions for time and quantity specify.

Jelly without added pectin is done when it reaches 8 degrees F. above boiling; usually, under good conditions at 1,000 feet or less, this is 220 F.

220 F. and a full rolling boil in all its glory: the jelly is done.

If you have no jelly thermometer, use the Sheet Test. Dip a cold metal spoon in the boiling jelly and, holding it 12 to 18 inches above the kettle and out of the steam, turn it so the liquid runs off the side. If a couple of drops form and run together and then tear off the edge of the spoon in a sheet, the jelly is done.

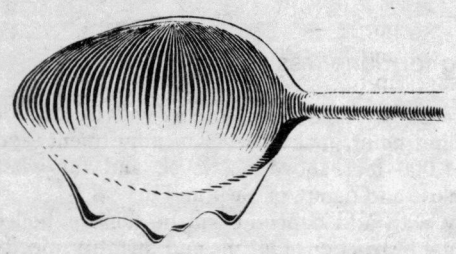

Or use the Refrigerator Test. Remove the kettle from the heat (so it won't raise Cain while your back's turned) and pour a tablespoon of jelly into a saucer. Put the saucer in the ice-cube compartment of your refrigerator for a minute or two: if it has become tender-firm, your jelly is ready to pour and seal.

Pouring & sealing, labeling & storing

The moment your jelly tests done, remove the kettle from heat and skim off the foam, working deftly so as not to stir any fluff down into the jelly. Ladle boiling-hot jelly immediately into clean sterilized glasses (see how to sterilize them and their caps in "About Jars and Cans" in the Canning chapter). With a clean cloth wrung out in boiling water, wipe any dribbles from the rim of the container and from the inside above the level of the jelly. Seal each jar at once, because airborne spoilers can settle on the cooling surface and produce mold during storage.

SEALING WITH PARAFFIN

If you seal with wax, leave ½ inch of headroom, wipe the inside lip of the glass, and cover the jelly with a ⅛-

inch layer of hot paraffin. Melt the wax in a double boiler arrangement (you can have a scary fire on top of your stove if you melt paraffin in a saucepan over high heat); or melt it in an old metal teapot or pitcher set in a larger pan of simmering water, and you have a good pourer for the liquid wax.

Use only fresh, clean paraffin, because re-used or dusty wax has acquired impurities that can cause spoilage. Wax that's too hot—a haze will rise from the surface if it is— can have little breaks in it after it cools. With a sterilized darning needle prick any air bubbles on the surface of the paraffin, because these are likely to cause holes in the wax as it cools, and there goes your seal.

A too-thick layer of wax usually will fail to seal, so stick to the ⅛-inch rule; and don't try to cover a faulty thin layer by topping it with more wax: seal again from scratch.

When the paraffin is cold, cover it with the metal cap or snap-on plastic lid that comes with most containers; such covers merely protect the wax seal, though, and are not sealers in themselves.

SEALING WITH MODERN LIDS

If you are using the modern jelly jars that have 2-piece screwband lids, leave only ⅛ inch of headroom. Wipe dribbled jelly from the rim of the jar, put on the sterilized caps and screw the bands tight. When each jar is tightly capped, invert it for a moment and right it again. A perfect vacuum seal will form as the hot jelly cools.

LABELING & STORING

During the 24 hours before you store your jelly, check it for loose texture and faulty seals (see "Jelly, etc., Failures and What to Do" on the next page). Carefully clean away any stickiness from the glasses/jars, giving extra attention to the tops around the closures where outside mold can attack the seal. Label each container with the kind and the date; it's a good idea to indicate the batch number if you made more than one lot of the same sort that day. And if you used a method or recipe different from your usual one, note this fact too: it could help to pinpoint reasons for trouble, as well as for outstanding successes.

Storage in a "cool, dark, dry place" is the same as for canned goods: ideally between 32 and 50 degrees Fahrenheit—certainly not where the contents will freeze (which can break seals), certainly not at normal room temperature (which encourages growth of spoilers). Dark, because the pretty colors can turn brown or fade in the light; otherwise you put them in cartons or wrap them in paper or cover the whole shebang with an old blanket or the like. And dry because humidity can corrode metal caps and lids and lead to broken seals.

Arrange the newly made jellies so that the last ones in will be the last ones out, unless you have indicated special priorities for their use.

Jelly, etc., Failures and What to Do

In theory, we'd all have perfect jellies, jams, marmalades, preserves, conserves and fruit butters if we used prime ingredients, if we measured carefully, and if we followed procedures conscientiously. But things can go wrong, even when we mean to be careful. Therefore here is a rundown of the symptoms and causes of common failures, listed now (rather than at the end of all the recipes, as other publications do), so you can keep them in mind as you work along.

We'll start by saying that you shouldn't stash away any of these Preserving Kettle products until they have stood handy by for 24 hours. Aside from allowing you a wonderful gloat, this day of grace before storing them will let you check the seals in time to do them over again if you find any poor ones. With paraffin it is not enough merely to add another layer of melted wax: you must remove the wafer of old wax, wipe the inside lip of the glass with a scrupulously clean cloth wrung out in boiling water, and pour on ⅛ inch of fresh hot wax.

Now for the problems that can be dealt with *safely*—provided that the seal is intact, and that there is no mold or fermentation in the contents.

Too stiff, tough. Too much pectin in proportion, or cooking no-added-pectin products too long; sliced citrus rinds in marmalades not precooked before added to sirup. Nothing can be done for pectin-added things, and it's not feasible to do the others over with more liquid. They're still probably tastier than store-bought.

Too-soft jelly. Tilt the containers: if you can see the contents shift, the jelly is too soft. This condition can be caused by cooking too long (as when the batch was too big and so was boiled beyond the ideal time limit); or by cooking too slowly; or by too much sugar; or by too little sugar or pectin or acid; or by not cooking long enough. Sometimes you can salvage such jelly by cooking it over; not always—but it's worth a try. Work with only 4 cups of jelly at one time.

Without added pectin. Bring 4 cups of the jelly to boiling and boil it hard for 2 minutes, then test it for signs of gelling. Let it try to sheet from a cold spoon, or (having removed the kettle from heat) chill a dab of it; if it shows signs of improving, boil a minute or so longer until it tests done. Then take it off the heat, skim, pour into hot sterilized containers, and seal.

With added powdered pectin. For each 4-cup batch of jelly, measure 4 teaspoons of powdered pectin and ¼ cup water into the bottom of the kettle; heat the pectin and water to boiling, stirring to keep it from scorching. Add the jelly and ¼ cup of sugar, bring quickly to a full rolling boil for 30 seconds, stirring constantly. Remove from heat, skim, pour into hot sterilized containers, seal.

With added liquid pectin. Bring 4 cups of jelly to a boil quickly. Immediately stir in 2 tablespoons of lemon juice, ¾ cup of sugar and 2 tablespoons of liquid pectin. Bring it back to a full rolling boil, and boil it hard for 1 minute, stirring constantly. Remove the kettle from heat, skim, pour into hot sterilized containers, and seal.

Runny jam. Jam isn't supposed to be as firm as jelly, so if it's only a little bit looser than you'd like it to be, don't bother to remake it. If it's really thin, though, try one of the remedies for too-soft jelly. If a test batch won't turn out right, make sure all the seals are intact and that storage is good—cool, dark and dry—and mark

the remaining jars to be used as a sweet topping for
ice creams, puddings, pancakes, etc. A homey and
hearty dessert: make a simple yellow cake, split it,
slather the jam between layers and serve it in squares
with chilled boiled custard to pour over it.

Runny conserves and butters. Often simply cooking them
over again will help; try a small batch. Is your storage
too warm?

"Weeping" jelly. This is the partial separation of liquid
from the other ingredients, and it can come from too
much acid or from gelling too fast—or from storage
that's warm. So check the seals, make sure there's no
mold or fermentation, and move it to a cool, dark, dry
place; this should help keep it from getting worse.
Such jelly is still usable: decant it just before serving
and mop up the juice with clean absorbent toweling.
Even perfect jelly can get forlorn if it languishes in
its serving dish in a warm or humid room.

Mold. Imperfect seals, unsterilized containers and lids,
letting the product stand uncovered before sealing,
re-used wax, storage in a warm, damp place—take
your choice. We don't believe in "a-little-mold-won't-
hurt-you" (see the comments on mold in the first
chapter of this book and in the Canning section), so
we say: *Destroy all jellies, jams, marmalades, preserves,
conserves and fruit butters THAT HAVE MOLD IN
THEM.*

Once in a while you'll find that a glass of your
paraffin-sealed jelly has mold outside on the top of the
wax, and it could be growing not on seepage from
the contents, but on a little smear of jelly that had not
been wiped from the rim or the neck of the jar. Even
if the wax seems to be tight, lift it off and look
critically for mold on the underside or on the surface
of the jelly. If there is no sign of mold, wipe the rim and
the inside (lip or neck, whichever) of the container with
a clean cloth wrung out in boiling water, and seal the
contents again with fresh hot paraffin.

Fermentation. The stuff has spoiled. Heave it.

212 F. = 100 C. / 220 F. = 104 C. / 1 U.S. cup (½ U.S. pint) = .24 L. /
1 U.S. pint (2 cups) = .48 L. / 1 U.S. tsp. = 5 ml. (5 cc.) / 1 U.S. T. =
15 ml. (15 cc.) / ½ inch = 1.25 cm.

Apple Jelly with Diced Fruit

Contrasts in its texture and flavor make this a fine spread or a good companion for cold ham or duck.

4 cups prepared apple juice (about 3 pounds apples)	6 to 8 peaches (or nectarines, or 12 apricots, or 2 cups pineapple chunks)
3 cups sugar	

To prepare the juice, cut washed apples in quarters, discarding stem and blossom ends. In a large kettle just barely cover apples with cold water; set over moderate heat, cover, and boil slowly until apples are quite soft—about 30 minutes. Turn the cooked apples into a damp jelly bag and let it hang until it drains well (don't squeeze the bag or the jelly will be cloudy).

Measure 4 cups juice and bring it to the boil in a large kettle; boil for 5 minutes, then add the sugar, stirring to dissolve it; remove from heat while you prepare the fruit.

Peel and dice the peaches or other fruit, add it to the apple jelly, return the kettle to heat and bring to boiling. Boil rapidly until the jelly sheets from a cold spoon or the temperature reaches 8 degrees F. above the boiling point of water in your kitchen—about 30 minutes. Remove from heat, skim off foam, and stir for 3 minutes to prevent floating fruit. Fill hot sterilized ½-pint jars that have 2-piece screwband lids, leaving ⅛ inch of headroom. Seal immediately. Makes about 6 ½-pint jars.

Basic Apple Jelly (Crabapple or Quince)

4 cups prepared apple juice	3 cups sugar

To prepare apple (crabapple or quince) juice cut washed apples in quarters, discarding stems and blossom ends. In a large kettle just barely cover apples with cold water and set over moderate heat. Cover and cook slowly for about 30 minutes, or until apples are quite soft. Turn into a damp jelly bag and drain well. Do not squeeze the bag or your jelly will be cloudy. Measure out 4 cups of the juice and bring to boiling in a large kettle; boil about 5 minutes, then add the sugar and stir until dissolved. Boil rapidly until the jelly sheets from a cold spoon or temperature reaches 8 degrees F. above the boiling point of water in your kitchen. Remove from heat and skim off foam. Pour into hot sterilized glasses, leaving ½ inch of headroom for paraffin seal, ⅛ inch of headroom for 2-piece screwband lids; seal at once. Makes 5 to 6 medium glasses.

Minted variation: At midpoint in the boiling, wash about 6 8-inch sprigs of fresh-picked mint and, holding the bunch by the cut ends, swish the mint around in the boiling jelly for 10 seconds; remove and proceed as above. It's just a hint of mint that adds a delightful bright flavor to the jelly.

Concord (or Wild) Grape Jelly Without Added Pectin

For best results use Eastern Concord or wild grapes (the latter have a flavor especially good with meats and game), and they should be *slightly underripe* for a natural pectin content higher than in fully ripe fruit. Holding the juice overnight in a cool place, and then straining again, will remove the crunchy little slivers of tartrate crystals that form in grape juice.

4 cups grape juice (3½ to 4 pounds of grapes)	1 firm apple
	3 cups sugar

Wash and stem the grapes, put them in a large kettle and crush. Wash the apple and cut it in eighths *without peeling or coring*, and add it with ½ cup of water (to prevent sticking). Bring all quickly to a boil, stirring, then reduce the heat and let the fruit cook gently until it is soft—about 10 minutes. Turn the pulp and juice into a damp jelly bag and drain well without squeezing. Refrigerate the bowl of juice overnight, and in the morning strain again through a damp jelly bag to remove the tartrate crystals. Measure 4 cups of juice into a large kettle, stir in the sugar, and boil quickly until the jelly sheets from a cold spoon or its temperature reaches 8 degrees F.

above the boiling point of water in your kitchen. Remove from heat, quickly skim off the foam, and pour immediately into hot sterilized glasses, leaving about ½ inch of headroom; seal at once with ⅛ inch of hot melted paraffin. (Or pour it into hot sterilized ½-pint jars, leaving ⅛ inch of headroom, cap tightly with 2-piece screwband lids; invert for a moment, cool upright to complete the seal.) Makes 5 to 6 medium glasses.

Grape Jelly II

    4 cups grape juice (3½   ½ bottle (3 ounces)
        to 4 pounds grapes)   liquid fruit pectin
    7 cups sugar

Sort, wash and stem ripe Concord (or wild) grapes; crush, add ½ cup of water, and bring to a boil. Reduce heat and simmer for about 10 minutes. Turn into a damp jelly bag and drain well; do not squeeze. Hold the juice overnight in a cool place, then strain through 2 thicknesses of damp cheesecloth to remove the tartrate crystals that form in grape products. Measure 4 cups of juice into a large kettle, add the sugar and mix well; bring quickly to a full boil that cannot be stirred down. Add the ½ bottle of pectin, bring again to a full rolling boil and boil hard for 1 minute. Remove from the heat, quickly skim off the foam, and pour the jelly into hot sterilized glasses, leaving about ½ inch of headroom; seal immediately with ⅛ inch of hot melted paraffin. (Or seal in ½-pint jars with 2-piece screwband lids, leaving ⅛ inch of headroom.) Makes 8 to 9 8-ounce glasses.

Lemon-Honey Jelly

  2½ cups honey   1 tablespoon grated
   ¾ cup fresh lemon juice,   lemon rind
        strained of all pulp   ½ bottle liquid fruit
                            pectin (3 ounces)

Combine honey, lemon juice and grated rind. Stir over moderate heat until mixture reaches a full boil. Add pectin and bring again to a full, rolling boil, stirring constantly. Boil hard 1 minute. Remove from heat and stir 3 minutes. Seal in hot sterilized glasses. Makes about 2 pints.

Apple-Herb Jelly

2½ cups prepared apple juice
¼ cup dried herb (such as tarragon, sage, thyme or savory)
¼ cup white vinegar

4 cups sugar
few drops red or green coloring (optional)
½ bottle liquid fruit pectin (3 ounces)

Heat apple juice just to boiling. Pour over the herb and let stand about 20 minutes. Strain through 2 thicknesses of cheesecloth. Add vinegar and sugar. Bring to a full, rolling boil, stirring frequently. Add coloring to suit taste. Stir in the pectin and bring again to a full, rolling boil. Boil hard 1 minute, stirring constantly. Remove from heat and skim. Seal in hot sterilized glasses. Makes about 6 medium glasses.

Fresh Mint Jelly

1 cup fresh mint leaves and stems, firmly packed
½ cup apple cider vinegar
1 cup water

3½ cups sugar
4 drops green food coloring
½ bottle liquid fruit pectin (3 ounces)

Do *not* remove leaves from the stems. Wash the mint, drain and place in a saucepan. Bruise well with the bottom of a heavy glass tumbler. Add vinegar, water and sugar. Bring to a full, rolling boil over high heat, stirring until sugar melts. Add coloring and pectin and bring again to a full, rolling boil, stirring constantly. Boil hard for 30 seconds. Remove from heat and skim. Pour through a fine sieve into hot sterilized glasses; seal. Makes about 2 pints.

This is a traditional accompaniment to any lamb dish.

Red Currant Jelly Without Added Pectin

4 cups currant juice (about 2½ quarts currants)	3½ cups sugar

Pick over the currants, discarding overripe or spoiled ones; wash quickly but carefully, and drain off excess water. Measure the washed currants into a large kettle, and *add no more than ¼ as much water as currants*. Over moderate heat, cook the currants until they are soft and translucent, stirring as needed to ensure that they cook evenly—about 10 minutes. Strain the currants with their juice through a damp jelly bag; do not squeeze, lest the juice become cloudy (this classic jelly should always be sparkling clear and jewel-like). Measure 4 cups of juice into a large kettle, bring to boiling, and boil briskly for 5 minutes. Add the sugar, stirring to dissolve it, and boil rapidly until the jelly sheets from a cold spoon or the temperature reaches 8 degrees F. above the boiling point of water in your kitchen. Remove from heat and skim off the foam; pour immediately into hot sterilized glasses, leaving about ½ inch of headroom, and seal at once with ⅛ inch of hot melted paraffin. (Or fill hot sterilized ½-pint jars, leaving ⅛ inch of headroom, and cap with 2-piece screwband lids; invert for a moment and let cool upright to form the seal.) Makes 5 to 6 medium glasses.

An alternate method for preparing the juice is to crush first a shallow layer of currants in the bottom of the kettle to start the juice, then pile on the rest of the currants. Heat slowly over low temperature setting for 5 minutes, then increase to medium heat and boil gently until currants are soft and translucent, stirring as needed. Strain.

Wine Jelly

This jelly is good with meats; and, in small pretty containers, makes an attractive present. It is *not boiled*, lest the alcohol be cooked away. Work only with small batches.

2 cups white wine　　　　½ bottle liquid fruit
　　(Chablis is good)　　　　　pectin
3 cups sugar

Mix wine and sugar in the top of a double boiler; put over boiling water, stirring constantly until sugar is dissolved—about 3 minutes. Remove and, off heat, stir in the pectin, mixing well. Pour quickly into hot sterilized glasses; as each jar is filled, promptly seal it with ⅛ inch of paraffin (delaying with the paraffin allows a white skin to form on the exposed surface as the mixture gels). Makes about 5 3-ounce glasses.

Variations: Substitute ½ cup sherry for ½ cup of the white wine for more emphatic wine flavor; or use all sherry if you can afford it.

Red wines of comparable sweetness to the white wine may be used, and of course Port wine jelly is a classic. Very dry red table wines, however, have not always produced a satisfactory gel for us—perhaps because their acid-to-sweetness ratio is out of kilter for the pectin. Experiment.

Rhubarb-Orange Jelly

4 cups chopped un-　　　　orange juice,
　　cooked rhubarb,　　　　　thawed
　　not peeled　　　　　　1 (1¾-oz.) package
2⅓ cups water　　　　　　　powdered fruit
1 (6-oz.) can frozen　　　　　pectin
　　concentrated　　　　　4 cups sugar

Combine rhubarb and 2 cups of the water and cook over moderate heat for about 15 minutes. Strain in a sieve, rubbing pulp through. Measure 2 cups juice, add the orange concentrate, the remaining ⅓ cup water, and stir in the pectin. Set over high heat and bring again to a full boil, stirring frequently. Add the sugar and bring again to a full boil, stirring constantly. Boil 1 minute. Remove from heat and skim. Seal in hot sterilized glasses. About 2½ pints.

Tomato Jelly

1¾ cups canned tomato juice	2 teaspoons Tabasco sauce
½ cup strained fresh lemon juice	4 cups sugar
	½ bottle liquid fruit pectin (3 ounces)

Combine all ingredients except pectin. Stir over high heat until mixture reaches a full, rolling boil. Stir in pectin and bring again to a full, rolling boil. Boil 1 minute, stirring constantly. Remove from heat. Stir and skim for about 3 minutes. Seal in hot sterilized glasses. Makes about 6 medium glasses.

This "lively" flavored jelly is equally good served with meats or spread on hot bread.

Wild Plum Jelly

6 cups prepared juice   4½ cups sugar

Prepare juice by covering cut-up plums with cold water and slowly bringing to a boil. Cook until very tender, stirring frequently. Press through a coarse sieve, then turn pulp into a jelly bag. Drain well, but do *not* squeeze the bag or jelly will be cloudy. Measure 6 cups of the juice and combine with 4½ cups sugar. Set over high heat and bring to a full, rolling boil, stirring constantly. Boil hard until jelly sheets (220 F.), stirring frequently. Remove from heat and skim. Seal in hot sterilized glasses. Makes about 8 medium glasses.

"Make Do" Corn Cob Jelly

12 cobs of red field-corn water	powdered fruit pectin
1 (1¾-oz.) package	4 cups sugar

Remove kernels, boil cobs in water to cover for 20 minutes. Drain liquid through a jelly bag. Measure 3 cups strained liquid into a large saucepan; stir in pectin. Bring to a full, rolling boil, then add the sugar. Bring again to boiling, stirring until the sugar is melted; boil hard 1 minute. Remove from heat, skim, and seal in hot sterilized glasses. Makes about 5 medium glasses.

This is a clear jelly, tasting a little like mild honey. It can be tinted, if desired, with food coloring.

FREEZER (UNCOOKED) JELLIES

The fresh-picked flavor of summer fruits çan be retained virtually intact in jellies or jams that are made to be stored in the freezer, and therefore are not cooked. The same basics must be present along with the fruit juice for a satisfactory result: the fruit's natural acid (eked out with lemon juice or a citric-acid solution if necessary, as described earlier), pectin and sugar.

The general handling of these jellies differs from that for conventional ones in several ways. Not being sterilized through boiling, they must be stored in the freezer to prevent spoilage—although freshly made or defrosted jelly will keep well for up to 3 weeks in the refrigerator. Also, because their natural pectin is not activated by boiling, pectin must be added; and it is added after the sugar, regardless of whether it is liquid or powdered. And finally, the jellies must be packed in sterilized freezer-proof jars, with headroom to allow for expansion; and of course they must be sealed with sterilized tight-fitting lids, *not* with paraffin.

Filled and capped, the containers must stand at room temperature until the jelly is set—which can take up to 24 hours—before going into the freezer.

In the rule below, powdered pectin is used; tested, good recipes are included in the folders enclosed with the containers of commercial powdered or liquid pectin.

The juice for this jelly is made from *unheated* fresh fruit. However, it can be made with frozen juice *not heat-extracted or sweetened;* or with juice from berries that have been frozen, *without added sugar,* for making jelly later on. It *cannot be made successfully from canned fruit juices,* because there the natural pectin has been activated by the heat of processing (or its gelling property may even have been impaired by the relatively long processing); and anyway it won't have the flavor you're after when you make freezer jelly.

Freezer (Uncooked) Berry Jellies with Powdered Pectin

3 cups prepared juice (about 2 to 2½ quarts of fresh strawberries, black-berries or red rasp-berries)

6 cups sugar
1 package powdered pectin (1¾ ounces)
¾ cup water

Crush the berries and strain them through a damp jelly bag or four layers of damp cheesecloth; squeeze gently if necessary. Add the sugar to the measured 3 cups of juice; stir well and let stand for 10 minutes (a few sugar crystals may remain, but they will dissolve in the time it takes the jelly to set). In a small saucepan stir together the ¾ cup water and the pectin; bring the mixture to the boil and boil hard for 1 minute, stirring constantly. Remove from heat and add it to the sweetened juice; continue stirring for 3 minutes. Then pour into sterilized jars that are freezable and have tight-fitting lids or 2-piece screwband lids, leaving ½ inch of headroom; seal. Let stand at room temperature until set—up to 24 hours—then freeze. Makes 6 8-ounce jars.

Diet Jellies

If you are merely watching calories, it's simpler to cut down your consumption of jellies and the like (or, strenuously, forgo them entirely). If there is another reason for curtailing your intake of natural sweeteners, you can do any of three things: (1) buy one of the good com-mercial sugarless jellies (which contain a number of ingredients not readily available to the average house-holder, such as guar gum, carrageen, dextrin and certain sodium or potassium compounds); (2) make a jelly—or jam—using low-methoxyl pectin, which requires only about ⅕ as much sugar as is needed with ordinary com-mercial pectins; or (3) use an artificial sweetener to make the perishable gelatin-based product described below.

Low-methoxyl pectin is made mainly from citrus-fruit rinds by a fairly complicated process. Its virtue is that certain calcium compounds will cause it to set, so the sugar is needed only for flavoring—roughly only ⅕ the amount of sugar needed in a conventional recipe. Low-methoxyl pectin is gettable from major natural- or health-food suppliers, and instructions for using it come with each packet.

SUGARLESS JELLIES

These jellies use an artificial sweetener for flavor and plain gelatin for body. Because they lack the preserving benefit of sugar, they must always be stored in the refrigerator and used within 3 to 4 weeks; freezing permits longer storage. They may be made from store-bought unsweetened juices, or from juices you have canned or frozen yourself.

Note: The recipes that follow call for a bottled liquid sweetener to the tune of 2 tablespoons = 1 cup sugar. The labels of various liquid or dry artificial sweeteners give their equivalents to sugar: *so read the label* to arrive at the same sweetness in proportion as given below.

Other fruit juices may be substituted for those given: just keep to the ratio of 4 teaspoons of gelatin for 2 cups of juice.

Sugarless Apple Jelly

2 cups unsweetened apple juice	to equal 1 cup sugar (read the
4 teaspoons unflavored gelatin	label, especially if substituting a dry
1½ tablespoons lemon juice	artificial sweetening agent)
2 tablespoons liquid artificial sweetener,	1 or 2 drops red food coloring if you like

Soften the gelatin in ½ cup of the apple juice. Meanwhile heat the remaining 1½ cups juice to boiling; remove from heat and stir in the softened gelatin until it is dissolved. Add the artificial sweetener, the lemon juice, the food coloring if you like it. Return to heat and bring to the boil, then pour into hot sterilized ½-pint jars that have 2-piece screwband lids, leaving ⅛ inch of headroom; seal. Store in the refrigerator when cool, and use within 3 to 4 weeks. Makes 2 8-ounce jars.

Sugarless Grape Jelly

1½ cups unsweetened
    grape juice
4 teaspoons unflavored
    gelatin
½ cup water
2 tablespoons liquid
    artificial sweetener,
to equal 1 cup
sugar (read the
label, especially if
substituting a dry
artificial sweeten-
ing agent)

Soften the gelatin in the ½ cup water. Meanwhile heat the grape juice to boiling; remove from heat and add the softened gelatin, stirring until it dissolves. Add the liquid artificial sweetener, and bring again to the boil. Remove, pour into hot sterilized ½-pint jars that have 2-piece screwband lids; seal. Store in the refrigerator when cool, and use within 3 to 4 weeks. Makes 2 8-ounce jars.

Jams and Marmalades

STEPS IN MAKING COOKED JAM AND MARMALADE

Jams and marmalades call for the same care in select-ing and preparing fruit that jellies do (q.v.). However, in several important ways their cooking and sealing differs from the handling of jellies. The most notable of these differences is that they all—*except for the so-called Diet and Freezer (Uncooked) jams*—are given a finishing Boiling-Water Bath as insurance against spoilage. This

means that they are packed in regular canning jars—
usually ½-pints or pints, with the right caps/lids—and
are given ½ inch of headroom.

Sugar and pectin

The general proportions for substituting other *non-
artificial* sweeteners for called-for sugar are the same for
jams and marmalades as for jelly (q.v.).

With added pectin, the crystalline type is mixed with
the unheated prepared fruit. Liquid pectin is added to the
cooked fruit-and-sugar mixture after the kettle is removed
from heat. With either form of pectin the cooking time is
the same: 1 minute at a full boil.

Without added pectin, the cooking time is increased to
a range of from 15 to 40 minutes, depending on the
character of the fruit. Jam is more likely to scorch than
jelly is, so stir it often during cooking.

For how pectin and sweeteners act in *diet jam*, see
page 323.

Testing for doneness

Jams, etc., with added pectin will be done when they
are boiled according to the individual instructions for time
and quantity.

Without added pectin, jam is done when it reaches 9
degrees F. above boiling, usually 221 F. at 1,000 feet of
altitude or below.

No thermometer? Jam is ready when it begins to hold
shape in the spoon (the Sheet Test does not apply to
jam, etc.). Or use the Refrigerator Test for jelly, above.

Pouring, sealing, processing, storing

Remove the kettle from heat, skim carefully, and then
stir the jam gently for 5 minutes to cool it slightly and thus
prevent floating fruit.

Ladle the still-hot jam or marmalade carefully into hot
sterile canning jars, leaving ½ inch of headroom; wipe the

mouths carefully with a clean cloth wrung out in boiling water, cap, and process in a B-W Bath (212 F.) for 10 minutes. Complete seals if necessary. Cool, clean the containers, label and store.

Basic Rule for Berry Jam

Weigh hulled and washed berries. Measure out an equal amount of sugar and set it aside. Crush berries well and set over low heat; bring to boiling slowly, stirring frequently. Add sugar and bring again to boiling. Simmer until thick, stirring frequently to prevent scorching as the mixture thickens. Pack in hot sterilized canning jars, leaving ½ inch of headroom, and process in a Boiling-Water Bath (212 F.) for 10 minutes. Complete seals if necessary. Cool, clean and store.

Aunt Mabel Berry's Rose-Hip Jam

> 1 pound rose hips° sugar
> 1 cup water

Simmer rose hips and water until fruit is very tender. Rub it through a sieve and weigh the pulp. To each pound or fraction, add an equal weight of sugar. Return to heat and simmer until thick, stirring frequently. Pack in hot sterilized jars and process.

This jam makes an attractive present, being non-commercial (as well as having a high Vitamin C content).

° Rose hips should be gathered after the first frost of autumn. Take care not to gather them from bushes that have been treated with insecticides or fungicides.

Sun-cooked Strawberry* Jam

You need a blistering hot, still day to do this. Have a table set up in the full sun, its legs set in cans or small pans of water to keep crawling insects from the jam. To protect it from flying insects, have handy a large sheet of clean window glass, the means to prop it at a slant over the platters, and cheesecloth or mosquito netting to tape like a curtain around the three sides left open to the air. And work in *small* batches.

Wash and hull berries, and measure them to determine how much sugar you need. Put a layer of berries in the bottom of a big kettle, cover with an equal number of cups of sugar; repeat a layer of berries and cover it with sugar. Set aside for about 30 minutes to let the berries "weep" and the juice start drawing. Place over very low heat and bring slowly to simmering, stirring occasionally to prevent scorching, until the sugar is dissolved.

Pour sirupy berries ½ inch deep into large plates or platters. Set platters on the table in strong sun. Prop the glass over them with one edge on the table, the opposite edge raised 4 to 6 inches high (this allows any condensation to run harmlessly down the glass onto the table, instead of dripping back on the jam to slow the jelling process). Arrange netting around the open sides.

As the fruit cooks in the sun, turn it over with a spatula—2 or 3 times during the day. When it has obviously jelled enough, pour it into hot sterilized ½-*pint* jars, leaving ¼ inch of headroom. Adjust lids, and process in a Boiling-Water Bath (212 F.) for *15 minutes* after the water boils. Complete seals if necessary. (If pint jars are used, increase the headroom to ½ inch, and process in the B-W Bath for 25 minutes.)

If the sun is not strong or if there's wind, jelling can take 2 or 3 days. In that case, bring the platters indoors each night.

* Or cherries. Or raspberries—but *not washed* before layering.

Old-Time Chokecherry Jam

Remove stems from chokecherries and wash. Drain. Add 1 cup water to every 4 cups fruit. Place over low heat and simmer until fruit is very tender, stirring occasionally. Rub pulp through a medium sieve; measure, and add an equal amount of sugar. Place over moderate heat and stir until sugar has melted. Bring to a full, rolling boil and cook until mixture sheets (220 F.). Stir frequently. Pack and process. Three cups pulp will make about 3 half-pints.

Green Tomato Marmalade

2 quarts sliced, small,   4 lemons, peeled (save
    green tomatoes   the rind)
½ teaspoon salt   4 cups sugar

Combine tomatoes and salt. Chop lemon rind fine and add. Cover with water and boil 10 minutes. Drain well. Slice the peeled lemons very thin, discarding seeds but reserving all juice. Add lemon slices and juice and the sugar to the tomato mixture. Stir over moderate heat until sugar melts. Bring to boiling, reduce heat and simmer until thick—about 45 minutes. Stir frequently. Pack in hot sterilized jars and process. Makes about 2 pints.

This classic from long ago is especially good served with meat to add a "company dinner" touch.

Honeyed Marmalade

2 cups diced, peeled   2 tablespoons grated lime
    apples   rind
2 cups diced, peeled   juice of 2 large limes
    carrots   2 cups sugar
1 cup diced, peeled   1 cup honey
    peaches

Combine all ingredients in order, mixing well. Set over low heat and bring slowly to boiling, stirring frequently. Simmer until thick. Stir frequently to prevent scorching as the mixture thickens. Pack in hot sterilized jars and process. Makes about 3 pints.

Carrot Marmalade

4 cups cooked, sliced   2 oranges
    carrots   6½ cups sugar
2 lemons

Put carrots and seeded lemons and oranges through the coarse knife of a food grinder: be sure to save all juice.

Add sugar, and cook very slowly until thick. Pour hot into hot sterilized jars and process. Makes about 3 pints.

This marmalade is beautifully colored and of excellent flavor, and is as much a favorite today as it was many years ago.

Ripe Tomato Marmalade

Small yellow tomatoes make a pretty marmalade.

8 cups chopped ripe tomatoes (about 4 pounds whole)	1 orange
	2 3-inch sticks of cinnamon
4 cups sugar	1 tablespoon whole cloves
2 lemons	

Wash, scald and peel the tomatoes; slice off blossom and stem ends. Scoop out the seeds, working over a strainer set into a bowl to catch the juice. Chop the tomato pulp in small bits, measure and put into a large kettle. Add ½ as much sugar as there is pulp, and add the juice you have saved. Slice the lemons and orange very thinly, remove the pips and cut the slices into small bits; add them to the tomato pulp. Loosely tie the spices in a square of muslin or double thickness of cheesecloth, add the spice bag to the mixture in the kettle. Cook and stir over low heat until the sugar is dissolved; bring to a rapid boil and boil until it starts to thicken and its temperature reaches 9 *degrees* F. (1 degree more than jelly) above the boiling point of water in your kitchen. Ladle at once into hot sterilized ½-pint jars that have 2-piece screwband lids, leaving ½ inch of headroom; adjust lids and process in a Boiling-Water Bath (212 F.) for 10 minutes. Remove; complete seals if necessary. Makes about 8 8-ounce jars.

Classic Orange Marmalade

The Scots make highly prized marmalades, among them this one from Mildred Wallace, which is characterized by a darker color and slightly bitter flavor compared with the most popular supermarket brands in the United

States. The precooking prevents the peel from becoming tough when it is boiled with the sugar.

2 pounds Seville oranges (or other bitter variety, like Cala- mondin), left whole	2 large lemons, whole 8 cups water (about) 8 cups (4 pounds) sugar

Wash the oranges and lemons well, removing any stem "buttons," and put the clean washed whole oranges and lemons in a large kettle with enough water to cover them; put the lid on the kettle and bring to a boil, then simmer until a slender fork will easily pierce the fruit— about 1½ hours. Remove the fruit to cool, saving the liquid; when they are cool, cut them in half the long way, then cut the halves in *very thin* slices (your knife must be sharp!), and take out and save the pips. Return the pips to the juice in the kettle and boil for about 10 minutes (this contributes to the bitter flavor). Strain the juice and return it to the kettle. Add the fruit slices and heat to boiling. Add the sugar, stirring until it dissolves, and continue cooking at a fast boil—stirring only enough to prevent scorching—until it starts to thicken and its temperature reaches 9 *degrees F.* (1 degree more than for jelly) above the boiling point of water in your kitchen. Remove from heat, skim off any foam, pour at once into hot sterilized ½-pint jars with 2-piece screwband lids, leaving ½ inch of headroom. Adjust lids; process in a Boiling-Water Bath (212 F.) for 10 minutes. Makes 5 to 6 ½-pint jars.

Rhubarb-Carrot Marmalade

6 cups diced, peeled, raw rhubarb 3 cups ground, peeled, raw carrots	2 medium oranges, unpeeled 4½ cups sugar

Combine rhubarb and carrots. Put oranges through the food grinder, using a medium knife. Discard seeds but reserve all juice. Add to rhubarb mixture, then add the sugar. Let stand overnight. Stir over low heat until boiling;

reduce heat and simmer until thickened—about 2 hours. Stir frequently. Pack in hot sterilized jars and process. Makes about 5 pints.

This is a very old rule indeed, and as good and honest as it is "out of the way."

Yellow Tomato Marmalade

3¼ cups chopped, peeled, ripe tomatoes	6 cups sugar
¼ cup fresh lemon juice grated rind of 1 large lemon	1 bottle liquid fruit pectin (6 ounces)

Place chopped tomatoes in small pan and set over low heat and cover. *Do not add any water.* Bring to boiling, reduce heat and simmer about 10 minutes, stirring frequently. Remove from heat and measure out 3 cups of the tomatoes and liquid. In a large kettle, combine the 3 cups tomatoes with the lemon juice, grated rind and sugar. Stir over moderate heat until boiling. Boil hard 1 minute. Turn off heat and add pectin. Stir for 5 minutes. Pack in hot sterilized glasses and process. About 1½ pints.

Applesauce Marmalade

4½ cups applesauce	2 small lemons, unpeeled
2 medium oranges, unpeeled	½ cup water
	4 cups sugar

Turn applesauce into a heavy saucepan. Put oranges and lemons through food grinder, using medium knife; discard seeds but reserve all juice. Add to the applesauce, then add water and sugar. Stir over low heat until boiling, stirring constantly. Reduce heat and simmer until thick—about 1 hour. Stir frequently: this marmalade will scorch easily during latter part of cooking time (an asbestos pad under the pan will help too). Pack in hot sterilized jars and process. About 3 pints.

Old-fashioned Pumpkin Marmalade

1 small pumpkin (*c.* 5 pounds, to give 4 pounds, cubed)	4 pounds sugar 3 lemons 1 orange

Peel pumpkin and cut in small cubes, discarding seeds and inner pulp. Add sugar and let stand overnight. Put lemons and orange through food grinder, discarding seeds but reserving all juice. Add to pumpkin. Stir over low heat until boiling. Simmer until thick and clear—about 3 hours. Stir frequently to prevent scorching. Pack in hot sterilized jars and process. About 4 pints.

The flavor of this prettily colored jam is predominantly citrus.

FREEZER (UNCOOKED) JAM

People seem to make more freezer jams than jellies, perhaps because there is more leeway to a jam's consistency than there is for jelly. Certainly the lovely garden-fresh flavor of berries and fruits is more pronounced in jams.

Do read the introductory comments for "Freezer (Uncooked) Berry Jellies," earlier: it explains the *Why* behind each variation in procedure from making cooked jams, and these differences are important for a successful product. Like their jelly counterparts, opened freezer jams must be refrigerated, and used within 3 weeks.

The recipe below is our favorite of all freezer jams. The folders that come with the containers of commercial powdered or liquid pectin have a number of good jams—and the proportions have been worked out after much testing. We recommend them.

"Best-Ever" (Frozen) Strawberry Jam

2 cups prepared fruit (about 1 quart ripe strawberries)	¾ cup water
	1 box powdered fruit pectin (1¾ ounces)
4 cups (1¾ lbs) sugar	

Thoroughly crush, one layer at a time, about 1 quart of fully ripe strawberries. Measure 2 cups of crushed berries into a large bowl. Add the sugar to the fruit, mix well, and let stand for 10 minutes; a few sugar crystals may remain but they'll dissolve as the jam sets. Mix water and pectin in a small saucepan, bring the mixture to a boil and boil for 1 minute, stirring constantly. Remove from heat and stir the pectin into the fruit; continue stirring for 3 minutes. Ladle quickly into sterilized freezable jars, leaving ½ inch of headroom. Seal immediately with sterilized tight-fitting lids, *not with paraffin*. Let jars stand at room temperature until the jam is set—which may take up to 24 hours—then freeze. Makes 5 to 6 8-ounce jars.

Diet Jams

For the *Why* of the following ingredients, method and storage, see the introduction to "Diet Jellies," earlier in this Preserving Kettle section, *and* "Artificial Sweeteners" on page 20.

The following recipe can be made just as well from raspberries or blackberries, which you may sieve if you don't like the seeds (allow about 4½ cups of whole berries to get the 2 cups of fruit pulp called for).

Gelatin will not gel with fresh, or frozen fresh, pineapple, by the way.

Sugarless Strawberry Jam

2 cups crushed straw-
  berries (about 1
  quart whole
  berries)
4 teaspoons unflavored
  gelatin
1½ tablespoons lemon
  juice (particularly
  if blackberries are
  substituted)

2 tablespoons liquid
  artificial sweetener,
  to equal 1 cup
  sugar (read the
  label, especially if
  substituting a dry
  artificial sweeten-
  ing agent)

Soften the gelatin in ½ cup of the juice from the crushed berries. Meanwhile heat the remaining 1½ cups of crushed berries to boiling; remove from heat and add the softened gelatin, stirring until it is dissolved. Add the lemon juice and the artificial sweetener. Return to heat and bring to the boil, then pour into hot sterilized jars that have 2-piece screwband lids, leaving ⅛ inch of head-room; seal. Store in the refrigerator when cool, and use within 3 to 4 weeks. Makes 2 8-ounce jars.

Preserves and Such

STEPS IN MAKING PRESERVES

Wash the fruit and remove stem and blossom parts. Peel peaches, pears, pineapples, quinces and tomatoes. Shred pineapple, less the core. Cut slits in tomatoes and gently squeeze out the seeds, cut large tomatoes in quarters, leave small ones whole. Pears and quinces are thinly sliced after halving and coring. Take the pits from sour cherries. Of course strawberries and raspberries are left whole.

To cook, carefully follow the specific recipe. Generally, dry sugar is added to the soft fruits to start the juice flowing. There should be enough juice to cook the fruit. Hard fruits are cooked in a sugar-and-water sirup. The recipe will tell you how long to cook each of the preserves.

Ladle hot preserves into hot sterilized canning jars, leaving ½ inch of headroom; wipe the mouths of the containers carefully with a clean cloth wrung out in boiling water; adjust lids, and process in a Boiling-Water Bath (212 F.) for 10 minutes; complete seals if necessary. Cool, clean the jars, label and store.

212 F. = 100 C. / 220 F. = 104 C. / 1 U.S. cup (½ U.S. pint) = .24 L. / 1 U.S. pint (2 cups) = .48 L. / 1 U.S. tsp. = 5 ml. (5 cc.) / 1 U.S. T. = 15 ml. (15 cc.) / ½ inch = 1.25 cm.

Southern Peach "Honey"

ripe peaches   sugar

Peel and halve peaches, discarding pits. Mash fruit thoroughly. Measure, and add 2 cups sugar to each cup of peach pulp. Set over low heat and bring to boiling, stirring constantly. Simmer until thick and clear—about 30 minutes. Stir frequently to prevent scorching. Pack in hot sterilized jars and process.

Some Southerners prefer this to all other spreads as a complement to hot biscuits.

Georgia Peach Conserve

4 cups coarsely chopped
　　peaches°
½ cup coarsely chopped
　　pitted prunes,
　　uncooked
⅓ cup seeded raisins

1 medium orange,
　　seeded and ground
½ cup water
1 cup light corn sirup
¼ cup chopped pecans
　　(or other nuts)

Turn peaches into a big heavy saucepan and, with the edge of a serving spoon, chop the fruit coarsely. Add the prunes, raisins, ground orange, water and the corn sirup.

Bring to a boil; reduce heat and simmer until dark and thick—about 1 hour. Add nuts and cook 5 minutes longer. Pack in hot sterilized jars and process. Makes about 2 pints.

* One quart of undrained canned peaches may be used, omitting the ½ cup of water and using only ¾ cup corn sirup.

Sweet-and-Sour Spiced Crabapples

3 pounds firm ripe crabapples (about)	3 dozen whole cloves
3 cups cider vinegar	4 to 6 3-inch sticks of cinnamon
3 cups water	6 short blades of mace*
2¼ cups sugar	

To prepare the crabapples, wipe the fuzz from the blossom ends, but leave stems on; wash well, then prick the skins with a large darning needle to keep the fruit from bursting while cooking. Tie the spices loosely in a square of muslin or double thickness of cheesecloth, and put the bag in a large enameled kettle with the vinegar, water and sugar; bring to boiling and boil together for 3 minutes. Add the crabapples and simmer until just tender—not mushy. (Test after 15 minutes by poking one deeply with a darning needle: there should be a little resistance.) Discard the spice bag, and pack the crabapples immediately in hot pint jars and cover them with the very hot sirup in which they were cooked, leaving ½ inch of headroom. Adjust the lids and process the jars in a Boiling-Water Bath (212 F.) for 10 minutes. Remove; complete seals if necessary. Makes about 5 pints.

* Or use 1 teaspoon ground mace or nutmeg, although it could come through the bag and make the liquid cloudy.

Sweet Cherry (or other) Preserves

4 cups pitted sweet cherries, tightly packed	3 cups sugar

In a 4-quart saucepan crush the cherries lightly to start the juice flow. Boil cherries and their juice about 10 minutes —or until fruit is tender. Add sugar to the cherries, stir well, and boil for 5 minutes more. Now cover the kettle and let the cherries stand for 2 minutes while they absorb more of the sugar. Stir the hot preserves to prevent floating fruit, then pour into hot sterilized jars, leaving ½ inch of headroom; process for 10 minutes in a B-W Bath (212 F.). Makes about 2 pints.

This is a basic rule for similar fruit preserves, and offers scope for variations of your own devising. Be individual!

Whole Crabapple Preserves

½ peck (4 quarts) ripe crabapples
4 cups water (cooking liquid + fresh water)
4 cups sugar (or combined with other sweeteners added to cooking liquid to make c. 6 cups of 50 percent, medium-heavy, sirup)

Choose uniform, blemish-free crabapples—fresh-picked and rosy-cheeked. Rub "whisker fuzz" from the blossom ends, remove stems or not; wash the fruit well.

Push a darning needle twice clear through each apple's "equator" to keep skins from popping during the cooking, ·and put the fruit in a large kettle (6- to 8-quart); add water just to cover. Boil gently until the apples are barely tender— test after 15 minutes by poking one deeply with a darning needle: there should be a little resistance.

Drain off and use the cooking water, plus any more of fresh water needed, to make 4 cups of liquid; add the sugar and bring to a boil. To the hot sirup add the apples, and cook them gently until they are tender and seem slightly transparent (apples with their skins are never as transparent as peaches).

Pack boiling-hot apples and sirup into clean, hot canning jars; leave ½ inch of headroom. Adjust lids. Process in a Boiling-Water Bath (212 F.) for 10 minutes. Remove; complete seals if necessary. Makes 5 to 6 pints.

For a variation, add a few red cinnamon candies to the sirup for color and flavor, or stick 3 or 4 whole cloves into each softened apple before cooking them in the sirup.

STEPS IN MAKING BUTTERS

Butters are nice old-fashioned spreads and they're good with meats. Their virtues are that they take about ½ as much sugar as jams from the same fruits (½ cup sugar to each 1 cup of fruit pulp), and they can be made with the sound portions of windfall and cull fruits that you'd probably not bother with for jelly or jam. Their one drawback is that they require very long cooking—and careful cooking at that, because they stick and scorch if you turn your back.

Butters are made from most fruits or fruit mixtures. Probably apple is the best-known ingredient, but apricots, crabapples, grapes, peaches, pears, plums and quinces also make good butters.

TO PREPARE THE FRUITS

Use prime ripe fruit or good parts of windfalls or culls. Wash thoroughly and prepare as follows:

Apples: Quarter and add ½ as much water or cider (or part water and part cider) as fruit.

Apricots: Pit, crush, add ¼ as much water as fruit.

Crabapples: Quarter, cut out stems and blossom ends, and add ½ as much water as fruit.

Grapes: Remove stems, crush grapes and cook in own juice.

Peaches: Dip in boiling water to loosen skins; peel, pit, crush and cook in their own juice.

Pears: Remove stems and blossom ends. Quarter and add ½ as much water as fruit.

Plums: Crush and cook in their own juice. The pits will strain out.

Quinces: Remove stems and blossom ends. They're hard, so cut in small pieces and add ½ as much water as fruit.

MAKING THE PULP

Cook the fruits prepared as above until their pulp is soft. Watch it—it may stick on.

Put the cooked fruit through a colander to rid it of the skins and pits, then press the pulp through a food mill or sieve to get out all fibers.

SUGAR AND COOKING

Usually ½ cup of sugar to each 1 cup of fruit pulp makes a fine butter. It's easiest to use at one time not more than 4 cups of fruit pulp, plus the added sugar.

Let the sugar dissolve in the pulp over low heat, then bring the mixture to a boil and cook until thick, stirring often to prevent scorching.

When the butter is thick enough to round slightly in a spoon and shows a glossiness or sheen, pack while still hot into hot, sterilized ½-pint or pint canning jars, leaving ½ inch of headroom. Adjust lids and process the jars in a Boiling-Water Bath (212 F.) for 10 minutes. Remove jars, complete seals if necessary. Cool and store.

Alternative cooking method: Butters stick so easily when they are cooking on the stove top that it's a real chore to keep them from scorching. Some cooks put about ¾ of the hot uncooked purée in a large, uncovered, heatproof crockery dish or enameled roasting pan and cook it in a 275 to 300 F. oven until it thickens. As the volume shrinks and there is room in the dish, add the other ¼ of the purée. When the butter is thick but still moist on top, ladle it quickly into containers and process.

OPTIONAL SPICES

Any spices are added as the butter begins to thicken. For 1 gallon of pulp use 1 teaspoon ground cinnamon, ½ teaspoon ground allspice, and ½ teaspoon cloves. Ginger is nice in pear butter—1 to 2 teaspoons to 1 gallon of pulp. For smaller quantities of pulp reduce measures of spices proportionately. If the butter is to be light in color, tie whole (not ground) spices loosely in a cloth bag and remove the bag at the end of the cooking.

Old-style Apple Butter

½ peck unpeeled apples	1 tablespoon ground cinnamon
2 cups sweet cider	
about 5 cups sugar	½ teaspoon ground allspice
1 teaspoon ground cloves	

Cut up apples and put them in a heavy kettle. Add the cider and cover. Cook over low heat until very tender, stirring occasionally. Cool slightly, then rub all through a sieve (or use a food mill). Measure and combine with ½ as much sugar. Stir in the spices. Simmer over low heat until dark and thick—about 2 hours. Stir frequently, for *this scorches easily:* an asbestos pad under the kettle will help. Remove from heat and pour into hot sterilized jars, leaving ¼ inch headroom. Adjust lids, process pints or quarts in a Boiling-Water Bath for 10 minutes to ensure the seals. The butter will thicken as it stands. About 5 pints.

Pickles, Relishes, Chutneys, Sauces

Pickles and relishes are first cousins. Their major difference is that vegetables and/or fruits for relishes are chopped before being put with the vinegar mixture, and those for pickles are left whole or cut to size for the recipe.

Any firm-fleshed vegetable or fruit may be used. There are some that hold their shape and texture particularly well in pickles, such as the black-spine type of cucumber and the Seckel pear.

Pickle products and relishes are packed in regular canning jars, trapped air is removed (see page 74), and the jars are capped *and processed in a Boiling-Water Bath (212 F.) for a specified time to ensure sterilization and a good seal,* and are stored as any other canned food is—in a cool, dry, dark place.

ESSENTIAL INGREDIENTS

THE PRODUCE ITSELF

Fresh, prime ingredients are basic. Move them quickly from garden or orchard to pickling solution. They lose moisture so quickly that even one day at room temperature may lead to hollow-centered or shriveled pickles.

Perfect pickles need perfect fruits or vegetables to start with. The *blossom ends* of cucumbers must be removed

EQUIPMENT FOR PICKLES AND RELISHES

If your pickle-making is modest in amount, you will need only a Boiling-Water Bath canner and a deep enameled kettle in addition to your regular kitchen utensils. The one pickle characteristic to keep in mind is the interaction of the vinegar and salt with metals: use enameled, earthenware or glass containers to hold or cook these mixtures—*never use* anything that's galvanized, or copper, brass or iron. A large stone crock is best for brining (but often they're hard to come by); glass 1-gallon, or larger, jars or straight-sided containers are a good substitute.

Essential:

> Boiling-Water Bath canner
> 6- to 8-quart enameled kettle for short brining and cooking pickles
> Jars in prime condition, with lids/sealers/gaskets ditto
> Minute-time with warning bell to time processing periods
> Shallow pans (dishpans are fine)
> Ladle or dipper
> Long-handled wooden or stainless steel spoon for stirring
> Wide-mouth funnel for filling containers
> Jar-lifter
> Colander for draining
> Large measuring cups, and measuring spoons
> Squares of cheesecloth to hold spices
> Plenty of clean dry potholders, dish cloths and towels

Nice, but not absolutely essential:

> Household scales
> Stoneware crocks
> Large trays

(since any enzymes located there can cause pickles to soften while brining), but do leave ¼ inch or so of *stem*.

SALT

Use only pickling, dairy or kosher salt without additives. These are plain, pure salt, either coarsely or finely ground.

Do not use table salt. Although pure, the additives in it to keep it free-running in damp weather make the pickling liquids cloudy; the iodine in iodized salt darkens the pickles.

Do not use the so-called rock salt or other salts that are used to clear ice from roads and sidewalks: they are not food-pure.

Salt, as used in brining pickles, is a preservative. A 10-percent brine, about the strongest used in food preservations, is 1½ cups salt dissolved in each 1 gallon of liquid. Old-time recipes often call for a brine "that will float an egg"; translate this to "10-percent brine."

Brine draws the moisture and natural sugars from foods and forms lactic acid to keep them from spoiling.

Juices drawn from the food dilute the brine, weakening the original salt solution.

VINEGAR

Use a high-grade cider or white distilled vinegar of 4 to 6 percent acidity (40 to 60 grain). Avoid vinegars of unknown acidity or your own home-made wine vinegar. The latter develops "mother" that clouds the pickling liquid. Use white vinegar if you want really light pickles.

And *never reduce the vinegar* if the solution is too tart: instead, add more sugar.

SWEETENERS

Use white sugar unless the recipe calls for brown. Brown, or raw, makes a darker pickle. Sometimes a cook in the northern United States or in Canada may use maple sugar or sirup in her pickles for its flavor—but this is feasible only if she has lots of it to spare. (See "Sweeteners" in jellies, earlier.)

SPICES

Buy fresh spices for each pickling season. Spices deteriorate and lose their pungency in heat and humidity, so they should be kept in airtight containers in a cool place.

ADDITIVES FOR CRISPING PICKLES

To enhance the crispness of various cucumber and rind pickles, old cookbooks sometimes called for a relatively short treatment with *slaked lime* (calcium hydroxide) or *alum* (see also "Firming Agents" in Chapter 1). Such chemicals are *not necessary* for good-textured products if the ingredients are perfect—well grown, unblemished and perfectly fresh—and are handled carefully according to directions.

However, a crisper-upper also mentioned in heirloom recipes is a natural one that, so far as we can discover, is safe to use without restriction. One bygone rule we've seen says to cover the bottom of the crock with washed grape leaves and put a layer of them on top of the pickles; and a Southern homemaker says scuppernong leaves are best.

WATER

Water used in making pickles should of course be of drinking quality (because otherwise contaminants can increase the bacterial load that leads to spoilage). Also, water with above-average calcium content can shrivel pickles, and iron compounds can make them darker than we like.

See page 16 for ways to rid water of some excess minerals, and page 536 for a discussion of how "hard" or "soft" water affects texture.

METHODS

LONG-BRINE

Vegetables such as cucumbers are washed and dropped into a heavy salt solution (plus sometimes vinegar and spice) and left in a cool place to cure for 2 to 4 weeks. Scum *must be removed* from the brine each day. Following this the pickles are packed loosely in clean jars and covered with the same or freshly made brine and processed in a Boiling-Water Bath.

SHORT-BRINE

Vegetables are left overnight in a brine to crisp up. The next day they are packed in jars, covered with a pickling solution and processed in a Boiling-Water Bath (212 F.) for a suitable time.

COMPLETE PRECOOKING

Complete precooking is the rule for relishes and similar cut-up pickle mixtures in a sweet-sour liquid. Packed hot in regular canning jars, these products then have a short Boiling-Water Bath.

Pickle, etc., Troubles and What to Do

If you find any imperfect seals during the 24 hours between processing and storing your pickles, relishes and sauces, you can dump the contents into the preserving kettle, bring to boiling, pack into hot, sterilized canning jars, and process again in the Boiling-Water Bath (212 F.) for the required time. (Of course if only one seal is imperfect, it's easier to pop that jar in the refrigerator and eat the food within the next few days.)

The interim day before storing in a cool, dry, dark place is the only time that these foods can be salvaged by repacking in sterilized containers and processing over again from scratch.

If, after these foods are checked and put in the storage area, you find any of the following, DESTROY THE CONTENTS SO THAT THEY CANNOT BE EATEN BY PEOPLE OR ANIMALS; then wash jars and closures in

hot soapy water, and boil them hard for 15 minutes in clean
water to cover. (Throw away the sterilized sealing discs and
rubbers; if sound, the sterilized jars may be used again.)

—Broken seal.
—Seepage around the seal, even though it seems firmly
seated.
—Mold, even a fleck, in the contents or around the seal
or on the underside of the lid.
—Gassiness (small bubbles) in the contents.
—Spurting liquid, pressure from inside as the jar is
opened.
—Mushy or slippery pickles.
—Cloudy or yeasty liquid.
—Off-odor, disagreeable smell, mustiness.

If this sounds strict, it's on purpose. Our most shiver-
producing bedside reading, *Botulism in the United States,*
1899–1973, a handbook prepared by the Center for Disease
Control of the U.S. Public Health Service and issued in
June 1974, listed condiments, including tomato relish, chili
sauce and pickles, among the most common vehicles of
botulism poisoning in home-canned foods.

Careless and unclean handling of raw materials and containers, inadequate processing, and warm storage conditions
are the main causes of spoilage in the foods described in
this section. Old "open-kettle" canning without benefit of a
Boiling-Water Bath to sterilize and to ensure the seal is one
type of inadequate processing. Another is failure to keep at
least 1 inch of boiling water over the tops of the containers
in the B-W Bath, and not keeping the water at a full boil
for the full time.

Often, low-acid vegetables are spoiled by the scum that
naturally forms on top of the fermentation brine; the scum
should be removed faithfully. And it is not only the top
layer of pickles that is affected, for the scum (which contains wild yeasts, molds and bacteria) can weaken the acidity of the brine.

Also, hard water that contains a great deal of calcium
salts can counteract some of the acid, or keep acid from
forming well enough during brining, and thus interefere
with the process that is meant to make certain pickles safe
with an otherwise adequate Boiling-Water Bath.

And of course "knife out" air bubbles before capping.

NOT PERFECT, BUT EDIBLE

If jars have good seals, if there are none of the signs of spoilage noted above, and if the storage has been properly cool, you can have less-than-perfect pickles that are still O.K. to eat.

Hollow pickles. The cucumbers just developed queerly on the vine; you can spot these odd ones when you wash them: usually they float. So use them chopped in relishes. Or they stood around more than 24 hours after being picked. If you can't get around to doing them the day you get them, refrigerate.

Shriveled pickles. This can come from plunging the cucumbers into a solution of salt, vinegar or sugar that's too strong for them to absorb gradually (here's the reason why some recipes handle pickles in stages). Or they've cooked too fast in a sugar-vinegar solution. Or the water used was too hard (see also "Water," page 344).

Darkened pickles. Iron in hard water, or loose ground spices.

Bleached-looking pickles. With no signs of spoilage present, this could mean that jars were exposed to light during storage. Wrap the jars in paper or put them in closed cartons if the place they're stored is not dark.

212 F. = 100 C. / 220 F. = 104 C. / 1 U.S. cup ($\frac{1}{2}$ U.S. pint) = .24 L. / 1 U.S. pint (2 cups) = .48 L. / 1 U.S. tsp. = 5 ml. (5 cc.) / 1 U.S. T. = 15 ml. (15 cc.) / $\frac{1}{2}$ inch = 1.25 cm.

Pickles and Relishes

Sweet Mixed Pickles

1 quart unpeeled cucumber cubes ($\frac{3}{4}$-inch)

1 quart tiny pickling onions, peeled

1 medium head cauliflower, broken into flowerets

1 large sweet red pepper, seeded and chopped

$\frac{1}{2}$ cup salt

4 cups vinegar

2 cups firmly packed light brown sugar (or raw)

$\frac{1}{4}$ teaspoon turmeric

1 tablespoon mixed pickling spices

2-inch stick whole cinnamon

6 whole cloves

1 teaspoon mustard seed

Combine the four vegetables and sprinkle with the salt. Cover with cold water and let stand overnight. Drain; rinse in fresh water and drain again thoroughly. Combine vinegar, sugar and turmeric in a large enamelware kettle. Tie spices in a small cloth bag and add. Place over moderate heat and bring to boiling. Cook 10 minutes, then add vegetables. Bring again to boiling and cook 5 minutes. Put hot in hot canning jars, leaving ½ inch of headroom, for ½-pints or pints, and process in a Boiling-Water Bath (212 F.) for 10 minutes. Makes about 3 pints.

Sweet Pickle Chips

These are so delicious and so easy that you'll want to make several separate batches as the cucumbers come along.

4 pounds pickling cucumbers (3 to 4 inches long)

Brining solution:

1 quart distilled white vinegar	1 tablespoon mustard seed
3 tablespoons salt	½ cup sugar

Canning sirup:

1⅔ cups distilled white vinegar	1 tablespoon whole allspice
3 cups sugar	2¼ teaspoons celery seed

Wash the cucumbers, remove any blemishes, nip off the stems and blossom ends and cut them crossways in ¼-inch-thick slices. In a large enameled or stainless steel kettle, mix together the ingredients for the *brining solution;* add the cut cucumbers. Cover and simmer until the cucumbers change color from bright to dull green (about 5 to 7 minutes).

Meanwhile have ready the *canning sirup* ingredients heated to the boil in an enameled kettle. Drain the cucumber slices and pack them, while still piping hot, in hot 1-pint canning jars, and cover them with very hot sirup, leaving ½ inch of headroom. Remove air bubbles, and adjust lids. Pack and add the hot sirup to one jar at a time, returning the sirup kettle to low heat between filling and capping each jar, so the sirup doesn't cool. Process filled and capped jars in a Boiling-Water Bath (212 F.) for 10 minutes. Remove jars and complete seals if necessary. Makes 4 to 5 pints.

Bread-and-Butter Pickles

This recipe is an especially good one, from Isabelle Downey's *Food Preservation in Alabama*.

6 pounds medium cucumbers	4½ cups sugar
1½ cups sliced onions	1½ teaspoons turmeric
2 large garlic cloves, left whole	1½ teaspoons celery seed
⅓ cup salt	2 tablespoons mustard seed
2 trays of ice cubes or crushed ice	3 cups white vinegar

Wash the cucumbers thoroughly; drain; cut unpeeled cucumbers into ¼-inch slices. In a large bowl, combine the cucumber slices, onions, garlic and salt; cover with the crushed ice or ice cubes, mix thoroughly and let stand for 3 hours. Drain off the liquid and remove the garlic. Combine the sugar, spices and vinegar and heat just to a boil. Add the cucumber and onion slices; simmer together 10 minutes. Pack loosely in clean, hot pint jars, leaving ½ inch of headroom; remove air bubbles. Adjust lids; process in a Boiling-Water Bath (212 F.) for 10 minutes. Remove, complete seals if necessary. Makes 7 pints.

Note: Save the pickling sirup left after you've opened and used a jar of these or Watermelon Pickles (q.v.). Pour it over drained cooked or canned sliced beets; cool, cover, store in the refrigerator for a day or two to develop the flavor, and serve.

Sweet Pumpkin Pickle

6 cups prepared pumpkin	2 large sticks whole cinnamon
2 cups vinegar	
2 cups sugar	

Prepare pumpkin by peeling and cubing flesh, discarding seeds and inner pulp. Place pumpkin cubes in a colander and set over boiling water: make sure water does *not* touch the pumpkin. Cover and steam until just tender. Drain.

Simmer vinegar, sugar and cinnamon for 15 minutes. Add pumpkin cubes and simmer 3 minutes. Set aside for 24 hours. Heat and simmer 5 minutes more. Remove cinnamon. Pack boiling hot in hot canning jars leaving ½ inch of headroom, adjust lids and process in a Boiling-Water Bath (212 F.) for 10 minutes. Makes 3 pints.

This pickle compares favorably with that made of cantaloupe.

Little Cucumber Crock Pickles

This is an old-time rule, producing small, crisp, whole pickles with good flavor. They take 4 to 5 weeks to make; and if they're put in brine as they come along in season, and kept in a cool place, they should last well into winter.

1 gallon cider vinegar (regular 5 percent)	(pure, no fillers) + salt to add later
½ cup sugar (or 1 teaspoon powdered straight saccharin)	optional: 4 fresh dill heads; or more, if you like stronger dill
1 cup whole mustard seed	3- to 4-inch pickling cucumbers (about 10 pounds total, or a scant peck)
1 cup pickling salt	

Thoroughly scrub a 5-gallon earthenware crock with hot water and soap, rinse well, then scald with boiling water; be energetic about it, because any residue of fat or milk from a previous use will ruin the pickles. In the crock mix together the vinegar, sweetening, mustard seed and salt; lay dill heads on the bottom if you like them. Keep the crock in a constantly cool place (40 to maximum 50 F.).

As they're gathered, wash the little cucumbers well, rub off the blossom ends (where enzymes are concentrated), and drop the cucumbers into the brine. Push newly harvested ones toward the bottom of the container as it fills, so the last ones in will not be the first ones out. Hold the pickles beneath the brine with a weighted plate (a pint jar filled with water weighs enough), and cover the crock with a layer of clean cheesecloth or muslin.

If you put all the cucumbers in at the same time, after three days add 1 cup more pickling salt, laid on the plate where it will dissolve slowly downward (the extra salt counteracts weakening of the brine as the natural juice is drawn from the cucumbers). One week later, put ¼ cup more salt on the submerged plate; and continue adding ¼ cup salt in this manner each week until the pickles are ready. At the end of a month, test by cutting a pickle crossways: if it is firm, and clear throughout with no white center, the pickles are ready to eat.

If you harvest your cucumbers piecemeal over a period of, say, two weeks, lay ½ cup pickling salt on the plate when the crock is half full, and add another ½ cup salt when the crock is filled; thereafter add ¼ cup salt each week until the pickles are ready.

A gray-white film will appear on the surface of the brine after the cucumbers have been in the pickling solution a few days: skim it off, and keep removing it as it forms. The film is to be expected as a natural part of the brining process, but if allowed to stay on the pickles it will hurt the acidity of the pickling solution, and your pickles will spoil.

When there's no more film, and your pickles test evenly clear to the center, start enjoying them. Always replace the weighted plate after taking any out, and cover the crock to keep the contents clean.

Alternate containers. Use a scrupulously clean glass crock (gettable in many housewares departments), or a graniteware (heavily enameled) preserving kettle whose inside is free from even tiny nicks or scratches that lay bare the metal. Or, cutting down proportionally on ingredients, use commercial 1-gallon jars whose mouths are large enough for your hand (these big jars are often gettable from a restaurant or institution that buys in quantity). For such jars, large *leakproof* plastic bags partly filled with water make good weights; they plop down to shape themselves in a good fit on top of the pickles.

Storage for glass containers should be dark as well as dry and cool, otherwise the light will fade your pickles.

Canning. If the conditions for storing your crock of pickles are not good, or if you foresee that you can't eat them all within their storage life of several months—can them.

Take all the pickles from the brine, and fit them vertically in clean pint or quart jars, leaving ½ inch of headroom. From the pickling solution remove any dill heads (and the mustard seed, if you like); bring the solution to boiling and pour it over the pickles, leaving ½ inch of headroom. Re-

move trapped air with the blade of a table knife, adjust lids
with their clean fresh rubbers or sealers, and process in a
Boiling-Water Bath—10 minutes for pints, 15 minutes for
quarts. Remove jars; complete seals if necessary.

Watermelon Pickles

8 cups prepared water- melon rind	4 teaspoons whole cloves
½ cup pickling salt	4 cups sugar
4 cups cold water	2 cups white vinegar
	2 cups water

Choose thick rind. Trim from it all dark skin and remains
of pink flesh; cut in 1-inch cubes. Dissolve salt in cold water,
pour it over rind cubes to cover (add more water if needed);
let stand 5 to 6 hours. Drain, rinse well. Cover with fresh
water and cook until barely tender—no more than 10 minutes (err on the side of crispness); drain. Combine sugar,
vinegar and water, add cloves tied in a cloth bag, and bring
to boiling; reduce heat and simmer for 5 minutes. Pour over
rind cubes, let stand overnight. Bring all to boiling and cook
until rind is translucent *but not at all mushy*—about 10 minutes. Remove spice bag, pack cubes in hot sterilized pint
jars; add boiling sirup, leaving ½ inch of headroom; adjust
lids. Process in a Boiling-Water Bath for 10 minutes. Remove jars and complete seals if necessary. Makes about 4
pints.

Ripe Cucumber Pickle

9 large ripe cucumbers	2 tablespoons mixed
⅓ cup salt	whole pickling
4 cups vinegar	spices
4 cups sugar	2 tablespoons mustard
	seed

Pare cucumbers and cut in half lengthwise. With a silver spoon, scrape out seeds and pulp, then cut in pieces about 1 × 2 inches. Sprinkle with the salt and cover with cold water. Let stand overnight. Drain; rinse with fresh water, and drain again. Combine vinegar and sugar in a large enamelware kettle. Tie spices in a small cloth bag and add. Bring to boiling and cook 5 minutes. Add drained cucumber slices and bring again to boiling. Reduce heat and simmer until cucumber is tender and appears translucent. Discard the spice bag; ladle into hot pint jars, leaving ½ inch of headroom, and process for 10 minutes in a Boiling-Water Bath (212 F.). Makes about 6 pints.

Sometimes this old recipe included a firming chemical in the overnight soak, but it's really not necessary if you're careful not to overcook.

Golden Glow Pickle

3 quarts prepared ripe cucumbers	¼ cup salt
	3 cups vinegar
6 medium onions, peeled and diced	3½ cups sugar
	15 whole cloves
2 green peppers, seeded and diced	2 tablespoons mustard seed
2 sweet red peppers, seeded and diced	1 teaspoon celery seed
	1 teaspoon turmeric

To prepare cucumbers: peel and cut lengthwise; discard seeds and pulp, and cut in ¾-inch cubes. Combine with diced onions and peppers. Sprinkle with the salt and let stand overnight. Drain; rinse in fresh water, and drain again thoroughly. Combine vinegar, sugar and spices in a large enamelware saucepan. Bring to boiling, then add the drained vegetables. Bring again to boiling, reduce heat and simmer 20 minutes: the cucumber cubes *should be just tender,* not mushy. Ladle hot into hot canning jars, leaving ½ inch of headroom, and process in a Boiling-Water Bath (212 F.) for 10 minutes for pint jars. Makes about 6 pints.

Zucchini Pickle

2 quarts thin slices of unpeeled, *small* zucchini squash	2 cups sugar
	1 teaspoon celery seed
	2 teaspoons mustard seed
2 medium onions, peeled and thinly sliced	1 teaspoon turmeric
¼ cup salt	½ teaspoon dry mustard
2 cups vinegar	

Combine zucchini and onions. Sprinkle with the salt, cover with cold water and let stand 2 hours. Drain; rinse with fresh water, and drain again. Combine remaining ingredients in an enamelware kettle and bring to boiling. Cook 2 minutes. Add zucchini and onions, remove from heat, and let stand 2 hours. Bring again to boiling and cook 5 minutes. Ladle hot into hot pint jars, leaving ½ inch of headroom, and process in a Boiling-Water Bath (212 F.) for 10 minutes. Makes about 4 pints.

Sweet Mustard Pickle

1 quart small green tomatoes, quartered	3 green peppers, seeded and diced
1 quart *small* unpeeled cucumbers (about 2 inches)	2 cups green beans, cut in 1-inch slices
	1 cup salt
1 quart unpeeled *medium* cucumbers	1 cup flour
	⅓ cup dry mustard
1 quart tiny pickling onions	2 teaspoons turmeric
	2 cups sugar
1 small head cauliflower, broken into flowerets	2 quarts vinegar

Combine vegetables and sprinkle with the salt. Cover with cold water and let stand overnight. Place over moderate heat and bring just to boiling point, then drain thoroughly. Combine remaining ingredients smoothly. Stir over moderate heat until smooth and thick. Add well-drained vegetables and bring *just to boiling point:* they should never be overcooked and mushy. Ladle hot into hot canning jars, allow ½ inch of headroom, and process in a Boiling-Water Bath (212 F.) for 10 minutes for pints or quarts. Makes about 4 quarts.

Dill Cucumber Pickles (Short-brine)

17 to 18 pounds of pickling cucumbers (3 to 5 inches)

2 gallons of 5 percent brine (¾ cup pickling salt to each 1 gallon of water)

6 cups vinegar

¾ cup salt

¼ cup sugar

9 cups water

2 tablespoons whole mixed pickling spices

14 teaspoons whole mustard seed (2 teaspoons go in each quart jar)

7 to 14 cloves garlic (1 to 2 cloves go in each quart jar)

21 dill heads (3 heads go in each quart jar)
OR

7 tablespoons dill seed (1 tablespoon to each quart jar)

Put washed and brush-scrubbed cucumbers in a noncorroding crock or kettle and cover with the 5 percent brine. Let stand overnight, then drain and pack cucumbers in clean, hot quart jars. Add the mustard seed, dill and garlic to each jar.

Combine vinegar, salt, sugar and water; tie pickling spices loosely in a clean, thin, white cloth and drop it into the mixture. Bring to a boil. Take out the spice bag and pour boiling liquid over cucumbers in jars, leaving ½ inch of headroom. Adjust the lids and process in a Boiling-Water Bath (212 F.) for 20 minutes. Makes about 7 quarts.

Green Tomato Pickles

7½ pounds green tomatoes (about 30 medium)	1 tablespoon whole cloves
6 good-sized onions	1 tablespoon dry mustard
¾ cup pickling salt	1 tablespoon peppercorns
1 tablespoon celery seed	½ lemon
1 tablespoon whole allspice	2 sweet red peppers
1 tablespoon mustard seed	2½ cups brown sugar
	3 cups vinegar (c. 5 percent acidity)

Wash tomatoes well, cut off blossom ends, blemishes and stems. Slice thin crossways. Peel and slice onions in thin rings. Sprinkle salt over alternate layers of sliced tomatoes and onions in an earthenware dish, and let stand in a cool place overnight. Drain off the brine, rinse the vegetables thoroughly in cold water and drain well. Slice the lemon thinly and remove the seeds; wash the peppers well, remove stems and seeds, slice thinly crossways. Tie all the spices loosely in muslin or a double layer of cheesecloth, add the spice bag and the sugar to the vinegar in a large enamelware kettle; bring to a boil. Add the tomatoes, onions, lemon and peppers. Cook for 30 minutes after the mixture returns to a boil, stirring gently to prevent scorching. Remove the spice bag, pack the pickles in hot jars and cover with boiling-hot liquid, leaving ½ inch of headroom. Adjust lids. Process in a Boiling-Water Bath (212 F.) for 10 minutes. Remove jars; complete seals if necessary. Makes about 6 pints.

RELISHES AND SAUCES

Piccalilli

6 medium-size green
tomatoes
6 sweet red peppers,
seeded
6 medium onions,
peeled
1 small cabbage

¼ cup salt
2 cups vinegar
2½ cups light brown
sugar (or raw)
2 tablespoons mixed
pickling spices

Put vegetables through the food grinder, using a coarse knife. Sprinkle with the salt, cover and let stand overnight. Drain; then cover with fresh water, and drain again. When thoroughly drained, put into a large kettle and add vinegar and sugar. Tie spices in a small cloth bag and add. Bring to boiling, then reduce heat and simmer about 20 minutes, stirring frequently. Remove the spice bag and turn the hot piccalilli into hot jars, leaving ½ inch of headroom; adjust lids and process in a Boiling-Water Bath for 10 minutes. Makes about 4 pints.

Beet Relish

12 cooked medium beets
(fresh or canned)
1 medium onion,
peeled and
chopped
2 cups finely chopped
cabbage
1 sweet red pepper,

seeded and finely
chopped
1½ teaspoons salt
¾ cup sugar
1½ cups vinegar
2 tablespoons prepared
horseradish

The vegetables may all be put through the food grinder, using a coarse knife. Combine all ingredients in a large saucepan. Bring to a boil, then reduce heat and simmer about 15 minutes, stirring frequently. Ladle hot relish into hot pint jars, leave ½ inch of headroom, and process in a Boiling-Water Bath for 10 minutes. Makes about 3½ pints.

Quick Pickled Beets, a side dish that's not canned, is given in the Vegetables Recipes.

❧

Corn Relish

4 cups corn kernels (about 9 ears' worth) *	2 teaspoons salt
	1½ teaspoons dry mustard
1 cup diced sweet green peppers	1 teaspoon celery seed
1 cup diced sweet red peppers	¼ teaspoon Tabasco sauce
1 cup finely chopped celery	½ teaspoon turmeric, for color (optional)
½ cup minced onion	2 tablespoons flour, for thickening (optional)
1½ cups vinegar	
¾ cup sugar	

Prepare corn by boiling husked ears for 5 minutes, cooling, and cutting from cob (do not scrape). In an enameled kettle combine peppers, celery, onion, vinegar, sugar, salt, celery seed and Tabasco sauce; boil 5 minutes, stirring occasionally. Dip out ½ cup hot liquid, mix it with dry mustard and turmeric, and return it to the kettle. Add the corn. (If you want the relish slightly thickened, blend the 2 tablespoons flour with ¼ cup cold water and add to the kettle when you put in the corn.) Boil for 5 minutes, stirring extra well if the relish has been thickened, so it won't stick or scorch. Immediately fill clean hot pint jars within ½ inch of the top, adjust lids, and process in a Boiling-Water Bath for 15 minutes. Complete seal if necessary; cool and store. Makes 3 pints.

* You can use frozen whole-kernel corn that's been thawed slowly: 3 10-ounce packages will equal 4 cups of fresh kernels.

Spiced Pears, see under Canning.

Indian Chutney

The rule for this fine Calcutta-style chutney was given us by Frances Bond, who lived twenty years in India before moving to Vermont, and she has tailored it for ingredients easy for the North American housewife to come by. It's ideal with budget-stretching curries or pilau, with hot or cold meats, and it makes a delightful present packed in decorative ½-pint canning jars.

For best results, the fruit—whether apples, peaches or pears—should be firm varieties, or slightly underripe. The fruit, raisins and crystallized ginger are added after the sirup ingredients have cooked together for 30 minutes, to let them keep their identity in the finished product: they should be tender but recognizable in the sirup, which is thick and a rich brown in color. The chutney improves after a couple of months in sealed jars.

juice, pulp and peel of 1 lemon, finely chopped
2 cups cider vinegar
2½ cups dark brown sugar (1 pound)
1 clove garlic, minced
pinch of cayenne pepper (⅛ teaspoon)
pinch of chili powder (⅛ teaspoon)
1½ teaspoons salt

5½ cups coarsely chopped firm apples, peeled and cored (about 3 pounds), or peaches or pears*
¾ cup crystallized ginger**—cut small but not minced (about 3 ounces)
1½ cups raisins, preferably seeded (½ pound)

Chop the lemon, removing seeds and saving the juice (a blender is good here), and put it in an open, heavy enameled kettle with the sugar, vinegar, minced garlic, salt, cayenne pepper and chili powder. Boil the mixture over medium heat for 30 minutes, stirring occasionally. Meanwhile prepare the apples (or peaches or pears), and add them to the sirup with the raisins and ginger. Boil all slowly, stirring to prevent sticking and scorching, until the fruit is tender but not mushy and the sirup is thick—about 30 to 45 minutes longer. Ladle the boiling-hot chutney into sterilized pint or ½ pint jars, filling to ⅛ inch of the top, and cap each jar immediately with a sterilized 2-piece screwband lid. Invert each jar for a moment after it is capped. Cool topside up and store. Makes 3 pints, or 6 ½-pints.

*Caught without fresh fruit in a chutney-making mood, Mrs. Bond substituted 5½ cups of coarsely chopped canned, drained pears, but added them in the last 10 minutes of cooking. Results: heavenly.

**Ground ginger contributes only flavor without texture, and reconstituted dry cracked ginger is usually woody, so don't substitute with either of them.

Tomato Ketchup

Because of the extra acidity from the vinegar and the long cooking in an open kettle, this good ketchup can be processed safely in a 10-minute B-W Bath. Some cooks add the sugar later with the vinegar to reduce the chance of scorching, but we think that adding the sugar earlier—with the spices—enables the flavors of the spices to develop in a pleasant way.

1 peck ripe tomatoes (8 quarts, or c. 50 medium tomatoes)	1 tablespoon whole allspice
2 cups finely chopped onions (c. 3 large)	1 tablespoon whole cloves
1 cup chopped sweet red peppers (c. 2 large)	1 tablespoon peppercorns
1 clove garlic, finely chopped	2 teaspoons mustard seed
1 tablespoon salt	1 bay leaf
1 tablespoon celery seed	¾ cup brown sugar
	2 cups cider vinegar

Wash the tomatoes, but don't bother to peel; cut them small, saving all the juice. In a heavy kettle combine the tomatoes, onions, peppers, garlic and salt, and simmer the mixture until soft—about 25 minutes. Press through a sieve or food mill to remove seeds and skins. Tie the spices and bay leaf in muslin or double-thick cheesecloth and add the bag and the brown sugar to the mixture. Cook quickly at medium boil, stirring frequently, until the mixture is reduced to ½ its original volume—about 1 hour. Remove the spice bag. Add the vinegar; simmer 10 to 15 minutes longer, stirring often, until the mixture thickens again. Pour while boiling into hot ½ pint or pint jars, leaving ½ inch of headroom. Run the blade of a table knife around the inner side of the jar to release any trapped air; adjust lids. Process ½ pints and pints in a Boiling-Water Bath (212 F.) for 10 minutes. Remove, complete seals if necessary. Makes 6 pints.

Chili Sauce

Because of extra acid from the vinegar and the long cooking, this can be finished safely in a B-W Bath.

4 quarts chopped ripe tomatoes (about 9 to 10 pounds)	1 3-inch stick of cinnamon, broken in pieces
5 large onions, peeled and chopped small	1 tablespoon mustard seed
4 sweet red peppers, seeded and chopped	2 teaspoons celery seed
	1½ teaspoons ground ginger
2 cups cider vinegar	1 teaspoon ground nutmeg
1 cup brown sugar, packed firmly	
2½ tablespoons salt	1 teaspoon peppercorns°

Peel, core and chop tomatoes; peel and chop onions; seed and chop peppers. Put them in a heavy enameled kettle, add the vinegar, sugar, salt; tie the spices in a double thickness of muslin or four thicknesses of cheesecloth (extra density of cloth will hold the ground spices better) and add the bag to the ingredients in the pot. Bring the mixture quickly to boiling, stirring so it won't scorch, then reduce the heat and cook at a slow boil until the sauce is thick—from 3 to 4 hours. It wants to be a little thicker than ketchup but not so thick as jam; and it will scorch if it's not watched and stirred, especially toward the end. Remove the spice bag and pack hot in hot canning jars, leaving ½ inch of head-room. Adjust lids and process in a Boiling-Water Bath (212 F.) for 15 minutes. Remove, complete seals if necessary. Makes about 5 to 6 pints.

*For "hotter" sauce, substitute 1 teaspoon crushed dried *hot* red pepper pods. This is one of those recipes whose seasonings can be tinkered with according to the family's taste.

MINCEMEAT

Mincemeat

1 pound boiled lean beef	2 cups sweet cider
½ pound beef suet	¾ teaspoon ground cinnamon
2½ cups seeded raisins	¾ teaspoon ground mace
¼ pound chopped citron	¾ teaspoon ground cloves
3 cups coarsely chopped apples	¼ teaspoon ground nutmeg
2 cups dried currants	¼ teaspoon ground allspice
2¼ cups light brown sugar	¼ teaspoon salt
3 tablespoons light molasses	1 cup brandy

Put beef, suet and raisins through the food grinder, using a coarse knife. Put citron and apples through the grinder. Combine all in a heavy kettle and add remaining ingredients in order, *except the brandy*. Bring to boiling, stirring con-stantly. Reduce heat and simmer about 1 hour, stirring fre-quently. *The mixture will scorch easily, so use an asbestos pad under the kettle.* Remove from heat and stir in brandy. Ladle hot into hot pint canning jars, allowing ½ inch head-room, and process at 10 pounds pressure (240 F.) for 20 minutes. Makes 5 pints, enough for 5 nine-inch pies.

Green Tomato Mincemeat

3 quarts prepared green
  tomatoes
3 quarts prepared apples
1 cup ground suet
1 pound seedless raisins
2 tablespoons grated
  orange rind
2 tablespoons grated
  lemon rind
5 cups well-packed light
  brown sugar (or
  raw)

¾ cup vinegar
½ cup fresh lemon juice
½ cup water
1 tablespoon ground
  cinnamon
¼ teaspoon ground cloves
¼ teaspoon ground all-
  spice
2 teaspoons salt

To prepare tomatoes: put through the food grinder, using a coarse knife. Peel and core apples and put through
grinder. Repeat with the suet. Combine all ingredients in
order in a large kettle, and bring to boiling, stirring frequently. Reduce heat and simmer until dark and thick—
about 2½ hours. Stir occasionally. An asbestos pad under
the kettle will help prevent scorching. Pour boiling hot into
pint jars, allowing ½ inch headroom, and process in a B-W
Bath (212 F.) for 25 minutes. Complete seals if necessary.
Makes 8 pints, enough for 8 nine-inch pies.

Pressure-processing is not needed in this old recipe to
ensure sterilization, because of the very long cooking time.

Drying 5

Drying as a way of putting food by goes back into prehistory, and it is still a prime method of preserving staples in hot, dry areas of the world where other means are lacking. Partly because of this very storybook quality, partly because it is basically a simple process under ideal conditions, partly because it creates a lightweight, highly concentrated food supply—for all these reasons drying at home has enjoyed a surge of popularity in recent years.

It is easy to be lured by its glamour, though, so start small. Try a bit of sun-drying first (if your climate permits), experiment with doing air-dried herbs in your attic, or with doing sliced apples in your oven. Even when you have the procedures down pat, it will make sense to preserve meats and fish and a majority of vegetables and many fruits by canning, freezing, root-cellaring or curing, or whatever.

WHAT DRYING DOES

The purpose of drying is to take out enough water from the material so that spoilage organisms are not able to grow and multiply during storage (see "Why Put-by Foods Spoil" in Chapter 1). The amount of remaining moisture that is tolerable for safety varies according to whether the food is strong-acid or low-acid raw material, or whether it has been treated with a high concentration of salt—and, to some degree, with the type of storage.

In addition, properly home-dried fruits and vegetables, uncooked, have roughly $\frac{1}{6}$ to $\frac{1}{3}$ the bulk and only around 10 to 20 percent of the water of their original fresh state.

Note: Although "drying," "dehydrating" and "evaporating" are often used casually as meaning the same thing, the USDA Research Service's fine *Agriculture Handbook No. 8, Composition of Foods: Raw, Pro-*

Complete metric conversion tables are in Chapter 1, but the following apply particularly to this section: 130 Fahrenheit = 54 Celsius / 140 F. = 60 C. / 150 F. = 66 C. / 175 F. = 79.5 C. / 212 F. = 100 C. / 1 U.S. teaspoon = 5 milliliters (5 cubic centimeters) / 1 U.S. tablespoon = 15 ml. (15 cc.)

cessed, Prepared lists as dehydrated those foods containing only 2.5 to 4 percent water—the other 96+ percent having been removed by highly sophisticated processes that we can't hope to equal at home. It lists as dried those foods still containing roughly 10 to 20 percent water (the amount depending on whether they're vegetables or fruits). We can take out all but this much moisture with the equipment and methods described in this section—and we'll call it *drying*.

GENERAL PROCEDURES IN DRYING

Choose perfectly fresh food in prime condition—just as you do for every food you put by—and handle it quickly and with absolute cleanliness at every step. Peel it, pit it, cut it up: whatever is required. Small or fairly thin pieces usually make the best product.

Many foods require some sort of treatment before drying to preserve color and nutrients, prevent decomposition, ensure even drying and prolong storage life. Depending on the type of food, these treatments are usually coating with an anti-oxidant, blanching, sulfuring, and, usually in the case of fish and meat, salting.

Good drying is done as rapidly as possible, so long as it doesn't actually cook the material and thereby spoil its looks and texture: the idea is to have the drying process outstrip decomposition. For most foods the heat is increased as the drying proceeds.

The specific appearance of proper dryness is described in the instructions for drying each food. (See also "Tests for Dryness in Produce.")

Some foods are better for being subjected to pasteurizing temperatures for a short time after they test dry.

Many foods require a conditioning period after they test dry: they are stored temporarily in bulk to distribute remaining moisture more evenly; periodic stirring and examination disclose which pieces are not dry enough for storage, and these are removed.

Properly dried foods are refrigerated (in the case of meats and fish), or are stored in a dry place, and in containers safe from insects and animals.

EQUIPMENT FOR DRYING

Keep everything simple, even rudimentary, in the beginning: aside from saving money it's a lot more fun in this hypertechnical age to return to elementals. You can always branch out with more sophisticated gear when you get your technique down pat for one type of food and start with a new one that requires a different treatment. Try for as much uniformity in size as possible, though, so you can swap equipment from one system to another.

Trays first

Shallow wooden—*never metal*—trays are necessary whether you dry outdoors in sun or shade, or indoors in a dryer or an oven. They should have slatted, perforated or woven bottoms to let the air get at the underside of the food. Don't make them of green wood—which weeps and warps; and don't use pine, which imparts a resinous taste to the food; and don't use oak or redwood, which can stain the food. Ingenuity will turn up many suitable materials: the following ideas are just a sampling.

The simplest frames to make would be those cut from wooden crates that produce comes in: saw the crates in several sections horizontally, rather as you'd split a biscuit.

If you're making frames from scratch, you can use 1-inch × 1-inch material of the kind you're likely to be using anyway for vertical cleats at the corners of a dryer or for bracing. It will give your trays only 1 inch of depth for holding food; this could be a disadvantage for sun-drying—which requires that food be protected by netting of some sort stretched over the top of the tray—but this extreme shallowness doesn't matter much in a dryer.

Tempered hardboard (this is not underlayment) would be good. It is strong despite its thinness and comes in 4-foot

× 8-foot sheets that is fine material for making dryers and sulfuring boxes. One sheet will give you a dryer 14 inches wide, 24 inches deep and 36 inches high, plus frames for 6 trays to fit inside, and some usable trimmings left over. For each tray with sides 2 inches high, cut two sides and two ends and fasten them together in a rectangle by nailing them, not to each other, but to 1-inch × 1-inch cleats; four-penny box nails will do the job well. Diagonal cross-bracing of the cleat material to form an X, with the end of each arm nailed to the corner cleat, will strengthen the frame, and will also provide valuable extra air space when trays are stacked; put it on after the bottom of the tray is attached.

HOW BIG?

Each 1 square foot of tray space will dry around 1½ to 2½ pounds of prepared food.

Loaded trays shouldn't be too large to handle easily, and they should be uniform in size so they stack evenly. The flimsier the construction, the smaller they should be; but even well-built ones for sun-drying are better if they're not more than 2 feet by 2 feet.

However, since you can have an emergency that means you will need to finish off in an oven or dryer a batch you've started outdoors, it makes sense to have the trays smaller, and rectangular. Make the trays narrow enough for clearance when you slide them inside, and 3 to 4 inches shorter than the oven or dryer is from front to back: you'll want to stagger the trays to allow air to zigzag its drying way up and over each tray as it rises from the intake at the bottom to the venting at the top.

Consider having the trays 1 to 2 inches deep, 12 to 16 inches wide, 16 to 20 inches long—but first having found the inner dimensions of the oven or dryer (less the fore-and-aft leeway for staggering the trays).

NO METAL FOR THE BOTTOMS EITHER

Don't use metal screening for the bottom. Aluminum discolors and, more important, it corrodes easily. Copper destroys Vitamin C. And galvanized screen has been treated with zinc and cadmium—and cadmium is dangerous stuff indeed to mix with food (when old-time instructions ask for "hardware cloth" they mean galvanized screen, by the way).

Steer clear of fiberglass mesh: minute splinters of fiberglass can be freed easily and impregnate the food.

Vinyl-coated screen in beguiling ¼-inch and ½-inch mesh looks like the answer at first glance, but what will it do at 140 F., the average heat in a dryer—melt? peel? And how will it react with the food after awhile? Certainly you could test a bit of it in sun-drying.

Any cloth netting will do if its mesh isn't larger than ½-inch. Two layers of cheesecloth work, as does mosquito net, etc.—but they're hard to clean without getting frazzled. Old clean sheets let less air up through but they're stouter. (In a pinch you can dry food on sheets laid flat in the direct, hot sun.) When cutting cloth for tray bottoms, allow 2 inches more all around so you can fold it over itself on the outside of the frame; then staple it in place.

We've seen good trays with bottoms of hay-bailing twine strung back and forth and then cross-hatched the other way. Draw the twine tight and flat, staple each loop to the outside of the frame-strip before you turn around and go back, keeping the strands ½ inch apart.

Strong, serviceable bottoms are made by nailing ½-inch wood strips to the bottom of the frame ½ inch apart; the strips run in only one direction. More finished—but worth it, because they're smooth and easy to clean—are ¼- or ½-inch hardwood dowels; these are nailed inside the frame with small box nails driven through from the outside, and they also go in only in one direction.

One thickness of cheesecloth laid over bottoms will keep sugar-rich food from sticking to them while it dries; so will a thin coating of oil. Even a few recent publications suggest mineral oil for lubricating the trays—it doesn't impart flavor and doesn't get rancid—but if you're leary of its effect on your body's vitamin absorption, use any fresh, low-flavored vegetable oil. You'll be scrubbing your trays anyway, regardless of what oil you use.

Simple dryers and drying aids

As with trays, there's somehow more satisfaction in using unfancy dryers, especially if you're just starting out. But the USDA *Farmers' Bulletin No. 984, Farm and Home Drying of Fruits and Vegetables* has explicit directions for making several types and sizes of dryers, ranging from a portable one that sits spang on the top of the cookstove (not recommended nowadays, particularly in view of the alternatives) to an elaborate affair suitable for a co-operative project. Unfortunately this very good pamphlet is out-of-print, but probably it can be seen at your County Agent's. The Extension Services in states where much drying is done also have material containing plans.

AN INDOOR BOX DRYER

The 14-inch × 24-inch × 36-inch dryer mentioned in the directions for making trays, above, will hold from 17 to 28 pounds of prepared food on its 6 trays if they're spaced a generous 4 inches apart, and 4 inches are allowed at the bottom and 6 inches at the top for circulation of air. This should be large enough for a small family who are going in for drying seriously. The sketch indicates the general construction of a drying box supported on legs tall enough to accommodate underneath it an upright space heater or a stand for raising an electric hot plate, etc., functionally close to the bottom of the dryer. Consider what you'll be heating with before you start to build: it makes sense to enclose the heat source, so, in order to have the 6 to 8 inches of clearance necessary all around an uninsulated radiant heater (such as an old metal laundry stove), you may have to increase the dimensions of the box or of the area inside the supports. See "Miscellaneous Furnishings" below for suggestions on heaters.

A simple box like this one is portable, a virtue for a modest operation. With a heat source that's not hard to move or that merely depends on a safe extension cord to a safe electrical outlet, the dryer can be set up in any handy dry area with good cross-ventilation—pantry, unused room or large passageway, shed, protected porch, even in the kitchen itself out of the traffic pattern. It should stand free and well away from walls at back and sides, with easy access to the front so you can regulate the heater, take frequent looks at the food and shift the trays during drying.

Make the dryer of the same tempered hardboard mentioned earlier, or of ⅜-inch *exterior* type plywood. Use 2-inch × 2-inch stock for the frame, to which the panels and top are nailed, and for the supporting posts, and 1-inch × 2-

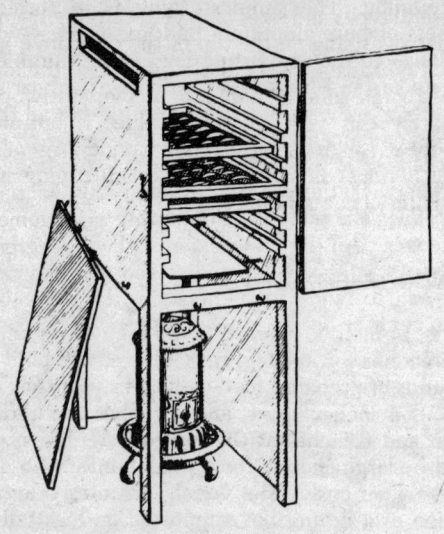

inch material for bracing and the door frame. Runways for the trays are of 1-inch × 1-inch material, and are placed 6 inches from the top and 6 inches from the bottom of the box, and spaced at 3-inch intervals (which gives 4 inches from the top of one runway to the top of the one below).

Cut the side and back panels 3 inches less than the box is tall, and nail them to the framing so they leave a 3-inch space on three sides for the warm moisture-laden air to escape at the top. Cover these ventilators with fine screen (non-galvanized metal screening is O.K. here, being away from the food).

Leave the bottom of the box open; but, 3 inches below the lowest tray runway, screw in partially—but substantially —several long screws on the interior of each side to support a heat-spreader that will be several inches smaller all around than the interior of the dryer. (The spreader, used whenever the type of heat is likely to be too intensely localized, should be metal; maybe we're hyper-finicky, but we'd steer clear of galvanized material here too—sheet aluminum would work all right; a beaten-up tin cookie sheet, cut down to size if need be, would be fine.)

To prevent heat loss below the dryer and control the intake of air at the bottom, enclose the sides and back of the underpinning. The simplest thing is to staple a double thickness of pure aluminum building paper—which is fire-retardant—to the supporting legs. More durable is skirting of hardboard or plywood protected on the heat side by two thicknesses of the aluminum paper; either nail the panels to the legs or hang them with hooks and screw-eyes snugly from the base of the dryer (or from the top horizontal cross-braces between the legs if, in order to contain an uninsulated metal stove, the table-like supporting arrangement is wider and deeper than the dimensions of the drying box proper). The fourth, and front, panel is shorter, leaving about 12 inches of clearance from the floor; and removable, so it's easy to tend the heater.

DRYING OVER A FLOOR REGISTER—NO

Drying food over a floor register leading from the furnace is not recommended: too much dust—even with scrupulously clean air filters; and there are usually fumes from the combustion—even though they're so mild as to be unnoticeable in the room.

A SOLAR DRYER

For small-scale drying outdoors with plenty of sunshine but higher humidity than is desirable for drying in direct sun, you can put together a dryer that looks, and acts, much as a coldframe does (see sketch).

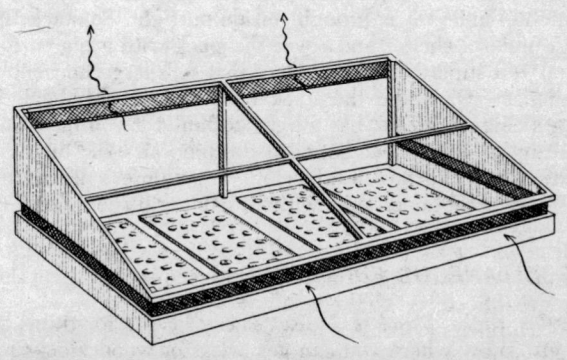

The tilted glass panel—one or more pieces of storm sash are fine—intensifies the heat from the sun, and this rise in temperature inside lowers the relative humidity correspondingly, so that drying occurs faster than is possible outside the dryer. The ample screened venting allows circulation of air.

This dryer is not effective on overcast days.

PROTECTIVE COVERINGS

Food dried in the open, whether outdoors or in a warm room, needs protection from insects and airborne gurry. If the trays have high enough sides (*c.* 2 inches) so the metal can't touch the food, fine-mesh non-galvanized screen does the job. If the trays are so shallow that the covering is likely to sag on to the food, a layer of cloth netting, stretched tight, is better. The problem is not so much what to use, as how to fasten it adequately and still have it easy to remove and replace at the times when the food is stirred or turned over to hasten drying.

Many people cut the covering 2 inches larger all around than the tray it's intended for, bend it over, and thumbtack the overlap to the sides of the tray. Or sometimes it's easier to stretch cheesecloth, etc., over several trays laid side by side, and fasten the cloth to the surface the trays rest on. Or you can use the framed screen from a window: its weight is usually adequate to keep it lying on top of a drying tray. And there's always simply laying a cloth net over the trays and weighting it down all around outside the trays.

Food that's drying outdoors must be protected from dew at night—unless it is brought inside outright. So stack the trays under a shelter and cover the stack with a big carton found at a supermarket or a store that sells large household appliances. Or drape the stack with a clean old sheet. If there's chance of rain, use a light tarpaulin if you have one (putting a clean sheet between the food and the tarp); a plastic shower curtain or tablecloth; or a painter's dropcloth, plastic or not, but certainly without paint or turpentine on it.

MISCELLANEOUS FURNISHINGS

Trestles, racks, benches. No set sizes or types for these, so just know where you can get bricks or wood blocks for raising the first course of trays off the ground; scrap lumber for building rough benches or racks to hang drying food from; smaller stuff to use as spacers. And not all at once and none of it fancy.

Sulfuring box. We're going to suggest sulfuring in certain instances, and we'll tell how to make and use a sulfuring box in a minute.

Auxiliary heaters. A wood-burning "globe" laundry or chunk stove small enough to go inside the skirts under the dryer throws fine radiant heat and is easy to regulate and pleasant to live with; *don't* use coal- or oil-burners, though: the fumes from freshly added coal "gassing off" or from burning oil are to be avoided in this enclosed little space (directly under food to boot). A cooking-gas ring is O.K.

An upright electric heater, thermostatically controlled and with a built-in fan, does well too (but any front-blowing heater laid on its back to direct the heat upward is a fire hazard).

An electric hot plate with low-medium-high settings can do an adequate job.

A cluster of 150- or 200-watt light bulbs (totaling around 750 watts for the smallish dryer we've been talking about) gives good, even heat—sometimes *too* even: unless there's an automatic control, you must unplug the whole thing or unscrew some of the bulbs to reduce the heat when you need to. Because the bulbs overheat themselves in a cluster, especially in such a confined space, they should be in porcelain sockets; and get extended-service bulbs with a heavy-duty element for longer bulb life.

The wiring of all electrical heating and blowing units must meet all safety criteria.

Electric fan. To boost the natural draft in an indoor dryer or to augment a cross-draft when drying in an open room or outdoors. It needn't be large; it should be directable, and *it must have a safety grill covering the blades.*

Thermometers. Even with a dryer or oven having a thermostat, you'll need a food thermometer—a roasting, candy or dairy type will do—to check on the heat of food being

processed; plus the most inexpensive kind of oven thermometer to move around between the top and bottom
trays to keep track of the varying temperatures.

Scale. Not vital but a great help is a scale weighing in
pounds (25 goes high enough, with quarter- and half-
pound gradations); use it for judging water-loss by
weight, per-pound treatments before drying.

Blanching kettle. Your preserving kettle or Boiling-Water
Bath canner will do. Enamelware is best; and with a
close-fitting cover. Plus a rack or basket—or even a
cheesecloth bag—to hold the food above the steam.

Assorted kitchen utensils. Dishpan, colander, crockery or
enameled bowls; stainless sharp knives for cutting and
paring; apple-corer and a melon-ball scoop; cutting-
board; vegetable slicer or a coarse shredder; spoons—
some wooden, at least one slotted; also plenty of clean
towels and paper toweling, and an extra packet of
cheesecloth.

Materials for storing. Several large covered crocks for conditioning dried food before storing—or strong cartons,
moisture-proofed with a lining of plastic sheeting; plastic or paper bags (not big) for packaging dried food in
small quantities; mouse-proof, sealable containers for
the packages. And cool, dark, dry storage when you're
done.

The Drying Methods

Basically, home-drying combines sustained mild heat
with moving air to accomplish its purpose—heat adequate
to extract moisture but moderate enough so it doesn't cook
the material, accompanied by currents of air dry enough to
absorb the released moisture and carry it off. These conditions can occur outdoors naturally, or they can be repro-
duced indoors in dryers.

SUN-DRYING

Successful outdoor drying is possible only in sun-
drenched regions with prolonged low humidity, where foods

are processed either in direct sun—in open air or in the solar "coldframe" described a moment ago—or in the shade. Because the food is covered or brought inside after dark so nighttime condensation won't undo the sun's work, sun-drying takes a minimum of several days, and generally longer.

Where to sun-dry?

In North America, the interior of California and the Southwestern states possess the ideal climate for sun-drying: predictably long periods of hot sun and low humidity. Next come the wide Plains east of the Rockies in the United States and Canada, where occasional showers are not a great problem if the food hasn't got wet and drying can be resumed in sunshine the next day. Despite their heat, the very humid areas of the South are less good; but sun-drying can be done in a limited way in parts of the Northwest, South and Northeast that are away from the moisture-laden air of the seacoasts.

Locate the drying area near enough to the house so you can tend the food several times during the day and put it under cover easily at night—but keep away from places where dust can be stirred up or where animals are quartered or pastured.

Don't dry food outdoors IF: you're in a smog belt; you're in urban sprawl with superhighways surrounding your community; or even if you're rural but have a well-traveled secondary road within a thousand yards of your home. (Dr. Henry A. Schroeder cites the case of a cow which aborted her calf after eating hay cut from a vacant lot in a small New England town: the hay was heavily polluted with lead from the exhausts of cars using the street, which is a numbered highway.)

What to sun-dry

The happiest-sounding source we've found for step-by-step procedures is the USDA Federal Extension Service's *Sun Dry Your Fruits and Vegetables* (1958), apparently compiled for groups like the Peace Corps to use abroad, and drawing on material from Greece and the Philippines as well as from universities in our Southwestern and South Central states. This pamphlet is the credential for the following list of sun-dryable produce—but please keep in mind that the list is based on (*a*) completely ideal conditions of continued sunshine, high temperatures and low humidity, and (*b*) the seeming lack of facilities for drying in the shade or with artificial aids.

It is also important to mention that most of the fruits were exposed to extensive sulfuring before drying—the length of the sulfur dioxide treatment presumably compensating for the long time required for total drying in the sun.

Fruits easier to sun-dry—Apples, apricots, cherries, coconut, dates, figs, guavas, nectarines, peaches, pears, plums and prunes.

Fruits harder to sun-dry—Avocados, bananas, blackberries, breadfruit, dewberries, grapes, Loganberries and mameys (tropical apricots).

Vegetables easier to sun-dry—Mature shell beans and peas, lentils and soybeans in the green state, chili (hot) peppers, sweet corn, sweet potatoes, cassava root, onion flakes, and soup mixture (shredded vegetables, and leaves and herbs for seasoning).

Vegetables harder to sun-dry—Asparagus, beets, broccoli, carrots, celery, greens (spinach, collards, beet and turnip tops, etc.), green/string/snap beans ("leather britches" to old-timers), green (immature) peas, green/sweet peppers, okra, pimientos, pumpkin, squash, and tomatoes.

MEAT (JERKY)

Fresh meat—the lean muscles of beef, lamb or deer—sometimes is dried in the sun with moderate success only in extremely dry sections of North America that have good air drainage, such as the high plains of the Rocky Mountain region. This classic jerky was never salted. The usual treatment, however, is to salt the meat before drying it in the sun, or in shade with slow fire to aid the process (see the individual instructions). Refrigerate jerky to store it; better still, freeze it.

The usual treatment, however, is to salt the meat before drying it in the sun, or in shade with slow fire to aid the process. See the individual instructions.

UNDER MORE TEMPERATE CONDITIONS

Because micro-climates and facilities vary from one sun-drying area to another, it's worth noting that authorities who speak only for conditions in the North Temperate Zone agree that vegetables dried in the sun are often inferior in quality to sun-dried fruits and are more likely to spoil during processing. Furthermore, even with some of the "easier" fruits consensus seems to be that color and flavor are better when they finish drying in stacks in the shade—if humidity is low enough—or are brought indoors to a dryer if it isn't.

In general, then, you might start with sun-drying herbs and flavoring leaves, slices of apples and large-stone fruits, kernel corn and slivers of pumpkin and squash and peppers, before undertaking more.

WHAT TO DRY IN SHADE

If high temperatures are constant, humidity is low even at night, and there is a good breeze, produce capable of being dried in direct sun may be dried well (and sometimes better) in the shade—if it has been given one of the pre-drying treatments and "started" in strong sun or in artificial heat, and provided that the stacks of trays are well ventilated.

Fish are not dried in hot sun, and much of their moisture is removed by thorough salting beforehand. See the individual instructions later.

Drying produce in the sun

Wash, peel, core, etc., and pre-treat according to individual instructions. Because vegetables must have more of their water removed than fruits do for safe drying, cut vegetables smaller than you cut fruits so they won't take too long to dry (being low-acid, vegetables are more likely to spoil during drying).

Spread prepared food on drying trays one layer deep
(½ inch, or depending on size of the pieces); put over it a
protective covering as described above; place trays in direct
sun on a platform, trestles, or merely raised from the ground
on stones or wood blocks—on any sort of arrangement that
allows air to circulate underneath them. The trays may be
laid flat or tilted by means of an extra support under one
end; often a slanting low roof is a good place.

If you use clean sheets or the like to hold the food, a
table, bench or shed roof is a good place. (Naturally you
lose the benefit of air circulating under the food in this
case.)

Stir the food gently several times each day to turn it over
and let it dry evenly.

Before the dew rises after sundown, bring the trays indoors or stack them in a sheltered spot outdoors. If the night
air is likely to remain very dry, the outdoor stack need not be
covered; otherwise wait a little until the warmth of the sun
has left the food, then drape a protecting sheet over the
stack. Return the food to the direct sun the next morning.

At the end of the second day, start testing the food for
dryness after it has cooled. If it doesn't test dry (see "Testing for Dryness" below), put it out in the sun again until it
does. Then it's ready for conditioning and packaging and
storing (see "Treatment After Drying," also below).

Stack-drying produce in shade

This variation of sun-drying relies on extremely dry air
having considerable movement, and therefore shouldn't be
attempted in muggy areas even though the sun is hot. But
where you have hot, dry breezes this method gives a more
even drying with less darkening than if the food was done
entirely in direct sun; apricots, particularly, retain more of
their natural color when shade-dried.

Prepare the food, cutting it in small pieces; put the trays
in direct sun for one day or more—until the food is ⅔ dry.
Then stack the trays out of the sun but where they'll have
the benefit of a full cross-draft, spacing them at least 6
inches apart with chocks of wood or bricks, etc. After several days the dried food is conditioned and packaged for
storing.

You can also stack-dry small quantities of food on an open porch by using an electric fan to boost the movement of the air and thus increase its evaporating power. Separate the trays of partly dried food with 6-inch blocks and set the fan at one end of the stack, directing its blast across the trays (which you'll have to shift around a good deal, since the food nearest the fan dries first). Condition and store.

Using a solar dryer

Because of the conditions it is designed to overcome, a solar dryer (sketched earlier) accommodates only a relatively small amount of material at one drying session, and the food "started" in it usually cannot be finished off by stack-drying. On the plus side, however, are the facts that it dries faster than would be possible in the open air, and that it costs nothing to operate.

Prepare the food as for sun-drying and spread it one layer deep on the trays. Check the material every hour or so, stirring it gently so it dries evenly; if convenient, turn the dryer several times so it faces direct sunshine. As soon as the interior has cooled when the sun gets low, cover all the ventilators. Remove the covering the next morning and set the dryer to get the full benefit of the sun's heat.

If the sun is hot enough and there is no exposure to humidity from the outside, the food should be dried in about two days. Condition and store.

INDOOR DRYING

Almost every food that sun-dries satisfactorily can make a better product dried indoors with applied heat and a natural or forced draft of air; and for some foods, especially low-acid vegetables, processing in an indoor dryer is recommended even though outdoor drying conditions are reliable during the harvest months.

Herbs dry best in the natural heat and draft of a well-ventilated room.

Depending on the water content and size of the prepared food, and whether the dryer is loaded heavily or skimpily, good drying is possible within 12 hours in an indoor dryer—

TESTS FOR DRYNESS IN PRODUCE

According to *Composition of Foods,* none of the fruits we'll be telling how to home-dry has less than 80 percent water in its fresh raw state, and the average comparable water content of the vegetables is not less than 85 percent.

The safe maximum percentages of water to leave in home-dried produce are: no more than 10 percent for vegetables, and no more than 20 percent for fruits. Commercially dried fruits often contain more water—especially when they're "tenderized"—but also they may contain additives other than simple sulfur dioxide to protect against spoilage from the higher content of moisture. But we don't have the food industry's highly refined means of testing for and controlling moisture, so we rely on appearance and feel to judge dryness.

Fruit generally can be considered adequately dry when no wetness can be squeezed from a piece of it when cut; and when it has become rather tough and pliable; and when a few pieces squeezed together fall apart when the pressure is released. "Leathery"—"suède-like"—"springy"—these are descriptions you'll see in the individual instructions. Several, such as figs and cherries, also are slightly "sticky."

Vegetables are generally "brittle" or "tough to brittle" when they're dry enough; an occasional one is "crisp." Again, instructions for specific vegetables will tell you what to look for.

When they are very nearly dry, some foods will rattle on the trays; this is another thing to check on when they're in natural heat/draft or in a dryer.

And then there's the ⅔ *dry* judgment. Without using refined correlations of drying rate against percentage of natural water and weight of solids before drying, we offer this rough rule-of-thumb: Compare the weight of a fresh sample of produce with its weight at some point during the total drying period—if it has lost ½ of its original weight, consider it ⅔ dry for your purposes.

Finally, foods still warm from the sun or hot from the dryer will seem softer, more pliable, more moist than they actually are. *So cool a test handful a few minutes before deciding it's done.*

and some materials, if cut small, may be dried in as little as 3 hours.

On a smaller scale, the oven of a cookstove can be made to perform as a dryer; the processing time is about the same.

Meats, fresh or lightly salted, dry better in an oven (which is an indoor dryer of sorts) than they do on trays in a regular drying box.

Salt fish is best done by shade-drying, since a breeze outdoors on a sunny day is preferable to the limited ventilation afforded by a dryer or an oven. But if the weather turns poor, and you don't mind the aroma indoors, you can finish off a batch in the dryer, or do a small amount in the oven.

Using a dryer

Here you're increasing the speed of drying by use of temperatures higher than those reached outdoors in the sun, so be prepared to regulate heaters and shift trays around if you want the best results.

HEAT—WHY/HOW MUCH/WHEN

There's no single across-the-board temperature ideal for box-drying everything, as you'll see from the individual instructions. Generally speaking, however, a steady 140 degrees Fahrenheit is a good heat to fall back on if your dryer is cranky and you don't mind some extra fiddling with trays. Preheat the dryer.

The *How* of drying by artificial heat is simple if you keep in mind a few *Why's*.

(1) In the usual home dryer—which has the heat source directly under the box, and relies on natural draft sometimes augmented by a small fan—the temperature at the top of the stack is less than the temperature at the bottom, near the heater.

So always keep track of the temperature at the lowest tray, so you can use this heat as the base for judging the temperature higher up.

And rotate the trays up or down every ½ hour to correct this difference and ensure even drying.

(2) Fresh food won't dry well if it is exposed to too much heat too soon: either it will case-harden—its exterior being seared and toughened, so natural moisture can't exude and be evaporated—or it will cook, with the cells rupturing and creating an unattractive product. (As a general thing,

the juicier the food, the more likely it is to rupture, and the lower is the temperature at which it should be started in the dryer.)

So if you add fresh food to a batch already drying, always put it in at the top of the dryer.

If the whole load is fresh, have the heat inside the cabinet 20 or so degrees below the maximum temperature you'll hold it at later. After probably not much more than 1 hour, start increasing the heat by 10 degrees each ½ hour till you reach the recommended maximum.

(3) But for the majority of its total drying time the food must have enough heat to kill the growth cells of some spoilers, as well as to remove moisture that lets other ones thrive (see "Dealing with the Spoilers" again, Chapter 1). This means that, no matter how low the temperature at which you start food in order to prevent case-hardening, etc., you have to raise the heat to a killing level and hold it there long enough to make it effective.

For more than half the total drying time, the heat must be sustained at at least 140 F.

(4) The first ⅔ of drying is accomplished at a faster rate than the last ⅓ is, and at this point—having been exposed for some time to maximum drying temperature, and having less protection through evaporation—the food can get too hot for the state it's in, and scorch. And just a hint of scorch will hurt the flavor.

When the food has reached the ⅔-dry stage, tend it with extra care to make sure it won't scorch. Keep rotating trays away from the heat source. If you need to, during the last 1 hour reduce the heat by 10 degrees or so, but stick to the 140-F. rule just above.

HANDLING FOOD IN A DRYER

Line or oil the trays—see the comments on tray bottoms under "Equipment for Drying"; spread prepared food on them one layer deep if it's in large pieces, not more than ½ inch deep if it's small. Place halved, pitted fruit with the cut side up (rich juice will have collected in the hollows if it was sulfured).

At this point newcomers to drying by artificial heat may want to weigh a sample tray of food for comparison with the weight of the same trayful after it's been in the dryer for a while, as a help in judging how much moisture has been lost.

Stagger the trays on the slides: one pushed as far back as possible, the next one as far forward as possible, etc. (as in the sketch earlier).

Check the food every ½ hour, stirring it with your fingers, separating bits that are stuck together. Turn over large pieces halfway through the drying time—but wait until any juice in the hollows has disappeared before turning apricots, peaches, pears, etc. Pieces near the front and back ends of the trays usually start to dry first: move them to the center of the trays.

If you add fresh food to a load already in progress, put the new tray at the top of the stack.

Make needed room for fresh food by combining nearly dry material in deeper layers on trays in the center of the dryer; it can be finished here without worry, but keep stirring it.

Using an oven

As far as you can, use an oven as you would a dryer, following general procedures and the specific instructions for each food.

Leave the upper (broiling) element of an electric oven turned off, and use only a low-temperature Bake setting for drying. With some electric ranges this broiling element stays partially on even with a Bake setting: if yours does this, simply put a cookie sheet on a rack in the uppermost shelf position to deflect the direct heat from the food being dried.

Most gas ovens have only one burner (at the bottom) for both baking and broiling. If your gas oven has an upper burner, don't turn it on.

Gas ovens are always vented, but electrics may not be. If your electric oven isn't vented, during drying time leave the door ajar at its first stop position.

If your oven isn't thermostatically controlled, hang an oven thermometer where you can see it on the shelf nearest the source of heat, leave the door ajar—and be prepared to hover more than usual over the food that's drying.

Preheat the oven to 140 F. *If the oven cannot be set this low, skip the lowest slide you would otherwise be using: keep the bottom tray at least 8 inches from the heat source.*

Don't overload the oven: with limited ventilation (even with a fan aimed toward the partly opened door) it can take as much fuel to dry a batch too big as might be used to dry two fairly modest batches.

Room-drying

Also done indoors is what can be called, for simplicity, room-drying. By this method food is hung in a warm room—the kitchen or the attic—for the days required to dry the material. Old-timers would suspend racks of drying food above the big wood-burning range, finish off a flitch of beef near by, and festoon strings of apple or pumpkin rings near the ceiling. Herbs are still usually dried in attics or the kitchens of country houses, hung either in the open or in paper bags to protect them from dust.

In extremely dry areas it may be feasible to stack-dry certain fruits and vegetables indoors, following enough time in sun or dryer to get them better than halfway along. Stack the trays with 6 inches of space between them, open windows to allow a free circulation of air, and force a draft across the trays with an electric fan. Shift the trays end for end occasionally and turn the food to ensure even drying.

PRE-DRYING TREATMENTS FOR PRODUCE

Before being dried at home by any method, fruits make a better product if they undergo one or more of the treatments given hereafter, while all vegetables are treated to stop the organic action that allows low-acid foods to spoil.

And, still speaking generally, a pre-drying treatment for fruits is optional for safety, but the pre-drying treatment for vegetables is a must.

The optional treatments for fruit involve (1) temporary anti-oxidants, to hold their color while they're peeled/ pitted/sliced; (2) blanching in steam or in sirup as a longer-range means of helping to save color and nutrients; (3) very quick blanching—either in boiling water or steam (lye is not recommended)—to remove or crack the skins; (4) sulfuring as longest-range protection for some nutrients and for color. We'll sum up for or against these fruit treatments as best we can after we describe them. Then it's up to you to use them or not.

The treatment for vegetables is steam-blanching. The quick dunk in boiling water that's used in freezing is not adequate to protect them against spoilage in drying; and the much longer boiling time needed here would waterlog the material, in addition to leaching away a number of its nutrients.

The following descriptions are given in the order that the treatments are likely to occur in handling produce for drying: they're not necessarily in order of importance.

Temporary anti-oxidant treatment

Even though you intend to blanch or sulfur certain fruits to protect them during the long haul of drying and storage, chances are that you'll want to do something to prevent their darkening piecemeal while you're actually cutting a batch up.

THE GOOD ASCORBIC-ACID COAT

Pure ascorbic acid is our best safe anti-oxidant, and is used a lot in preparing fruits for freezing (q.v.). Use it here too. But with the difference that the solution will be somewhat stronger, and thus food coated with it can hold its color in transit in the open air for a longer time.

One cup of the solution will treat around 5 quarts of cut fruit, so prepare your amount accordingly. Sprinkle it over the fruit as you proceed with peeling, pitting, coring, slicing, etc., turning the pieces over and over gently to make sure each is coated thoroughly.

For apples: dissolve 3 teaspoons of pure crystalline ascorbic acid in each 1 cup of cold water.

For peaches, apricots, pears, nectarines: dissolve 1½ teaspoons of pure crystalline ascorbic acid in each 1 cup of cold water.

If the variety of fruit you're working with is likely to become especially rusty-looking when the flesh is exposed to air, it's O.K. to increase the concentration of ascorbic acid as needed. The proportions above usually do the job.

The commercial anti-oxidant mixtures containing ascorbic acid don't work as effectively, volume for volume, as the pure Vitamin C does, but they're often easier to come by. Follow the directions for Cut Fruits on the package.

Blanching fruits in heavy sirup

It seems excessive to precook fruit in heavy sirup in order to help prevent discoloration while drying—particularly when we have better ways to do the job. But if you want to turn fruit into an extra-sweet confection, here is the sirup-blanch most often described for apples, apricots, figs, nectarines, peaches, pears, plums and prunes.

Make a sirup of roughly 1 part sweetener to 1 part water, using refined or raw sugar, or part corn sirup or honey (and see "Sugar Sirups for Canning Fruits," Chapter 2); heat it to 212 F., and in it simmer the fruits for 10 to 15 minutes, depending on the size of the pieces. Remove the kettle from the heat and let the fruit stand in the hot sirup for about 15 minutes more. Then lift the fruit out and drain it well on paper toweling to remove as much surface moisture as you can. Save the sirup for the next batch.

Dry the fruit by whichever method you like. Take extra care that it doesn't stick to the trays, or scorch during the last stage of drying in artificial heat. Fruit treated this way is more attractive than usual to insects, so cover it well during sun-drying. And package it well for storage in a really cool, dry place.

Treating fruit skins without lye

Not only is lye tricky to work with in its own right (do see especially "Warnings About Lye" in canning Hominy and Soapmaking, Chapters 2 and 9), but the alkali in soda compounds hurts many B vitamins and Vitamin C.

A very quick dip in boiling water, *quite apart from the steam-blanching that helps keep the color and nutrients of certain cut fruits*, works well instead of the lye treatment. And it's safer for you and for your food.

FOR CHECKING THE SKINS

Nature provides a wax-like coating on the skins of cherries, figs, grapes, prunes and small dark plums, and certain firm berries like blueberries and huckleberries, and they all dry better if this waterproofing substance is removed beforehand.

The chances of case-hardening and rupturing (q.v.) are also reduced if the relatively tough skins of such fruits and berries are cracked minutely in many places; this is called "checking," and it allows internal moisture to be drawn through to the surface and there to be evaporated.

Because the 30 to 60 seconds required for the de-waxing and checking operation is often too short to let live steam be effective for the contents of the blanching basket, the answer is a very quick dip in briskly boiling water, followed by a dunk in very cold water, and thorough draining. Use the method described in Chapter 2 for treating the skins of blueberries in the Raw pack variation, or use the one in Chapter 3 for blanching vegetables before freezing.

Length of the dip depends on the relative toughness of the fruits' skins (cranberries are tougher-skinned than currants, for instance). Lay absorbent toweling on the fruit to remove excess moisture from their surface, and continue with the next step in handling the specific fruits.

FOR PEELING

Routine peeling of any fruit except apples isn't necessary for making a good dried product (see "When to Peel" in Drying Fruits), but if you feel that you must remove the skins from peaches, and even from apricots and nectarines, simply dip them, a few at a time, in boiling water for 30 to 60 seconds—ample time for firm-ripe fruit—cool them quickly in cold water, and pull their skins off by hand.

Steam-blanching before drying

On the whole, vegetables to be dried are blanched in full steam at 212 degrees F. for longer time than they are blanched, either in steam or boiling water, before being frozen. The length of blanching time is given for each vegetable in the individual instructions.

Steam-blanching also is suggested for certain fruits as an aid to discouraging oxidation, and for softening berries, as well as for checking (cracking) the skins of grapes, prunes and figs instead of treating them with lye. In addition to stopping the decomposing action of the enzymes and helping to fix the color of the food, steam-blanching protects some of the nutrients and loosens the tissues so drying is actually quicker.

Put several inches of water in a large kettle that has a close-fitting lid; heat the water to boiling, and set over it—high enough to keep clear of the water—a rack or wire basket holding a layer of cut food not more than 2 inches deep. Cover, and let the food steam for half the time required; then test it to make sure that each piece is reached by steam. A sample from the center of the layer should be wilted and feel soft and heated through when it has been blanched enough.

In a pinch you can use a cheesecloth bag, skimpily loaded with food, and placed on the rack to steam. Be careful not to bunch the food so much that steam can't get at all of it easily.

Remove the food and spread it on paper toweling or clean cloths to remove the excess moisture while you steam the next load; lay toweling over it while it waits for further treatment or to go on the drying trays.

Sulfur, Spoilage and Nutrients

For many years sulfur has been used to preserve the color of drying fruits whose flesh darkens when exposed to air. The fruits generally treated with sulfur have been apples, apricots, nectarines, peaches and pears; light-fleshed varieties of cherries, figs, plums and prunes have also been treated with sulfur to prevent oxidation, though not so routinely. Meanwhile it was noted that unsulfured fruits appear more likely to sour, get moldy or be attacked by insects during prolonged drying or in storage than is the case with the sulfur-treated dried fruits.

Further, sulfur is a mineral essential for life, and therefore is not harmful *per se* in the quantities used in home-drying (which allow plenty of leeway). The sulfur forms sulfurous acid when it unites with the water in the fruits' tissues, and the sulfurous acid evaporates during prolonged sun-drying, during storage, and from cooking; the residue that is eaten is turned into an innocuous compound that is then excreted. The more strongly acid the food, the greater is the power of sulfurous acid to inhibit the growth of molds and the bacteria that cause souring.

Further, nutritionists have published the results of research showing that, among others, Vitamins A (carotene), B_1 (thiamine) and C (ascorbic acid) are destroyed by exposure to air, but that sulfur aids in retention of A and C when they're hit by air, but helps to destroy what's left of B_1 (which often is present in negligible amounts anyway, compared with A and C).

Summarizing. These points seem to be adequate support for the consensus of the country's leading authorities on home-drying—to sulfur certain cut fruits to ensure better storage, less darkening, and less loss of Vitamins A and C. Sulfuring is optional: the results of drying without sulfur may be disappointing as to color, flavor and some aspects of nutrition, but food safety relies mainly on how competently these fruits are dried.

A POOR WAY TO SULFUR (SOAKING)

There are two ways generally given for sulfuring at home, and we'll mention the quicker but unsatisfactory method first, to get it out of the way.

As in all soaks, cut fruits held in a sulfite solution lose some of their water-soluble nutrients and tend to get waterlogged; in addition, the sulfur compound may often penetrate the tissues unevenly.

Further, for us the available instructions offered too wide a spread for comfort. The substances were given variously as sodium sulfite, sodium bisulfite and potassium metabisulfite, with the amount to use in each 1 gallon of water ranging from 1½ teaspoons on up to 3½ tablespoons. And soaking times, seemingly not geared closely enough to the strength of the solutions, were from 15 to 30 minutes.

HOW TO SULFUR WELL (FUMES, IN A BOX)

We'll describe this at some length because, having decided that *sulfuring is O.K. in certain cases,* we experimented with the mechanics of the procedure in order to amplify the sketchy instructions found so far.

The main bother in burning pure sulfur for the dioxide treatment is that the raw fumes irritate eyes and breathing

passages, and therefore the business must be done outdoors in the open air. (All to the good, though, is the fact
that they are so irritating: one whiff and you duck away—
as you do from ammonia or activated lye, though perhaps
not as fast—so you're not likely to keep on breathing them
unknowingly.)

Otherwise, it's very simple and direct. And fun, if you
relax and think of sulfur as something the ancients knew
well, for it's the brimstone of the Bible. Anyway, sulfur
first melts—at around 240 F.—becoming a brown goo
before it ignites and burns with a clear blue flame that
produces the acrid sulfur dioxide that penetrates evenly
and is easy to judge the effect of. The usual amount to
use is 1 level teaspoon burned for each 1 pound of prepared fruit.

Local drugstores had several kinds of dry sulfur, but
we chose the "sublimed" variety—99½ percent pure, to be
taken internally mixed with molasses (the classic folk
tonic); it's a soft yellow powder with no taste and the
faintest of scents that's nothing like the rotten-egg odor of
hydrogen sulfide. The 2-ounce box was enough to do 16–
18 pounds of prepared fruit. Also, from the hardware store
we got a 4-ounce cake of 100 percent refined sulfur (with
a short wick to start it melting and burning, because it's
a fumigating candle to use in sealed rooms to get rid of
bugs in the woodwork). This cake we found much harder
than the powder to gauge the dose of: its weight figured
in teaspoons, the amount of sulfur needed for a batch of
fruit took longer to burn than the time required for exposure to the fumes.

Loading and stacking the trays. You'll be weighing your
prepared fruit to determine how much sulfur to use
in the treatment, and thus you can keep the batch
within limits of what your sulfuring box will handle
effectively. Nor should you sulfur more food in one
session than your drying arrangements will accommodate as soon as the sulfuring is done—6 trays' worth,
for example, if you use an indoor dryer the size of the
one described earlier. Spread the fruit one layer in
thickness on the trays (which *don't* have metal bottoms), placing any convex pieces with their hollow side
up in order to prevent loss of the rich juice that col-

lects during treatment. Limit each sulfuring batch to food of the same type and size: you won't be able to shuffle the food around to compensate for different exposure times, as you do in a dryer. And don't overload the trays or crowd the sulfuring box, because it's easier in the long run to deal with two short stacks, widely spaced, than with one stack too tall and inadequately spaced.

Put the bottom tray on blocks of some kind to raise it at least 4 inches above ground. Separate the trays above it with wooden spacers to hold them 3 inches or so apart. Allow for 6 inches of clearance between the top of the stack and the sulfuring box that will be inverted over it.

The sulfuring box. This you can make from the same materials used for the indoor dryer (q.v.); or you can cover

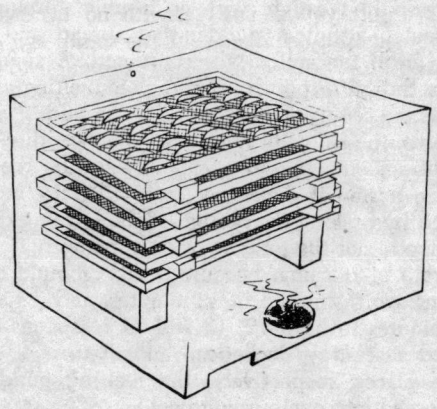

a slatted crate snugly with building paper; or you can use a stout, large carton of the sort that household appliances are shipped in. The box should be tall enough to cover an adequately spaced stack of up to 6 trays, and be about 12 inches longer than the trays from front to back so there'll be room for the sulfuring dish *beside* the stack.

At the bottom near one corner cut out a slot 6

inches wide by 1 inch high if the box is large, 3×1 inch if it's just the average supermarket carton; this will be the only air intake needed to keep the sulfur burning, and you'll cover the slot after the sulfur is consumed. Near the top of the side opposite the intake slot, make a hole the diameter of a pencil; this also will be covered when the sulfur has burned. (The need for this tiny upper hole is Norman Rogers's discovery: it wasn't mentioned in the various directions we started out with, but he found that, without it, the sulfur stopped burning prematurely and required relighting; and the escape of fumes through it was negligible.)

Lighting and burning. Experimenting with sulfur either loose in a small pan/dish or twisted in a bit of paper, Norman found that the powder spread in a shallow container burned better than the same amount in a paper spill: it caught with no trouble, forming under the match a brown puddle that spread and fed the flame until the sulfur was consumed. A spent match left in the sulfur prevented complete melting and burning, as did scraps of charred paper, he said.

Also, the sulfur burned best—i.e., steadily, without relighting, quickly and completely—when the powder was in a smooth layer *not more than ½ inch deep*, settled by tunking the bottom of the burning dish. And the depth, not the total amount, apparently determines the rate of burning, because a little mound of 2 teaspoons' worth burned in 12 minutes, and it took about 15 minutes to burn 12, 18 and 24 teaspoons of sulfur spread a scant ½ inch deep and lit in one, two, and three places, respectively; the melting puddles soon converged into one burning pool.

He said he wasn't bothered by the little sulfur dioxide that escaped when he tilted the box to snake out the sulfuring dish to check on the burning. And the brimstone smell had evanesced from his hands within an hour after he was through with his experiments.

The sulfuring container should be set at the side or in front of the stack: if it were underneath, the food on the bottom tray directly above it would discolor from concentrated exposure to even the small amount of smoke involved in the combustion.

Whatever you have for a burning dish, it should be perfectly clean before each use. The dish can be metal, enameled, or heavy crockery. Metal corrodes from the sulfur after a number of uses (one reason why tray bottoms of metal screening are never put in a sulfuring chamber), and aluminum becomes pitted most quickly; he simply made one-time-use dishes by molding a double thickness of household foil around the bottom of a flat-bottomed container of the appropriate size. He made the sides of his dishes a good 1 inch higher than the layer of sulfur—about 1½ inches all told—because if the sides are too high in proportion to the burning surface of the powder, the flame can smother for lack of air. A flat round dish 3 inches in diameter takes about 12 teaspoons of powdered sublimed sulfur smoothed ½ inch deep; a 4-inch one takes about 18, and a 5-incher about 24 teaspoons.

Be careful about using bright red or yellow ceramic dishes: certain of these pigments on glazed pottery can contain cadmium. Unless you know that the glaze of either color is cadmium-free, don't use such pottery for sulfuring. Or for anything else.

How much sulfur? Weigh the prepared fruit before spreading it in a single layer on the treatment trays, and use 1 teaspoon of powdered pure sulfur for each 1 pound of fruit. This 1-for-1 ratio is conservative but, so far as we can find out, it's adequate for the average small-family operation; much better to adjust sulfuring *time* to your method than to increase the sulfur you burn per pound.

How long to sulfur? Sulfuring time varies according to the texture of the fruit, whether it's peeled, how big the pieces are, and whether it's exposed for a long time in air—as in sun- and shade-drying outdoors—or has a relatively short exposure, as in a drying box or an oven. Specific times are given in the instructions for individual fruits.

Start to count sulfuring time *after* the sulfur has finished burning, which will take about 15 minutes (see above), and you have tightly closed off the air-intake slot at the bottom of the box and the tiny breathing hole at the top. The reason: the necessary amount of sulfur dioxide must first be created by total combustion of the sulfur which has been measured for the weight of the batch being treated; then the fumes must be given time to reach and penetrate the surfaces of the fruit on the stacked trays. With the sulfuring box made airtight, you simply leave it inverted over the stacked trays for the required period.

In general, apples, apricots, peaches and pears are sulfured twice as long for sun-drying as they are for an indoor dryer. And a practical rule-of-thumb is to sulfur small slices of these fruits for the box dryer for a minimum of 20 to 30 minutes, and larger pieces for comparatively longer. Therefore, since we *recommend sulfuring for sun-drying* because of the extended time in vitamin-destructive open air, these fruits would be sulfured for about 60 minutes if they're to be sun-dried as slices, with double that sulfuring when they're quartered, and more than two hours when they're halved.

On the other hand, home-drying done in an indoor box or an oven is often a small-batch affair—and in such cases it's not always feasible to get cranked up to for the sulfuring operation. Though not as effective over-all as sulfuring, there are other measures that can be substituted for all or part of the sulfur treatment for fruit to be dried by artificial heat in small batches; they're correlated in the "Pre-drying Treatments for Produce" earlier in this chapter.

Unloading the sulfur box. Stand to windward of the box so the fumes won't come your way, reach across the top, and tilt the whole thing toward you until the box rests on its side or end: it's like lifting the lid of a Pressure canner to let the steam escape away from your direction.

Remove the trays from the top, being careful not to spill any juice that has collected in the hollows of the fruit. If sulfuring was done on trays that fit a dryer, just slide them as is into the preheated drying box or oven; otherwise lift the fruit on to trays all ready for going immediately out in the sun or in artificial heat.

POST-DRYING TREATMENTS FOR PRODUCE

Even after a sample from each tray of food has shown no moisture when cut and pressed, and feels the way its test says it should, you can't take for granted that the whole batch is uniformly dry. And especially if it's been dried outdoors do you need to get rid of any spoilers—airborne micro-organisms or bugs you can see—that may have got to it somewhere along the line.

Conditioning

This makes sense particularly for food done in a dryer because there's often more chance of spotty results than in sun-drying, and you'll want moisture content equalized between under- and overdried pieces.

Cool the food on the trays, then pour it all into a large, open, non-porous container *that's not aluminum*—a big crock, enamel- or graniteware canner, even a washtub lined first with plastic and then with clean sheeting (washtubs are generally galvanized). Have the containers raised on trestles or tables, and in a warm, dry, airy, well-screened, animal-proof room.

Stir the food once a day—twice if you can manage to —for 10 days or 2 weeks, depending on the size of the pieces. It's O.K. to add freshly dried food to the conditioning batch, but naturally not if the food in a container is almost ready to store.

Fruits, usually being in larger pieces (and therefore more likely to need finishing off) than vegetables, need more conditioning time.

Pasteurizing

Recommended strongly for sun-dried fruits and vegetables and for dryer-processed vegetables that have been cut small enough so their drying time perhaps hasn't been long enough to kill the spoilers.

Don't bother cranking up the dryer for this, and don't do large amounts at a time: use an oven with a thermometer in it, and time the process.

Preheat the oven to 175 F. Spread the food loosely not more than 1 inch deep on the trays; don't do more than two trays' worth at the same time. Heat brittle-dried vegetables, cut small, for 10 minutes at 175 F.; treat fruits —cut larger and therefore needing more time—for 15 minutes at 175 F.

Remove each pasteurized batch and spread it out to cool on clean toweling, etc. Cover lightly with cheesecloth to keep dried food clean. Package one batch while other batches are pasteurizing.

PACKAGING AND STORING

Have dried, treated produce thoroughly cooled before putting it up in small amounts (if packaged warm it will sweat, especially in plastic bags).

Use small paper or plastic bags; twist the tops and secure the closures with string, wire tape, rubber bands, etc.; store packets in large critter-proof containers with heavy lids in a dry, cool place (in humid climates, store in glass jars). Label if necessary.

Drying Fruits

The following instructions are merely individual applications of the principles, methods and treatments described up till now in this section, and newcomers to drying are likely to have better results if they look at the introductory *Why-How* (and sometimes pro/con) before they tackle specific fruits.

In general below, sulfuring is strongly recommended for all cut fruits that are *sun-dried,* in order to help save important vitamins, prevent insect infestations and hold color; steam-blanching is recommended for fruits processed in a *dryer;* a temporary ascorbic-acid coating is suggested for certain fruits that oxidize readily, whether dried in the sun or in artificial heat outdoors; some whole fruits are "checked" with boiling water to crack their skins; pasteurizing is recommended for sun-dried fruits, to kill off any bugs that have got to them. All these treatments are optional: they are included because they result in a more nutritious and attractive product.

It is assumed that all the fruits are firm-ripe, without blemishes, and have been washed carefully in cold water as a preliminary to preparation.

See "Tests for Dryness in Produce," earlier.

COOKING DRIED FRUITS

Pour boiling water over them in a saucepan *just to cover*—no more now: they shouldn't be drowned, and you can always add more if you need to—and simmer the fruit, covered, for 10–15 minutes, depending on the size of the pieces. Remove from heat and let cool, still covered. Sweeten to taste at the very end of cooking, or when removed from heat (sugar tends to toughen fruit fibers in cooking). For best flavor, chill the fruit overnight before serving.

If the fruit is to be "reconstituted" to use in a cooked dish (a pie or a cream dessert, say), put it in a bowl, add boiling water just to the top of the fruit; cover; and let it soak up the water for several hours, or until tender. Add water sparingly and only if the pieces seem still to be tough, because the liquid is full of good things and should be included in the recipe as if it were natural juice.

130 F. = 54 C. / 140 F. = 60 C. / 150 F. = 66 C. / 175 F. = 79.5 C. / 212 F. = 100 C. / 1 U.S. tsp. = 5 ml. (5 cc.) / 1 U.S. T. = 15 ml. (15 cc.)

APPLES

Best for drying are late-autumn or early-winter varieties, including: Baldwin, Ben Davis, Northern Spy, Spitzenburg; then Winesap, Jonathan, Greening, Rome Beauty, both Delicious, the Russets.

Prepare. Peel, core, slice in ⅛-inch rings. As you go, coat slices with strong ascorbic-acid solution to hold color temporarily.

Dryer. Steam-blanch 5 minutes; remove excess moisture. Begin them at 130 F.; raise gradually to 150 F. after the first hour; when nearly dry, reduce to 140 F. Test dry. Condition. Package; store. Average total drying time: up to 6 hours, depending on size of slices.

Sun-drying. Prepare as above, using ascorbic-acid coating. If not steam-blanched for 5 minutes, sulfur for 60 minutes; if blanched, sulfur for 30 minutes. Proceed with drying. Test dry. Pasteurize. Package; store.

Room-drying. Prepare as above. Steam-blanch 5 minutes *and* sulfur 30 minutes; or sulfur only for 60 minutes. Thread on clean string, and festoon near the ceiling of a warm, dry, well-ventilated room (attic) or above the cookstove (kitchen). Test dry. Pasteurize.

The slices may also be dried on stacked trays with an electric fan blowing across them.

DRY TEST. Leathery, suède-like; no moisture when cut and squeezed.

APRICOTS

Pick before they are so ripe they drop from the tree.

Prepare. Halve and stone. Hold against oxidizing with ascorbic-acid coating.

Dryer. Steam-blanch halves 15 minutes, slices 5 minutes. Remove excess moisture and start in the dryer at 130 F., raise gradually after the first hour to 150 F. Reduce to 140 F. for last hour or when nearly dry. Test dry. Condition; store. Average total drying time: up to 14 hours for halves, up to 6 hours for slices.

Sun-drying. Prepare as above with ascorbic-acid coat. If steam-blanching (above, as for a dryer) sulfur slices 30 minutes, halves 90 minutes. If not blanched, sulfur slices 1 hour, halves 2 hours. Remove halves carefully to drying trays so as not to spill juice in the hollows, and place cut-side up in the drying trays. Turn when all visible juice has disappeared. Test dry. Pasteurize; store.

DRY TEST. Leathery, pieces fall apart after squeezing; no moisture when cut.

BERRIES, FIRM

Prepare. Check (crack) the skins of blueberries, huckleberries, currants and cranberries, etc. by dipping for

15–30 seconds (depending on toughness of skin) in rapidly boiling water. Plunge into cold water. Remove excess moisture.

Dryer. Start at 120 F., increase to 130 F. after one hour, then to 140 F.; they will rattle on the trays when nearly dry. Keep at 140 F. until dry. Test dry. Condition; store. Average total drying time: up to 4 hours.

Sun-drying. Check the skins as above for the dryer. Remove excess moisture and put on trays one layer deep in the sun. Test dry. Pasteurize; store.

DRY TEST. Hard. No moisture when crushed.

BERRIES, SOFT

There are so many better ways to use these—canned, frozen, in preserves—that there's not much use in drying them. Strawberries are especially blah and unrecognizable when dried.

CHERRIES

If not pitting cherries, check their skins with a 15–30-second dunk in boiling water; cool immediately. Some people sirup-blanch (q.v.) before drying cherries.

Dryer. Remove excess moisture from checking treatment. Start at 120 F. for one hour, increase gradually to 145 F. and hold there until nearly dry. Reduce to 135 F. the last hour if danger of scorching. Test dry. Pasteurize. Cool and store. Total drying time: up to 6 hours.

Sun-drying. Pit. Sulfur for 20 minutes. Dry. Test dry. Pasteurize. Cool and store.

DRY TEST. Leathery and sticky.

FIGS

Prepare. Small figs or ones that are partly dry on the tree may be dried whole. Large juicy figs are halved.

Dryer. Check skins by a quick dunk in boiling water for

30–45 seconds. Cool quickly. If cut in half, steam-blanch for 20 minutes. Some people sirup-blanch (q.v.) whole figs before drying. To dry, start at 120 F., increase temperature after the first hour to 145 F. When nearly dry, reduce to 130 F. Test dry. Condition. Cool and store. Total average drying time: up to 5 hours for halves.

Sun-drying. Check the skins as above if drying whole. Sulfur light-colored varieties (like Kadota) for 1 hour before drying. If figs are to be halved, do not check the skins—instead, steam-blanch the halves for 20 minutes and then sulfur for 30 minutes. Test dry. Pasteurize. Cool and store.

DRY TEST. Leathery, with flesh pliable; slightly sticky to the touch, but they don't cling together after squeezing.

GRAPES

Use only Thompson or other seedless varieties for drying.

Prepare. Check the skins by dipping 15–30 seconds in boiling water and cooling immediately. Proceed as for whole Cherries.

Dryer. Proceed as for Cherries. Total average drying time: up to 8 hours.

Sun-drying. Handle like Cherries, but don't sulfur.

LEATHERS (PEACH, etc.)

These sheets of pliable dried pulp may be made from virtually all fruits and berries, with peaches, apples and wild blackberries leading the field. The following is a general rule, so experiment with only small batches until you get the texture and fresh, tart flavor you like.

Added sweetening helps to bind the texture but increases the chance of scorching, so add a little water as needed, and stir often (the amount of sweetener suggested may not be enough for very tart plums). The pectin in apple skins helps texture: remove bits of cooked skin by straining.

Prepare. Use fully ripe fruit. Peel or not, core/stone, cut small; coat with an anti-oxidant (q.v.) if you like, but the brief precooking should prevent some darkening. Measure prepared fruit, and add 1½ tablespoons sugar or honey for each 1 cup of cut fruit. Bring just to boiling, cook gently until tender; remove from heat, and when the fruit is cool enough to handle either put it through a fine sieve or food mill, or whirl it briefly in a blender.

Lay long sheets of foil or plastic *freezer*-wrap on large cookie pans, allowing extra at ends and sides, and oil it well. Pour enough fruit pulp in the center of each sheet to allow it to be spread very thinly—no more than ¼ inch deep—to within 2 inches of the rims.

Dryer. Start at 130 F.; raise to 145 F. after the first hour and hold there until the surface is no longer tacky to the touch, or for 45 minutes. When nearly dry, reduce heat to 135 F. Test dry. Cool.

Drying time depends on juiciness of the fruit— usually about 2 to 3 hours. To make heavier leather, spread fresh pulp thinly on a layer that has lost all tackiness: building up on a nearly dry layer is better than working with a too-thick original layer.

To store, leave each sheet of leather on the plastic wrap on which it was dried and roll it up, tucking in the sides of the wrap as you go along. Overwrap each roll for further protection against moisture. Refrigerate until used, up to a couple of months; freeze for long-term storage.

Sun-drying. Cover from dust and insects with cheesecloth held several inches away from the fresh fruit pulp, and place in direct sun. Bring inside at night. Protective cover can be left off when the leather is no longer tacky to the touch. Finish with a pasteurizing treatment at 145 F. for 30 minutes. Total sun-drying time about 24 hours.

DRY TEST. Pliable and leathery, stretches slightly when torn; surface slick, with no drag when rubbed lightly with the fingertips.

NECTARINES

Prepare as for Apricots.

Dryer. Steam-blanch halves 15–18 minutes, slices for 5 minutes. Dry with the same heat sequence as for Apricots. Test dry. Condition; store. Drying time will be roughly the same as for Apricots.

Sun-drying. Prepare and handle like Apricots. Test dry. Pasteurize. Cool and store.

DRY TEST. Pliable, leathery.

PEACHES

Yellow-fleshed freestone varieties are the best for home-drying.

Prepare. Commercially dried peaches are dried in halves, and next to never peeled. For home-drying in slices, however, peel either by a knife as you are working along, or dip whole fruit in boiling water for 30 seconds, cool quickly and strip off skins.

Halve and stone the fruit; leave in halves or cut in slices. Scoop out any red pigment in the cavity (it darkens greatly during drying). Treat slices or halves with ascorbic-acid coat as you go along to hold color temporarily.

Dryer. Steam-blanch slices 8 minutes, unpeeled halves 15–20 minutes. Start drying at 130 F., increase gradually to 155 F. after the first hour. Turn over halves when all visible juice has disappeared. Reduce to 140 F. when nearly dry to prevent scorching. Average total drying time: up to 15 hours for halves and up to 6 hours for slices.

Sun-drying. Prepare as for the dryer. If steam-blanching slices and halves as above, sulfur slices 30 minutes, halves for 90 minutes. If not blanched, sulfur 60 minutes and 2 hours, respectively. Be careful not to spill the juice in the hollows when transferring the halves to drying trays, where they're placed cut-side up. Proceed as for Apricots. Test dry. Pasteurize. Cool and store.

DRY TEST. Leathery, rather tough.

PEARS

Best for drying is the Bartlett. (Kieffer is better used in preserves.) For drying, pears are picked quite firm and before they are ripe. They are then held at not more than 70 F. in boxes in a dry, airy place for about 1 week—when they are usually ripe enough for drying.

Prepare. Split lengthwise, remove core with a melon-ball scoop, take out the woody vein. Leave in halves or slice (pare off skin if slicing); coat cut fruit with ascorbic acid as you work, to prevent oxidizing.

Dryer. Steam-blanch slices 5 minutes, halves 20 minutes. Start at 130 F., gradually increasing after the first hour to 150 F. Reduce to 140 F. for last hour or when nearly dry. Test dry. Condition; cool and store. Average total drying time: up to 6 hours for slices, 15 hours for halves.

Sun-drying. Sulfur as for Peaches; dry like Peaches. Test dry. Pasteurize; cool and store.

DRY TEST. Suède-like and springy. No moisture when cut and squeezed.

PLUMS AND PRUNES

These may be dried whole, halved or sliced.

Prepare. Check the skins with a 30–45-second dunk in boiling water. Cool immediately.

Dryer. Steam-blanch 15 minutes if halved and stoned, 5 minutes if sliced. Start *slices* and *halves* at 130 F., gradually increase to 150 F. after the first hour; reduce to 140 F. when nearly dry.

Start *whole*, checked fruit at 120 F., increase to 150 F. gradually after the first hour; reduce to 140 F. when nearly dry. Test dry. Condition; cool and store. Average total drying time for slices: up to 6 hours; halves, up to 8 hours; whole, up to 14.

Sun-drying. Check the skins of whole fruit. Sulfur whole fruit for 2 hours. Sulfur slices and halves for 1 hour. Test dry. Pasteurize. Cool and store.

DRY TEST. Pliable, leathery. A handful will spring apart after squeezing.

Drying Vegetables and Herbs

See the over-all introduction to Chapter 5 for a description of the techniques involved in the directions given below.

Vegetables are precooked by steam-blanching before being dried. *This treatment is NOT optional:* it helps stop the spoilers occurring in low-acid foods.

Except for corn dried on the cob, all vegetables are pasteurized if their processing heat has not been high enough, or prolonged enough, to destroy spoilage organisms. Pasteurizing is particularly important for sun-dried vegetables.

Vegetables are cut smaller than fruits are, in order to shorten the drying process—for the faster the drying, the better the product (so long as the food isn't *cooked*). The approximate total drying times in a dryer are not given below, but they range from around 4 to 12 hours, depending on the texture and size of the pieces.

See "Tests for Dryness in Produce" just before Fruits section.

COOKING DRIED VEGETABLES

Before being cooked, all vegetables except greens are soaked in cold water just to cover until they are nearly restored to their original texture. Never give them any more water than they can take up, and always cook them in the water they've soaked in.

Cover greens with enough boiling water to cover and simmer until tender.

BEANS—GREEN/SNAP/STRING/WAX
(LEATHER BRITCHES)

Prepare. String if necessary. Split pods of larger varieties lengthwise, so they dry faster. Steam for 15–20 minutes.

Dryer. Start *whole* at 120 F. and increase to 150 F. after the first hour; reduce to 130 F. when nearly dry. For *split beans*, start at 130 F., increase to 150 F. after first hour, and decrease to 130 F. when nearly dry. Test. Condition; pasteurize. Cool and store.

Sun-drying. Handle exactly as for the dryer. Test dry. Pasteurize certainly. Cool and store.

Room-drying. Prepare as above but do not split. String through the upper ⅓ with clean string, keeping the beans about ½ inch apart. Hang in warm, dry, well-aired room. Test. Pasteurize certainly. Cool and store. (Old-timers would drape strings near the ceiling over the wood cookstove; they gave the name "leather britches" to these dried beans—probably because they take so long to cook tender.)

DRY TEST. Brittle.

BEANS, LIMA (AND SHELL)

Allow to become full-grown—beyond the stage you would when picking them for the table, or for freezing or canning—but before the pods are dry. Shell. Put in very shallow layers in the steaming basket and steam for 10 minutes. Spread thinly on trays.

Dryer. Start at 140 F., gradually increase to 160 F. after the first hour; reduce to 130 F. when nearly dry. Test dry. Condition; pasteurize. Cool and store.

Sun-drying. Not as satisfactory for such a dense, low-acid vegetable as processing in a dryer is. However, follow preparation as for a dryer. Test dry. Condition if necessary; pasteurize certainly. Cool and store.

DRY TEST. Hard, brittle; break clean when broken.

BEETS

Prepare. Choose beets small enough so they have no woodiness. Leave ½ inch of the tops (they will bleed during precooking if the crown is cut). Steam until cooked through, 30–45 minutes. Cool. Trim roots and crowns. Peel.

Slice crosswise no more than ⅛ inch thick; OR shred them with the coarse blade of a vegetable grater: the smaller, thinner pieces dry more quickly, but of course their use in cooking is more limited.

Dryer. Put slices in at 120 F. and increase to 150 F. after first hour; reduce to 130 F. when nearly dry.

Put finer shreds in at 130 F. Increase gradually to 150 F. after first hour; turn down to 140 F. when nearly dry. Test dry. Condition; pasteurize. Cool and store.

Sun-drying. Prepare as for dryer, but shreds are recommended here instead of slices because they dry more quickly. Condition if necessary; pasteurize certainly. Cool and store.

DRY TEST. Slices very tough, but can be bent; shreds are brittle.

BROCCOLI

Prepare. Trim and cut as for serving. Cut thin stalks lengthwise in quarters; split thicker stalks in eighths. Steam 8 minutes for thin pieces, 12 minutes for thicker pieces.

Dryer. Start at 120 F., gradually increasing to 150 F. after the first hour; reduce to 140 F. when nearly dry. Test dry. Condition; pasteurize. Cool and store.

Sun-drying. Prepare as for dryer. Test dry. Condition if necessary; pasteurize certainly. Cool and store.

DRY TEST. Brittle.

CABBAGE

Most people put it by in a root cellar (q.v.) or as Sauerkraut (see under Salting); but some of it dried can be handy for soup.

Prepare. Remove outer leaves. Quarter; cut out core, and shred with the coarse blade of a vegetable grater, about the size for cole slaw. Steam 8–10 minutes. Cabbage—as do all leaf vegetables—packs on the trays during drying, so spread it evenly and not more than

½ inch deep. At a time, you'll dry only about half as much by weight of leaf vegetables as you do with other types of food, because they need more room to prevent matting.

Dryer. Start at 120 F., increase gradually to 140 F. after the first hour; reduce to 130 F. when nearly dry: the thin part of the leaves dries more quickly than the rib, and therefore is more likely to scorch and turn brown. Keep lifting and stirring the food on the trays to keep it from matting. Test dry. Condition if necessary; pasteurize. Cool and store.

Sun-drying. Follow procedure for dryer. Test dry. Pasteurize certainly. Cool and store.

DRY TEST. Extremely tough ribs; the thin edges crumble.

CARROTS

Like beets, above, they keep so well in a root cellar (q.v.) that it seems a shame to dry them—unless it's a choice between dried carrots and no carrots at all.

Choose crisp, tender carrots with no woodiness. Leave on ½ inch of the tops.

Prepare. There's no need to peel good young carrots: just remove whiskers along with the tails and crowns— which is done *after* they're steamed. Steam until cooked through but not mushy—about 20–30 minutes, depending on the size. Trim off tails, crowns with tops, and any whiskers. Cut in ⅛-inch rings, or shred.

Dryer. Proceed as for Beets, either sliced or shredded. Test dry. Condition; pasteurize. Cool and store.

Sun-drying. Proceed as for Beets, either sliced or shredded. Test dry. Condition if necessary; pasteurize certainly. Cool and store.

DRY TEST. Slices very tough and leathery, but will bend; shreds are brittle.

CELERY

For drying *leaves*, see Herbs.

Prepare stalks. Split outer stalks lengthwise, leave small center ones whole; trim off leaves to dry as herb seasoning. Cut all stalks across in no larger than ¼-inch pieces. Steam for 4 minutes.

Dryer. Start at 130 F., increase to 150 F. after the first hour; reduce to 130 F. when nearly dry. Test dry. Condition; pasteurize, because the maximum heat may not be long enough to stop spoilers. Cool and store.

Sun-drying. Prepare as for the dryer. Test dry. Pasteurize certainly. Cool and store.

DRY TEST. Brittle chips.

CORN-ON-THE-COB

Use popcorn and flint varieties for this. (Flint corn was the food-grain of the Colonists, who were taught by the Indians to use it. It is different from "dent" corn, which shrinks as it dries.) The kernels of both flint and popcorn remain plump when hard and dry.

These varieties are allowed to mature in the field and become partly dry in the husk on the stalks. Both are usually air-dried in the husk. However, in some hot countries the husks are peeled back from the partly dried ears and braided together or tied together. In northern Italy, in the fall, whole sides of brick-lattice buildings are golden with corn being finished in the sun. Both popcorn and flint corn (which is ground into meal) rub easily from the cob when the kernels are dry. Save the cobs for smoking home-cured meats (q.v.).

DRY TEST for popcorn: rub off a little and pop it. If the result's satisfactory, then immediately put it into moisture-vapor-proof containers with tight closures, to prevent it from getting too dry to pop (the remaining moisture in the kernel is what makes it explode in heat).

DRY TEST for flint corn: brittle—it cracks when you whack it. Store in sound air- and moisture-proof barrels; but if you must hold it in large cloth bags, invert the bags every few weeks: this prevents any moisture from collecting on the bag where it touches the floor.

CORN, PARCHED

Correctly dried sweet corn is more than a stop-gap for the many people who consider it superior in flavor to canned corn. Any variety of sweet corn will do.

Prepare. Gather in the milk stage as if it were going straight to the table. Husk. Steam it on the cob for 15 minutes for more mature ears, 20 minutes for quite immature ears (the younger it is, the longer it takes to set the milk). It's a good idea to separate the corn into lots with older/larger and younger/smaller kernels so you can handle them uniformly. When cool enough to handle, cut it from the cob as for canning or freezing whole-kernel corn (q.v.). Don't worry about the glumes and bits of silk: these are easily sifted out after the kernels are dry.

Dryer. Spread shallow on the trays. Start at 140 F.; raise to 165 F. gradually after the first hour; reduce to 140 F. when nearly dry, or for the last hour. Stir frequently to keep it from lumping together as it dries. Test dry. Condition. (Pasteurizing is not necessary following processing in a dryer *if the temperature has been held as high as 165 F. for an hour*.) The silk and glumes will separate to the bottom of the conditioning container; but if you don't condition, shake several cupfuls at a time in a colander whose holes are large enough to let glumes and silk through. Best stored in moisture-vapor-proof containers in small amounts.

Sun-drying. Prepare exactly as for the dryer. Stir frequently to avoid lumping. Pasteurize certainly. Shake free of glumes and silk. Package and store.

To cook. Rinse in cold water, drain; cover with fresh cold water and let stand overnight. Add water to cover, salt to taste, and boil gently until kernels are tender—about 30 minutes—stirring often and adding a bit more water as needed to keep from scorching. Drain off excess water, season with cream, butter, pepper.

DRY TEST. Brittle, glassy and semi-transparent; a piece cracks clean when broken.

GARLIC

It keeps so well when it's conditioned after harvesting that it's seldom dried at home. If you do want to dry it, though, treat it like Onions.

HERBS

This category includes celery leaves as well as the greenery from all aromatic herbs—basil, parsley, sage, tarragon: whatever you like.

All such seasonings are *air-dried* at temperatures never more than 100 F. (higher, and they lose the oils we value for flavor); and as much light as possible should be excluded during the process.

For the best product, dry only the tender and most flavorful leaves from the upper 6 inches of the stalk. Especially with bag-drying, check after the first week and every few days thereafter: the stalks should not be brittle-dry, lest you have trouble separating tiny pieces of stick from the fully dried leaves.

Gather and prepare. Cut them on a sunny morning after the dew has dried, and choose plants that have only just started to bloom; cut them with plenty of stem, then strip off tougher leaves growing lower than 6 inches on the stalk, and remove blossom heads. Hold in small bunches by the stems and swish the leaves through cold water to remove any dust, or soil thrown up from a rain. Shake off the water and lay on absorbent toweling to let all surface moisture evaporate.

Bag-drying. Collect 6 to 12 stems loosely together, and over the bunched leaves put a commodious brown-paper bag—one large enough so the herbs will not touch the sides. Tie the mouth of the bag loosely around the stems 2 inches from their ends, and hang the whole business high up in a warm, dry, airy room. When the leaves have become brittle, knock them from the stems and package them in air-tight containers and store away from light. You can pulverize the leaves by rubbing them between your hands; then store.

Tray-drying. Prepare as for drying in bags. Cut off the handle-stems, spread the leafed stalks one layer deep on drying trays. Put the trays in a warm, dark room that is extra well ventilated (if you use a fan, don't aim it *on* the trays—the herbs could blow around—instead, "bounce" the forced draft off a wall, so it will be gentler). Turn the herbs several times to ensure even drying. Test dry. Remove from stems, and package as above.

Celery leaves. Cut out the coarsest midribs. Tray-dry as for any leaf herb.

DRY TEST. Readily crumbled.

MIXED VEGETABLES

These are never dried in combination: drying times and temperatures vary too much between types of vegetables. Dry vegetables and seasoning separately, *then* combine them in small packets to suit your taste and future use.

MUSHROOMS

Only young, unbruised, absolutely fresh mushrooms should be dried.

Prepare. Wash quickly in cold water if necessary; otherwise wipe with a damp cloth. Remove stems (they're denser than the caps as a rule, so shouldn't be dried in the same batch with the tops). Slice the caps in ⅛-inch strips—cut stems across in ⅛-inch rings—and treat them with the ascorbic-acid coating if it's important to you to keep them from darkening as you work. Steam for 12–15 minutes.

Dryer. Start at 130 F., increase gradually to 150 F. after the first hour; reduce to 140 F. when nearly dry. Test dry. Condition; pasteurize. Cool and store.

Sliced caps and stems process at the same temperature sequences, but stem pieces usually take longer.

Sun-drying. Prepare as for the dryer. Test dry. Condition if necessary; pasteurize certainly. Cool and store.

DRY TEST. Brittle.

ONIONS

Prepare. Peel; slice in rings about ⅛-inch thick. Uniformity is important here, because slices too thin can brown and scorch; it's better to have evenly thicker pieces than some ⅛ inch and others paper-thin. No steaming is necessary.

Dryer. Put them in at 140 F. and keep them there until nearly dry, watching carefully that thinner pieces are not browning. Reduce to 130 F. for the last hour if necessary. Test dry. Condition. Cool and store.

Sun-drying. Prepare as for the dryer. Test dry. Pasteurize. Cool and store.

DRY TEST. Light-colored, but brittle.

PEAS, BLACK-EYED, treat like Beans (Shell), above.

PEAS, GREEN

Choose young, tender peas as you'd serve them fresh from the garden. From there on, treat them like Shell Beans, above.

DRY TEST. Shriveled and hard; shatter when hit with a hammer.

PEPPERS, HOT (CHILI)

Choose mature, dark-red pods. Thread them on a string through the stalks, and hang them in the sun on a south wall. When dry, the pods will be shrunken, dark, and may be bent without snapping.

PEPPERS, SWEET (GREEN OR BELL)

Prepare. Split, core, remove seeds; quarter. Steam 10–12 minutes.

Dryer. Start at 120 F., gradually increase to 150 F. after the first hour; reduce to 140 F. when nearly dry (if any are thin-walled, reduce to 130 F. toward the end, and keep stirring them well). Test dry. Condition. Cool and store.

Sun-drying. Prepare as for the dryer. Test dry. Condition if necessary; pasteurize certainly. Cool and store.

DRY TEST. Crisp and brittle.

POTATOES, SWEET (AND YAMS)

Only firm, smooth sweet potatoes or yams should be used.

Prepare. Steam whole and unpeeled until cooked through but not mushy, about 30–40 minutes. Trim, peel; cut in ⅛-inch slices, or shred.

Dryer. Proceed as for sliced or shredded Beets. Test dry. Condition; pasteurize. Cool and store.

Sun-drying. Prepare as for dryer. Test dry. Condition if necessary; pasteurize certainly. Cool and store.

DRY TEST. Slices extremely leathery, not pliable; shreds, brittle.

POTATOES, WHITE (IRISH)

These root-cellar too well to bother drying. But dry like Turnips, below.

PUMPKIN

Deep-orange varieties with thick, solid flesh make the best product. There's not much use in drying in chunks, because they're to be mashed after cooking.

Prepare. Take them directly from the garden (they shouldn't be conditioned as for root-cellaring). Split in half, then cut in manageable pieces for peeling and removing seeds and all pith. Shred with the coarse blade of a vegetable grater (less than ⅛ inch thick). In shallow layers in the basket, steam for 6 minutes.

Dryer. Proceed as for shredded Beets, above. Test dry. Condition; pasteurize if length of maximum processing heat isn't enough to stop spoilers. Cool and store.

Sun-drying. Prepare as for the dryer. Test dry. Pasteurize certainly. Cool and store.

DRY TEST. Brittle chips.

Historical note. In olden days, pumpkins were often halved at their equators, then cut in rings about 1 inch thick; rind and seeds and pith were removed from each ring. The rings were hung on a long stick and dried slowly in front of a fire until they were like tough leather.

SPINACH (AND OTHER GREENS)

Prepare. Use only young, tender, crisp leaves. Place loosely in the steaming basket and steam for 4–6 minutes, or until well wilted. Remove coarse midribs; cut larger leaves in half. Spread sparsely on drying trays, keeping overlaps to a minimum (leaves tend to mat).

Dryer. Start at 140 F., increase to 150 F. after the first hour; if necessary, reduce to 140 F. when nearly dry, to avoid browning. Test dry. Condition. Cool and store.

Sun-drying. Prepare as for the dryer. Test dry. Pasteurize certainly. Cool and store.

DRY TEST. Easily crumbled.

SQUASH (ALL VARIETIES), treat like pumpkin.

TURNIPS AND RUTABAGAS

Like white (Irish) potatoes, which root-cellar extremely well if handled right (q.v.), turnips and rutabagas are seldom dried.

If you're compelled to dry them, though, quarter and peel them, then shred—and steam and dry as for shredded Carrots.

DRY TEST. Brittle chips.

Drying Meat and Fish

For 99 percent of North America's householders, drying meat and fish should be considered "playing with food" since they lack either the appropriate conditions or the need to do it—and climate and necessity are what make the remaining 1 percent masters of the art.

We shan't give blow-by-blow instructions for making jerky as the Mountain Men did, or drying codfish with the expertise of a Newfoundland native. Here are the basic steps. Work in small batches, with complete sanitation; don't cut corners. Refrigerate or freeze the finished product: high-protein foods like these invite spoilage.

USING DRIED MEAT AND DRIED FISH

Jerky traditionally was shaved off (or gnawed off) and eaten as is, because it was a staple for overland wanderers who were traveling light and far from assured supplies of fresh meat. (Helpful ins-and-outs of concentrated journey food are to be found in Horace Kephart's *Camping & Woodcraft;* see "pemmican" especially.) Today several versions of it appear in stick form as snacks, for either the Long Trail or a cocktail party.

Dried salt fish—the type described below—is always freshened by soaking beforehand, either in cold water or fresh milk; the soaking liquid is discarded here, because of the extremely high salt content.

Such fish were standard fare even in the hinterlands far from salt water. A small roadway in our wooded Green Mountains is still called by old-timers "Codfish Alley"— so named by homesteaders who went down to the flatboats coming up the Connecticut River to buy salt codfish, which they carried home in armloads like billets of stovewood. See Recipes.

130 F. = 54 C. / 140 F. = 60 C. / 150 F. = 66 C. / 175 F. = 79.5 C. / 212 F. = 100 C. / 1 U.S. tsp. = 5 ml. (5 cc.) / 1 U.S. T. = 15 ml. (15 cc.)

DRYING MEAT (JERKY)

Jerked meat is roughly ¼ the weight of its fresh raw state.

Preferred meats for jerking are mature beef and venison (elk is too fatty), and only the lean muscle is used. Cut lengthwise of the grain in strips as long as possible, 1 inch wide and ½ inch thick.

DRY TEST. Brittle, as a green stick: it won't snap clean, as a dry stick does. Be sure to test it *after* it cools, because it's pliable when still warm, even though enough moisture is out of it.

Unsalted Jerky

This does not mean unseasoned—there's a bit of salt for flavor—but the meat is not salted heavily to draw out moisture or to act mildly as a preservative.

Lay cut strips on a cutting-board, and with a blunt-rimmed saucer or a meat mallet, pound the following seasonings (or your own variations thereof) into both sides of the meat: salt, pepper, garlic powder, your favorite herb. Use not more than 1 teaspoon salt for each 1 pound of fresh meat, and the other seasonings according to your taste.

Arrange seasoned strips ½ inch apart on wire racks—cake-cooling ones, or the racks from the oven; put them in a preheated 150 F. oven, and immediately turn the heat back to 120 F. If your oven is not vented, leave its door ajar at the first stop position. After 5 or 6 hours turn the strips over; continue drying at the same temperature for 4 hours more, when you check for dryness. When dry enough, jerky is shriveled and black, and is brittle when cooled.

Wrap the sticks of jerky in moisture-vapor-proof material, put the packages in a stout container with a close-fitting lid, and store below 40 F. in the refrigerator (or freeze it). Reconstitute by simmering in water to cover.

Salted Jerky

Try this in the sun; or, if you're emulating the frontiersmen, over a very slow, non-smoking fire that's not much more than a bed of coals.

Prepare a brine of 2½ cups of pickling salt for 3 quarts of water, and in it soak the cut strips of meat for 1 or 2 days. Remove and wipe dry.

OVER COALS

Arrange a rectangular fire-bed and drive forked poles into the ground at each corner; the forks should come about 4 feet above the ground. Two hours before you're ready to begin drying the salted meat, start a fire of hardwood and let it burn down to coals.

Cut two fairly heavy poles to go from fork to fork on the long sides, and sticks as thick as a finger—and sharpened at one end—to lay at right angles over the sidepoles. The salted strips may be draped over the crosssticks; or they may be suspended from the sticks (pass the sharpened stick through the meat about 3 inches from one end) and spaced several inches apart.

Feed the fire with small hardwood so carefully that juice does not ooze out from the excess heat, or the meat start to cook.

Depending on conditions, drying could take 24 hours. Test for dryness; package and store in refrigerator or freezer.

IN THE SUN

This is nothing to try in a back yard—or indeed in any place near civilization. It needs clear, pure air, uncontaminated by animals or human beings: after all, it is a method used on old-time hunting trips deep into the High Plains.

Choose a time when you'll have good—but not roasting—sun, dry air day and night, and a gentle breeze. Hang the salted strips from a drying frame such as described above (of course with no fire), and leave them there until they become brittle-dry.

A BASIC PROCEDURE FOR DRYING FISH

Drying fish at home is not something to be undertaken lightly. The fish must be fresh-caught and handled scrupulously at every stage; it must undergo a long dry-salting period before it is put out to dry; and, since home-drying is best done outdoors in the shade, the procedure requires a trustworthy breeze, fairly low humidity, and critter-proof holding tubs and racks.

Dry any *lean* fish (cod of course is the classic), because fatty fish don't keep as well. It should be abundant, to make drying worth the bother.

Coat all surfaces of each fish liberally with pure pickling salt, using 1 pound of salt for 2 pounds of fish, and stack the opened fish flesh-side up on a slatted wooden rack outdoors. Don't make the stacks more than 12 layers deep, with the top layer skin-side up. Leave them stacked from 1 to 2 weeks, depending on the height of fish and the dryness of the air. Brine made by the salt and fish juices will drain away. Move the pile inside each night and weight it down to press out more brine.

Scrub the fish again to remove the salt, and put them on wooden frames outdoors to complete the necessary removal of moisture from their tissues. Hang or spread the fish on cross-pieces in an open shed with good ventilation; direct sun on the fish can start it to sunburn (cook) at only 75 degrees F. Bring fish in at night, re-piling to ensure even drying; re-spread on the racks more often with skin-side up.

To store, cut in manageable chunks if the fish are large; wrap in moisture-vapor-proof plastic; pack in tight wooden boxes, and store in a dry, cool place between 32 and 40 F.

DRY TEST. No imprint is left when the fleshy part of a fish is pinched between thumb and forefinger.

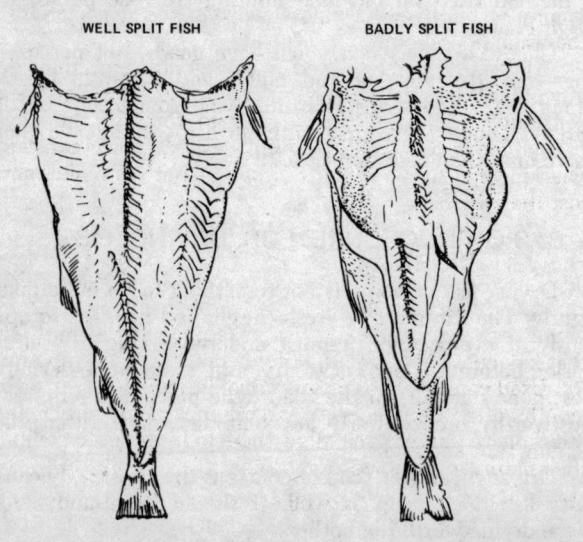

WELL SPLIT FISH　　　　　　　BADLY SPLIT FISH

Root-Cellaring 6

Of all the time-tested ways of putting food by, only wintering-over in cold storage is less satisfactory today than it was a century or more ago. And the reason is simple: all the technological advances we're so pleased with in construction and heating have given us cozy, dry basements instead of cool, damp cellars, and the chilly shed off the pantry has given way to a "utilities room" or a warm passageway between carport and kitchen.

So this section is telling how to turn back the clock and create conditions that several generations of North Americans have devoted themselves to improving. It includes some indoor areas that are warmer and drier than the traditional outbuilding or cellar with stone walls and a packed earthen floor, and it also includes some arrangements outdoors that are a good deal more rough-and-ready.

Several of the ones described may require outlays of money or effort or hardihood beyond the expectations of the usual householder.

And most of them need more maintenance than does any other type of storage discussed up till now. This means constantly watching the weather to forestall the effects of sudden extra cold/warmth/wetness, and constantly checking the food for signs of spoil—but it's maintenance just the same.

HOW IT WORKS

Root-cellar is a homemade verb embracing ordinary winter storage for fresh, raw, whole vegetables and fruits that have not been processed to increase their keeping quality. Used commonly, it means to hold these foods for several months after their normal harvest in a cold, rather moist atmosphere that will not allow them to freeze or to complete their natural cycle to decomposition.

419

The freezing points and warmth tolerances of produce vary. The range to shoot for generally, though, is 32 to 40 degrees Fahrenheit—the effective span for refrigeration —with only a couple of vegetables needing warmer storage to keep their texture over the months. In this range the growth of spoilage micro-organisms and the rate of enzymatic action (which causes overripening and eventual rotting) are slowed down a great deal.

Good home root-cellaring involves some control of the amount of air the produce is exposed to, since winter air is often let in to keep the temperature down. But fresh whole fruits and vegetables respire after they're harvested (some more than others: apples seem almost to *pant* in storage), so the breathing of many types is reduced by layering them with clean dry leaves, sand, moss, earth, etc., or even by wrapping each individually in paper. These measures of course aren't as effective as those of commercial refrigerated storage, which rely in part on drastic reduction of the oxygen in the air supply, but they work well enough for the more limited results expected from home methods.

THE HUMAN LIMITATIONS

We'll be describing a variety of storage arrangements in a minute, including a couple that are drier and sometimes warmer than the traditional ones (since we can't leave in limbo those foods that require something different from the classic old cellar treatment). But first a word about practicality.

The beauty of root-cellaring is that it deals only with whole vegetables and fruits and there are no hidden dangers: if it doesn't work, we know by looking and touching and smelling that the stuff has spoiled, and we don't eat it.

On the other hand it's something that sounds a lot more feasible than it may really turn out to be.

First, the householder must learn something about the idiosyncrasies of the fruits and vegetables he plans to store on a fairly large scale: for example, apples and potatoes— the most popular things to carry over through winter— can't be stored near each other, and the odor of turnips and cabbages in the basement can penetrate up into the living quarters, and squashes want to be warmer than carrots do.

Then he casts around for the right sort of storage. And the solution may cost more than its value to his over-all food program, especially if it's a structure more elaborate or permanent than the family's make-up warrants.

But aren't there the less pretentious outdoor pits, or the more casual barrels sunk in the face of a bank? Yes; and they're fun to use—except in deep-snow country when they can be a worry to get at.

Samuel Ogden of Landgrove, Vermont, organic gardener and noted Green Mountain countryman, warns the newcomer to cold-climate root-cellaring to avoid three things: counting too heavily on cold storage, having too much diversity, and having it inaccessible in bad weather. But for the family with a serious, long-term food program that depends in great degree on its own efforts, though, he recommends the Vermont experience of Helen and Scott Nearing described in *Living the Good Life:* the Nearings' last two root cellars are well-thought-out and substantial affairs—and represent total dedication to a way of life.

INDOOR STORAGE

The classic root cellar downstairs

There are fewer of these to be found as the years go by, even in the old houses in our part of the country. Usually in the corner of the original cellar-hole, they have two outside walls of masonry (part of the foundation), the floor is packed earth, and any partitions are designed more to support shelving than to keep out warmth from a nonexistent furnace. They incorporate at least one of the small windows that provide cross-ventilation for the whole cellar to keep overhead floor joists from rotting; propped open occasionally during the winter, it's the answer for regulating temperature and humidity.

Such a place can be ideal today, although the house is "restored" and now contains a furnace (protected from seepage in the springtime by a surrounding pit and an automatic sump pump). Just complete boarding-off the

---

EQUIPMENT FOR ROOT-CELLARING

Storage place, indoor or outdoor (see below)

Clean wooden boxes/lugs/crates or barrels; or stout large cardboard cartons (for produce that wants to be dry, not damp)

Plenty of clean paper for wrapping individually, or shredding

Plenty of clean dry leaves, sphagnum, peat moss or sand

A tub of sand to keep moistened to provide extra humidity if needed

Simple wall thermometer certainly; humidity gauge (optional)

---

two inside walls, cut and hang a door in one of them, and apply whatever is handiest for insulation against heat from the furnace.

To be perfect, it should have an inner partition to separate storage of fruits and vegetables; and its own electric light. For the rest, build stout shelves or put up trestles to hold boxes, crates and baskets off the floor on the sides away from the window. Reserve one well-drained corner for the vegetables that will be clumped upright with their roots set in soil or sand and moistened by hand to keep them fresh.

Darken the window(s)—potatoes turn green in light when they're stored, and this isn't good. If necessary, keep clearing snow from the areaway that's below ground leading to the window.

And check the whole thing for places where field mice can get in and feast on your crops during the lean winter months, and stop them up.

A modern basement store room

To some extent, in a closed-off corner of a modern basement you can copy the conditions of the old-time downstairs root cellar.

Choose a corner preferably on the north or east (where the temperature is likely to be most even); it should have no heating ducts or oil or water pipes running through it. And if it has a window you'll be saved having to figure out a system for governing temperature, ventilation and humidity.

The store room should be at least 6 × 6 feet if you're going to bother building it at all.

The existing right-angled outside walls of the basement will become the outside walls of the store room. Make two inner right-angled walls of fiberboard or ½-inch lumber nailed to 2 × 4 studding spaced 2 feet apart and secured to a footing. Leave open space on one side to frame and hang a door.

Insulate the new room from the inside—glass batts that include a vapor barrier go easily between the studs. Finish the inside with wallboarding if you want to, covering the seams with common lath. Insulate the door and give it a simple latch. *Insulate the inside walls only.*

Make an air-duct box to cover at least the lower ⅔ of the window when it's opened, and carry the box part way down the side of the wall (see detail of the sketch). The duct brings cold incoming air to the lower part of the room and lets warm air from the upper part be drawn outdoors through the upper ⅓ of the opened window.

Ideally you should have a fruit storage room too—or at least a part of the all-purpose store room blocked off for fruit. If you have a window in the fruit room too, build an air-duct box for it. Otherwise you'll regulate temperature/humidity by opening and closing the entrance to the fruit section.

Using a bulkhead

Many middle-aged houses have an outside entrance to the cellar: a flight of concrete steps down to the cellar wall, in which a wide door is hung to give access to the cellar. The top entrance to the steps—the hatch—is a door laid at an angle 45 degrees to the ground.

On the stairway, which probably is closed from the outside during the coldest months anyway, you can store barrels/boxes of produce. You could put up rough tempo-

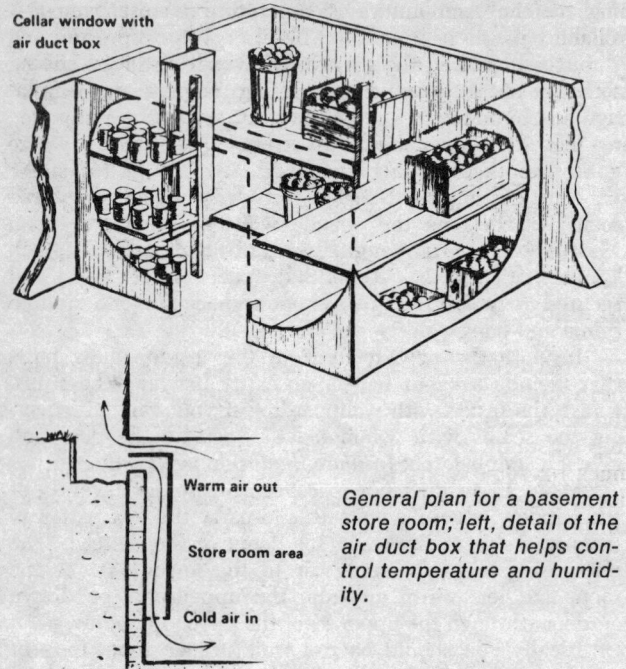

General plan for a basement store room; left, detail of the air duct box that helps control temperature and humidity.

rary wooden side walls along the steps; but certainly lay planks on the steps to set your containers on. Insulate the door into the cellar proper with glass batts. If you need to, keep a pail of dampened sand on one of the steps to add humidity. You're likely to be propping the hatchway door open a few inches from time to time to help maintain proper temperature on the steps. This means shoveling snow from the bulkhead, and piling it back on when this outside door is closed again.

A dry shed

This takes the place of the garage advocated by some people—but not by us: too much oil and gasoline odor (some produce soaks stray odors up like a sponge), and far too great a quantity of lead-filled emissions from running motors. And anyway temperature is often uncontrollable.

Instead of using the garage, partition off storage space in the wood-floored shed leading into the kitchen, if you have an old house in the country. Or segregate a storage area in a cold, seldom used passageway.

Storage areas like these are usually not fit for such long-term storage as the basement root cellar or store room is.

Up attic

An old-fashioned attic generally is the last place in the house to cool off naturally as cold weather sets in; and unless the roof is well insulated, the attic temperature rises on sunny winter days. This fluctuation doesn't matter much for some foods, however (see the chart), though it does for onions, say. The answer is to wall off, and ceil, a northeast corner for anything that needs maintained low temperature and dryness. Then you can put pumpkins and such near the stairway leading to the attic—and leave the hall door downstairs open whenever you need to.

In styrofoam picnic chests/hampers

Harvest carrots, beets, turnips, etc., late in the fall. The handling described for carrots works well for the others too.

Cut tops off carrots, leaving about ½ inch of stem. Wash away garden soil, then wipe fairly dry—"fairly dry" because you want a little moisture; but not wet, lest the vegetables mold. Sack them, 4 to 5 pounds at a crack, in durable, good-sized plastic bags: the garbage bags sold in supermarkets to line wastebaskets, etc., are fine. Press out excess air from each filled bag, twist the top and tie it tight with string, rubber bands, coated wire.

Pack the bags in a styrofoam chest of the inexpensive sort you use to carry picnic food on ice, and store the chest in any cold spot like an unheated off-corner of the cellar or an enclosed sunless porch. Keep the lid tightly on the chest except when removing meal-size amounts from a bag.

Carleton Richardson of Brattleboro, Vermont, who taught us this type of storage, sometimes puts bags of beets directly on the cool cement floor of his cellar—and they keep beautifully for 6 months.

Late apples in milk cans (or small trash cans)

High in Green Mountain ski country, Albert and Millie Dupell use four old 40-quart milk cans with tight-fitting lids to store two bushels each of Northern Spys and winter McIntoshes.

Wrap each perfect, fresh apple snugly in one standard-size page of clean newspaper, having balled and smoothed the paper beforehand so it will create tiny air pockets next to the fruit. Pack fruit in the cans firmly but without bruising, and store the cans anywhere that holds between 30 and 45 F. Don't open too often—take out a couple of meals' worth at a time, refrigerating the extras—and seat the lid very firmly afterward.

To use small metal trash barrels, you'll need to provide some insulation so they will equal the efficiency of milk cans (which are becoming "collectibles" for nostalgia buffs, and so are fairly scarce). Put 3 inches of dry sawdust in the bottom of the barrel and pack 1 inch of sawdust between the outside apples and the metal sides. A sheet of tough plastic draped over the top of the barrel under the lid will ensure that the cover fits as smoothly tight as a milk can's does.

SMALL-SCALE OUTDOOR STORAGE

There's a good deal of information around that contains ideas for full-dress outdoor buildings for root-cellaring. Of these we suggest two: the USDA *Home and Garden Bulletin No. 119, Storing Vegetables and Fruits in Basements, Cellars, Outbuildings, and Pits*—available from your County Agent—and the September 1971 issue of *Organic Gardening and Farming* (Rodale Press, Emmaus, Pennsylvania 18049), with more variety.

Use either one to take off from in designing an elaborate building within your climate zone and your means. We limit the discussion below to arrangements for small-scale outdoor storage.

Some mild-climate pits, etc.

The USDA bulletin and the other sources describe several easy-to-make and cheap outdoor storage facilities, all either on well-drained ground or sunk only several inches below the surface. See the chart for which produce likes the conditions they provide.

However, such arrangements can be counted on *only in places with fairly mild winters* that have no great extremes in temperature. At any rate, make a number of small storage places, fill them with only one type of produce to each space, and be prepared to bring the entire contents of a store-place indoors for short storage once the space is opened.

MILD-CLIMATE CONE "PITS"

Most instructions call a storage place like this a pit, but it's really a conical mound above ground. To make it,

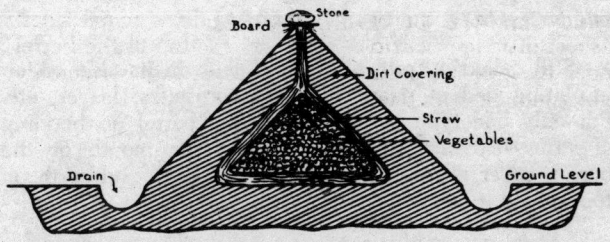

lay down a bed of straw or leaves, etc.; pile the vegetables *or* fruits (don't mix them together) on the bedding; cover the pile well with a layer of the bedding material. With a shovel, pat earth on the straw/leaf layer to hold it down, extending a "chimney" of the straw to what will be the top of the cone to help ventilate and control the humidity of the innards of the mound. Use a piece of board weighted by a stone to act as a cap for the ventilator. Surround the "pit" with a small ditch that drains away surface water.

As colder weather comes, add to the protective layer of earth, even finishing with a layer of coarse manure in January.

MILD-CLIMATE CABBAGE "PIT"

This is quite like the cone above, except that it's longer and allows stored food to be removed piecemeal.

Lay the uprooted cabbages head-down on a bedding of straw, etc., pack insulating straw/leaves around them, and cover all with earth. Cut a drainage carry-off on each side of the pit.

MILD-CLIMATE COVERED BARREL

Still called a "pit" is a barrel laid on its side on an insulating bed of straw, chopped cornstalks, leaves, etc. Put only one type of produce in the barrel on bedding of straw/leaves. Prop a cover over the mouth of the barrel; cover all with a layer of straw, etc., and earth on top to hold it down.

OUTDOOR FRAME FOR CELERY, ETC.

Dig a trench 1 foot deep and 2 feet wide, and long enough to hold all the celery you plan to store. Pull the celery, leaving soil on the roots, and promptly pack the clumps upright in ranks 3 to 5 plants wide. Water the roots as you range the plants in the trench. Leave the trench open until the tops dry out, then cover it with a slanted roof. This you make by setting on edge a 12-inch board along one side of the trench, to act as an upper support for the cross-piece of board, etc., that you lay athwart the trench. Cover this pitched roof with straw and earth.

Walter Needham's cold-climate pit

As he told in *A Book of Country Things*, he was raised in rural Vermont by a grandfather for whom a candle-mold was a labor-saving device. So whenever we want to learn about totally practical methods of pioneer living in the cold country, we turn to Walter Needham. He was the first to point out that the conical "pit" wouldn't do an adequate job in 20-below Zero weather. This is his alternative:

Choose the place for your pit on a rise of ground to avoid seepage. There, shovel out a pit about 1½–2 feet deep and 4 feet wide at the bottom, throwing dirt up all around to build a rim that will turn water away; dig a V-shaped drainage ditch around it for extra protection (see sketches). Take out any stones near the sides of the pit because frost will carry from one stone to another in rocky ground. The pit needn't go below the deep frost-line if such frost conductors are removed. Pack the bottom of the pit with dry mortar sand 2 to 3 inches deep: the loam, having retained moisture, will freeze; the sand holds the food away from the loam.

On the layer of sand make a layer of vegetables not more than 1 foot deep; cover the vegetables with more fine sand, dribbling it in the crevices, to fill the pit nearly to ground level. Cover the sand with straw or spoiled hay, mounded to shed the weather. Hold down this cover with a thin layer of sod—or, nowadays, plastic sheeting weighted down with 1 to 2 inches of earth. Cover one end of the mound with a door laid on its side and slanted back almost like a bulkhead entrance. In winter you'll move the door away to dig in for the vegetables, and, as they're taken out, move the door back along the mound.

This root-pit is best for beets, carrots, turnips and potatoes.

Covered barrel, above, and celery frame—both for a mild climate.

Walter Needham's sunken barrels

Again, these are for cold-winter areas with uneven temperature.

Into the face of a bank dig space to hold several well-scrubbed metal barrels with their heads removed—one barrel for apples, say; one for potatoes, one for turnips. Take out any large stones that would touch the barrels and conduct frost to them, and provide a bedding of straw/dry leaves, etc., for the barrels to rest on. Slant the open end of the barrels slightly downward, so water will tend to run out.

Put straw or whatever in the barrels for the produce to lie on, and fill the barrels from back to front, using dry leaves or similar material to pack casually around the individual vegetables or fruits if they need it.

Over the opening put a snug cover propped against it —a stout wooden "door" with a *wooden* handle (did you ever have the skin of your palm freeze on to metal in below-Zero weather?). Dig a shallow V-shaped drainage ditch to carry surface water away from the barrels (as in the top-view sketch on the next page).

The snow will be added protection in the deepest cold of the winter. Shovel it back against the door after removing food from the barrels.

Root-Cellaring Fruits and Vegetables

ROOT-CELLARING FRUITS

Only several of the most popular fruits root-cellar well; and of these, apples retain their texture and flavor longest, with several varieties of pears next in storage life.

Like vegetables, fruits to be stored over the winter should be harvested as late as possible in the season, and be as chilled as you can get them before they're put in their storage containers (it will take even a properly cold root cellar a good deal of time to remove the field heat from a box of warm apples).

Because they absorb odors from potatoes, turnips and other "strong" vegetables, fruits should have their own

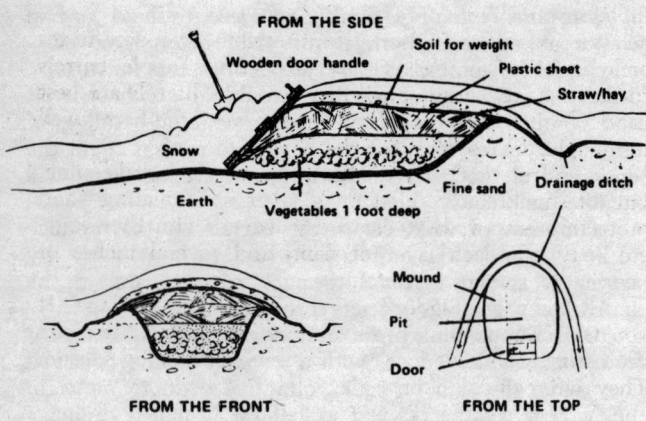

FROM THE SIDE

Wooden door handle

Soil for weight

Plastic sheet

Straw/hay

Snow

Earth

Vegetables 1 foot deep

Fine sand

Drainage ditch

FROM THE FRONT

Mound

Pit

Door

FROM THE TOP

Cold-climate storage in a pit, above, and in sunken barrels.

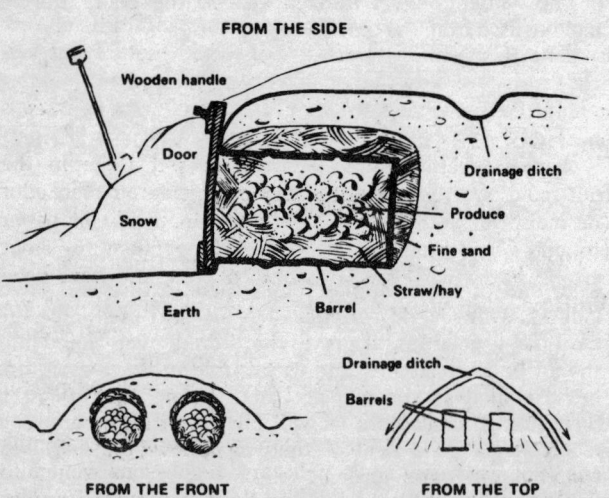

FROM THE SIDE

Wooden handle

Door

Snow

Drainage ditch

Produce

Fine sand

Earth

Barrel

Straw/hay

FROM THE FRONT

Drainage ditch

Barrels

FROM THE TOP

special section partitioned off in the root cellar if they are stored in quantity; otherwise put them in another area where the conditions simulate those of a root cellar (see the chart), or keep them as far from the offending vegetables as possible.

We recommend clean, stout cartons, wooden boxes or splitwood fruit baskets over the classic apple barrels for storing fruits for a small family. Metal barrels are best used for fruits in underground storage, with the barrel well insulated from frost in the earth.

Some fruits are individually wrapped for best keeping, but all should be bedded on a layer of insulating—and protecting—straw, hay, clean dry leaves, with the straw, etc., between each layer of fruit, and several inches of bedding on top of the container.

All fruits need checking periodically for spoilage. If you're afraid your fruit will deteriorate faster than you can eat it fresh, have a midwinter preserve-making session. They're fun on cold lowery days.

APPLES

Best keepers: *late* varieties, notably Winesap, Yellow Newton, Northern Spy; then Jonathan, McIntosh in New England, Cortland, Delicious. Pick when mature but still hard, and store only perfect fruit. Apples kept in quantity in home cold storage usually will be "aged" from Christmas on.

Apples breathe during storage, so put them in the fruit room of a root cellar so they don't give off their odor (or moisture) to vegetables. Wrap individually in paper (to cut down their oxygen intake); put them in stout cartons, boxes, barrels that can be covered, and have been insulated with straw, hay or clean dry leaves. If you use large plastic bags or liners for the boxes, etc., cut ¼-inch breathing holes in about 12 places in each bag.

They also may be stored in hay- or straw-lined pits or in buried barrels covered with straw and soil, etc.

Farmers in Vermont often store wrapped apples in milk cans (not used now since bulk-tank regulations went into effect) with tight-fitting lids; as they respire, the apples make their own humidity. Cans then may be held in a dry

Recommended Conditions for

Produce	Food Freezes At (F.)	Type of Storage
Fruits:		
Apples	29.0	RC-F
Grapefruit	29.8	RC-F
Grapes	28.1	RC-F
Pears	29.2	RC-F
Vegetables:		
Beans, dried	won't	DS; A
Beets	(*c.* 30)	BSR; P/B; RC
Cabbage	30.4	P/B; RC; DS
Carrots	(*c.* 30)	BSR; P/B; RC
Cauliflower	30.3	RC
Celery	31.6	BSR; frame; RC
Chinese cabbage	(*c.* 31.9)	BSR; frame; RC
Dried seed, live	won't	A
Endive	31.9	frame
Horseradish	(*c.* 30.4)	BSR; P/B; RC
Kale	(*c.* 31.9)	frame
Kohlrabi	(*c.* 30)	BSR; P/B; RC
Leeks	(*c.* 31.9)	BSR; P/B; RC
Onions	30.6	DS; A
Parsnips	30.4	BSR; P/B; RC
Peas, dried	won't	DS; A
Peppers	30.7	BSR; RC; DS
Popcorn	won't	A
Potatoes	30.9	BSR; P/B; RC
Pumpkins	30.5	BSR; A
Salsify	(*c.* 30.4)	BSR; P/B; RC
Squash	30.5	BSR; A; RC
Sweet potatoes	29.7	BSR; DS
Tomatoes, green	31.0	BSR; DS
Turnips	(*c.* 30)	P/B; RC
Winter radishes	(*c.* 30)	BSR; P/B; RC

A = attic : BSR = basement store room : DS = dry shed :
frame = coldframe : P = outdoor pit : B = buried barrel :
P/B = pit or barrel : RC = root cellar : RC-F = root cellar for

Over-the-winter Cold Storage

Ideal Temperature (F.)	Relative Humidity (%)	Air Circulation	Average Storage Life
at 32	MM: 80–90	moderate	4–6 months
at 32	MM: 80–90	slight	1–1½ months
at 32	MM: 80–90	slight	1–2 months
at 32	MM/M: 85–90	slight	2–7 months
32–50	D: 70	moderate	12+ months
32–40	M: 90–95	slight	4–5 months
at 32	MM/M: 85–90	slight	late F–W
32–40	M: 90–95	slight	6 months
at 32	MM: 80–90	slight	1½–2 months
at 32	MM/M: 85–90	slight	late F–W
32–34	VM: 95–98	slight	3–4 months
32–40	D: 70	slight	12+ months
at 32	MM/M: 85–90	moderate	2–3 months
at 32	M: 90–95	slight	4–6 months
at 32	VM: 95–98	moderate	1 month
32–40	M: 90–95	slight	2–3 months
at 32	MM: 80–90	moderate	1–3 months
at 32	D: 70	moderate	F–W
at 32	M: 90–95	slight	F–W
32–50	D: 70	moderate	12+ months
45–50	MM: 80–90	slight	½–1 month
to 75	D: 70	slight	12+ months
35–40	MM: 80–90	slight	F–W
at 55	MD/D: 70–75	moderate	F–W
at 32	M: 90–95	slight	4–5 months
at 55	VD: 50–70	moderate	F–W
55–60	MD/D: 70–75	moderate	F–W
55–70	MM: 80–90	moderate	1–1½ months
at 32	M: 90–95	slight	2–4 months
at 32	M: 90–95	slight	2–4 months

fruit : D = dry : VD = very dry : MD/D = moderately dry to dry :
M = moist : MM = moderately moist : MM/M = moderately moist
to moist : VM = very moist : F = fall : W = winter.

shed where they won't freeze, rather than in a root cellar. (See at end of "Indoor Storage.")

Check periodically and remove any apples that show signs of spoiling. See the chart for ideal conditions.

GRAPEFRUIT AND ORANGES

Store unwrapped in stout open cartons or boxes in the fruit room of a root cellar (see chart for conditions). Inspect often for spoilage, removing spoiled ones and wiping their mold off sound fruit they've touched.

GRAPES

Catawbas keep best, then Tokays and Concords. Pick mature but before fully ripe.

Grapes absorb odors from other produce, so give them their own corner of the root-cellar fruit room (see chart for conditions). Hold in stout cartons or boxes lined with a cushion of straw, etc., with straw between each layer; don't burden the bottom bunches with more than three layers above them, fitting the bunches in gently. Cover with a layer of straw. Check often for spoilage.

PEARS

Best keeper of the dessert varieties is Anjou, with Bosc and Comice popular among the shorter keepers. (Bartlett and Kieffer ripen more quickly and earlier: the former is especially good for canning, the latter for spicing whole or used in preserves; see Canning and The Preserving Kettle.)

Pick mature but still green and hard. Hold loosely in boxes in a dry, well-aired place at 50–70 F. for a week before storing. Then store them like apples. See chart for conditions.

Warning: Pears that have started ripening above 75 F. during the interim between picking and storage, or are root-cellared at too high a temperature, will spoil, often breaking down or rotting inside near the core while the outside looks sound.

ROOT-CELLARING VEGETABLES

Root-cellared vegetables freeze sooner than fruits do, as a rule; and, if you store a variety beyond the commonest root crops—beets, carrots, potatoes, turnips and rutabagas—you need several different kinds of storage conditions. See the chart again, and the individual instructions below.

Wooden crates and movable bins, splitwood baskets, stout cartons—all make good containers for indoor storage. Insulating and layering materials are straw, hay, clean leaves, sphagnum and peat moss, and dry sand. The moist sand suggested for certain vegetables shouldn't be at all puddly-wet: if it's cold to the touch and falls apart when squeezed, leaving a few particles stuck to your hand, it should be the right degree of dampness.

Don't fill containers so deeply that the produce at the bottom is ignored in the periodic examinations for spoilage. And forgo building permanent bins that can't be moved outside for between-season scrubbing and sunning—stout shelving for the containers at convenient heights off the floor is a much better use of storage space.

BEANS (SHELL), DRIED

Cool the finished beans and package in plastic bags which you then put in large, covered, insect- and mouse-proof containers. See chart for conditions.

BEETS

Harvest in late fall after nights are 30 F. (they withstand frosts in the field) but when the soil is dry. Do not wash. Leave tails and ½ inch of crown when removing the tops. Pack in bins, boxes or crates between layers of moist sand, peat or moss; or line containers with a large plastic bag that has ¼-inch breathing holes cut in about 12 places. See chart for conditions.

CABBAGE, LATE

Cabbage is not harmed by freezing in the field if it's thawed slowly in moist sand in the root cellar and not allowed to refreeze. Late cabbage can be stored effectively in several ways. (1) Roots and any damaged outer leaves are removed and the heads are wrapped closely in newspaper before being put in bins or boxes in an outdoor root cellar (the odor is more noticeable when they are wrapped than when covered with sand or soil). (2) With roots removed, the heads are covered with moist soil or sand in a bin in the root cellar. (3) In pit storage, stem and root are left on and they are placed head-side down. Straw, hay or clean, dry leaves may be packed between the heads for added protection and the whole business covered with soil. (4) The outer leaves are removed and cabbages are hung upside down in a dry place at normal room temperature for several days or until they "paper over." Then they are hung upside down in the root cellar.

Warning: Cabbages have one of the strongest odors of all vegetables, so don't store them where the smell can waft through the house. See chart for conditions.

CARROTS

Carrots may stay in the garden after the first frosts. After digging, handle like Beets. See chart for conditions.

CAULIFLOWER

Another hardy vegetable that can withstand early frosts. Cut off the root and leave plenty of protecting outer leaves; store in boxes or baskets with loose moist sand around and covering the heads. See chart for conditions.

CELERY

Celery should not be stored near turnips and cabbages, which taint its flavor.

Pull the plant, root and all; leave the tops on. Do not wash. Place the roots firmly in moist sand or soil, pressing it well around the roots. Water the covered roots occasionally to keep them moist *but do not water the leaves.*

The procedure for celery may be followed in a trench,

a coldframe-bed, or in a corner of the root-cellar-floor that has been partitioned off to a height of six inches. The closer the celery is stood upright, wherever it's stored, the better. See chart for conditions.

CHINESE CABBAGE

Pull and treat like Celery. See chart for conditions.

DRIED SEED (LIVE)

So long as it is kept quite dry, live seed won't germinate. Store in plastic bags that are then put in a large, mouse-proof covered container; or in canning jars that are wrapped in newspaper to keep out the light. It can't be hurt by natural low temperatures: see chart for conditions.

ENDIVE

Pull as for Celery. Do not trim, but tie all the leaves close together to keep out light and air so the inner leaves will bleach. Set upright and close together with moist soil around the roots, again as for Celery. See chart for conditions.

HORSERADISH

One of the three vegetables that winters-over beautifully in the garden *when kept frozen*. Mulch carefully until the weather is cold enough to freeze it, then uncover to permit freezing and, when it has frozen in the ground, mulch heavily to prevent thawing. For root-cellaring, prepare and handle like Beets. See chart for conditions.

KALE

Treat like Celery right on down the line. See chart for conditions.

KOHLRABI, handle like Beets; see chart.

LEEKS, see Celery and chart.

ONIONS

Pull onions after the tops have fallen over, turned yellow and have started to dry—but examine for thrips (which can cause premature wilting, etc.).

Bruised or thick-necked onions don't store well.

Onions grown from sets are stored in a cool, very dry place on trays made of chicken wire with the tops pointing down through the mesh.

Onions must be conditioned—allowed to "paper over" —in rows in the field; turn them several times so their outsides dry evenly. Smaller amounts may be surface-dried on racks in a dry, airy place under cover; or the tops may be braided and the bunches hung in a dry room. After they are conditioned, trim the tops and hang the onions in net bags or baskets in a dry, airy storage place. See chart for conditions.

PARSNIPS

Actually *improved* by wintering frozen in the garden (and not allowed to thaw), but may be root-cellared if necessary. Treat like Beets or Horseradish. See chart for conditions.

PEPPERS

Careful control of temperature and moisture is imperative in storing peppers (see chart): they decay if they get too damp or the temperature goes below 40–45 F.

Pick before the first frost; sort for firmness; wash and dry thoroughly—handling carefully because they bruise easily.

Put them one layer deep in shallow wooden boxes or cartons lined with plastic in which you cut about 12 ¼-inch holes; close the top of the plastic. Even under ideal conditions the storage life is limited.

POPCORN, see Beans (Shell), dried; see chart.

POTATOES, EARLY

Don't harvest after/during heavy rains, or on a hot day. Dig them carefully early in the morning when the temperature is no more than 70 F. Condition them for 2 weeks at 60–70 F. in moist air to allow any injuries to heal: early potatoes will not heal if they are conditioned in windy or sunny places. After conditioning, store at 60 F. for 4 to 6 weeks. These early varieties do not keep long, and spoil readily held at over 80 F.

POTATOES

Late potatoes are much better keepers than early varieties. Dig carefully. Hold them in moist air about 2 weeks between 60–75 F. to condition: do *not* leave them out in the sun and wind. Put them, not too deep, in crates, boxes or bins stored in a dark indoor or outdoor root cellar; cover to keep away all light (to prevent their turning green, which could mean the presence of selenium, not good in large doses).

After several months' storage, potatoes held at 35 F. may become sweet. If they do, remove them to storage at 70 F. for a week or so before using them. Potato sprouts must be removed whenever they appear, especially toward the end of winter. Early sprouting indicates poor storage conditions. See chart for conditions.

Warning: Potatoes make apples musty, so don't store these two near each other unless the apples are well covered.

PUMPKINS

Harvest before frost, leaving on a few inches of stem. Condition at 80 F. for about 2 weeks to harden the rind and heal surface injuries. Store them in fairly dry air at about 55 F. (see chart). Watch the temperature carefully: too warm, and they get stringy; and pumpkins (and squashes) suffer chill damage in storage below 50 F.— they're not for outdoor cellars or pits. Just because they are big and tough doesn't mean they can be handled roughly, so place them in rows on shelves, not dumped in a pile in a corner.

SALSIFY

The third vegetable (with parsnips and horseradish) that winters-over to advantage in the garden—so long as it remains frozen. If they must be stored, dig them when the soil is dry late in the season but before they freeze. Handle and root-cellar like Beets (and see chart).

SQUASHES, condition and store like Pumpkins—but dried (see chart).

SWEET POTATOES

If a killing frost comes before you can dig them, cut the plants off at soil level, so decay in the vines can't penetrate down into the tubers.

Sweet potatoes are really quite tender, so handle them gently: sort and crate them in the field. Condition at 80–85 F. for 10 days to 2 weeks near a furnace or a warm chimney, maintaining high humidity by covering the stacked crates (which have wooden strips between for spacing) with plastic sheeting or a clean tarpaulin, etc. Then store in fairly dry and warm conditions (see chart). Like Pumpkins and Squashes, they damage from chill below 50 F.

TOMATOES, MATURE GREEN

For storing, harvest late but before the first hard frost, and only from vigorous plants. Wash gently, remove stems, dry; sort out all that show any reddening and store these separately.

Pack no more than two layers deep with dry leaves, hay, straw or shredded paper (plastic bags with air-holes are more likely to cause decay). Sort every week to separate faster-ripening tomatoes. See the chart.

TURNIPS

These and rutabagas withstand fall frosts better than most other root crops, but don't let them freeze/thaw/freeze. Storage odor can penetrate up from the basement, so store them by themselves outdoors (see chart for conditions).

Handle like Beets; pack in moist sand, peat, etc.

WINTER RADISHES, handle like Beets; see chart.

Curing 7

Salting and Smoking

Curing—that is, impregnating with a heavy concentration of salt—is almost as old as drying is, and, because it isn't dependent on climate, it has been used even more widely throughout the world as a means of putting by certain foods.

"Certain foods" again— Here are the limitations.

Obviously, only those foods whose flavor is compatible with salt may be cured in the first place.

And all such foods are simply not edible unless most of the salt is washed out—along with a good deal of some of the nutrients—before they are cooked not enough salt is used to stop spoilage all by itself, the food must have something further done to it to make it safe to eat, or even make it seem appetizing. This additional handling involves simple refrigeration, or fermenting and refrigeration, or smoking and refrigeration, or canning or freezing, depending on how long you hope to keep it.

Salting

WHAT SALTING DOES

A concentrated brine—which is salt + juice drawn from the food by the salt (called "dry salting"), or salt + water if juice is limited or not easily extracted (called "brining")—cuts down the activity of spoilage microorganisms in direct relation to the strength of the solution. The following general proportions give the idea (with percentages reflecting ratios by weight of salt to water, not sophisticated salinometer readings; for more about Salt, see Chapter 1).

A 5 percent solution (1 pound, or 1½ cups, of salt to 19 pints of juice/water) *reduces* the growth of most bacteria.

A 10 percent solution (1 pound, or 1½ cups, of salt to 9 pints of juice/water) *prevents* the growth of *most bacteria.*

A solution from 15 percent (1 pound, or 1½ cups, of salt to 5½ pints of juice/water) to 20 percent (1 pound, or 1½ cups, of salt to 4 pints of juice/water) *prevents* the growth of *salt-tolerant bacteria.*

The amounts of salt given in the individual instructions are designed to give the necessary protection to the food being cured, provided that any further safeguards are followed as well. Sometimes a brine is added to make sure that enough liquid is present to carry out the curing process, because you can't add plain water without diluting the strength of the salt required to treat the particular food satisfactorily.

EQUIPMENT FOR CURING (SALTING)

Especially for vegetables:

Large stoneware crocks or jars (5-gallon size is good here)

OR

The biggest wide-mouth canning jars you can get— or ask the high-school cafeteria or your friendly neighborhood snack bar for empty gallon jars (wide-top) that their mayonnaise or pickles came in

OR

Sound, unchipped enamelware canner (if you can spare it) with lid

Vegetable grater with a coarse blade; large old-style wooden potato-masher

Safe storage area at *c.* 65–70 F. for fermenting vegetables; plus cooler—*c.* 38 F.—storage for longer term

(Continued on the next page)

Especially for meats and fish:

Large stoneware crocks (10-gallon or larger)
> OR
Wooden kegs or small barrels—new, or thoroughly scrubbed and scalded used ones (before curing in them, though, fill them with water to swell the staves tight together, so the containers won't leak when they're holding food)
Moisture-vapor-proof wrappings; plus stockinet—tubular cotton-knit—for holding the wrap tight to the meat after it's packaged
Safe, cold storage area (ideally 36–38 F.) for curing meats and fish—and for longer-term storage of meats and vegetables in their curing solutions

For both vegetables and meats, etc.:

Cutting-boards and stainless-steel knives (see Canning Meats)
Large enameled or glass/pottery pans or bowls for preparing the curing mixtures
Big wooden spoons, etc., for mixing and stirring
China or untreated hardwood covers that fit down inside each curing container: an expendable plate, a sawed round, etc.
Weights for these covers, to hold the food under the curing brine—a canning jar filled with sand is good; but nothing of limestone or iron, which mess up the curing solutions
Plenty of clean muslin (old sheets do beautifully) or double-weight cheesecloth
Glass measuring cups in 1-cup and 4-cup sizes
Scale in pounds (up to 25 is plenty, with ¼- and ½-pound gradations)
Good-sized working space, particularly for dealing with meats

Because salt draws moisture from plant and animal tissues in proportion to its concentration, heavy salting is often a preliminary step in drying or smoking high-protein foods.

USE PURE MEDIUM-COARSE PICKLING SALT FOR ALL CURING

NEVER USE "FREE-FLOWING" OR IODIZED TABLE SALT FOR CURING

Salting Vegetables

Unless you're fermenting vegetables—as for sour cabbage (sauerkraut), etc., below—there's only one reason for salting them: you have no other way to put them by, so you either salt your vegetables now or do without vegetables later.

Vegetables are brined whenever they don't release enough natural juice to form adequate liquid during their cure, i.e., when they are not cut small or when they have little natural juice to start with.

Juicy or finely cut vegetables are dry-salted. If a relatively small amount of salt—2½ percent by weight —is added to certain vegetables, they ferment to make a "sour" product (sauerkraut again); but if 25 percent salt by weight is added to these or other prepared vegetables, the high concentration of salt prevents the growth of the yeasts and bacteria that cause fermentation, and thereby preserves them.

DRY SALTING TO PRESERVE

Corn, green/snap/string/wax beans, greens, even cabbage and Chinese cabbage and a number of root vegetables may be dry-salted, but some are more interesting— and certainly more nutritious, since freshening isn't necessary before cooking—if they are fermented instead.

It's not good to add newly prepared vegetables to any already curing, so use a container that will hold the batch you're working with. Just be sure to allow enough headroom to keep the food submerged under the weight: 4 inches should be enough for a 5-gallon crock, less for a smaller one; and the weight can rise above the rim of the container.

Salted Sweet Corn

Select sweet corn in the milk stage as you'd choose it for serving in season as corn-on-the-cob. Husk, remove the silk, and steam it for 10 to 15 minutes over rapidly boiling water or until the milk is set. Cut it from the cob about ⅔ the depth of the kernels (as for whole-kernel processing); weigh it. Mix 4 parts of cut corn with 1 part salt—1 pound of pure pickling salt for each 4 pounds of corn; or 1 cup of salt to 4 cups of cut corn if you don't have a scale.

Pack the corn-salt mixture in a crock to within about 4 inches of the top, cover with muslin sheeting or a double thickness of cheesecloth, and hold the whole business down with a clean plate or board on which you place a weight. If there isn't enough juice in 24 hours to cover the corn, add a salt solution in the proportions of 3 tablespoons salt to each 1 cup of cold water; replace the weighted plate to submerge the corn.

Store the crock in a safe, cool place (about 38 F.). The corn will be cured in from 3 to 5 weeks. Remove meal-sized amounts by dipping out corn and juice with a glass or china cup (don't use metal). Change the cloth as it becomes soiled, and always replace the weighted plate. Keep the crock in cool storage.

To cook the corn, freshen it (soak in cold water a short time, drain, and repeat) until a kernel tastes sweet. Simmer until tender in just enough water to prevent scorching; serve with butter or cream and seasoning to taste.

Salted Green/Snap/String/Wax Beans

Use only young, tender, crisp beans. Wash, remove tips and tails; cut in 2-inch pieces or french. Steam-blanch 10 minutes and cool. Weigh the beans, and measure 1 pound (1½ cups) of pure pickling salt for every 4 pounds of beans. Sprinkle a layer of salt in the bottom of a crock, add a layer of beans; repeat until the crock is filled to within 4 inches of the top or until the beans are used; top with a layer of salt. Cover with clean muslin sheeting or doubled cheesecloth and hold down with a weighted plate. If not enough brine has formed in 24 hours to cover the beans, eke it out with a solution in the proportions of 3 tablespoons of salt to each 1 cup of cold water.

Proceed as for Salted Sweet Corn, above.

Salted Dandelion (or other) Greens

Green salads were a rarity with New England hill folk in the early nineteenth century; nor did they go in for leaf vegetables much, except for dandelions in early spring and beet or turnip tops from their gardens in late summer.

Sometimes they salted their greens according to the 1-to-4 rule. Nowadays we'd go them one better, though, and steam-blanch the washed, tender leaves until they wilt —from 6 to 10 minutes, depending on the size of the leaves. Cool the greens, weigh them, and layer them in a crock with 1 pound (1½ cups) of pure pickling salt for every 4 pounds of greens. Proceed as for Salted Sweet Corn and Salted Green Beans, above.

To cook, rinse well and freshen in cold water for several hours, rinse again, drain, and simmer gently in the water adhering to them. Season with small dice of salt pork cooked with them, or serve with vinegar.

Salted Rutabagas (or White Turnips)

Use young, crisp vegetables without any woodiness. Peel; cut in ½-inch cubes. Steam-blanch from 8 to 12 minutes, depending on size of the pieces; cool. Weigh the prepared turnips and proceed with the 1-to-4 rule—1 pound (1½ cups) of pure pickling salt for each 4 pounds of turnips—and handle thereafter like Salted Sweet Corn, above.

To cook, rinse and freshen for several hours in cold water, rinse again; then simmer until tender in just enough water to keep from scorching. Mash if you like, and serve with butter and seasoning to taste.

Salted Cabbage

Remove bruised outer leaves; quarter, cut out the core. Shred as you would for cole slaw. Steam-blanch for 6 to 10 minutes until wilted. Cool, weigh; follow the 1-to-4 rule for Salted Greens above, and continue with the cure.

To cook, rinse and freshen for several hours, rinse again; then simmer until tender in just enough water to prevent scorching. Season during cooking with 2 teaspoons of vinegar and ¼ teaspoon caraway; or drain and return to low heat for 3 minutes with crumbled precooked sausage or small dice of salt pork; or serve with butter and seasoning to taste.

DRY SALTING FOR "SAUERKRAUT" VEGETABLES

Most often fermented are cabbage (sauerkraut) and Chinese cabbage, and rutabagas or white turnips.

Generally speaking, the sweeter vegetables make a more flavorful product, while firmer ones provide better texture. Don't relegate tough, old, woody vegetables to the souring crock—use the best young, juicy ones you can get.

If you feel like experimenting with a small batch (5 pounds, say, in a 1-gallon jar, or less in a smaller container) you could add with the salt the traditional German touches of caraway or dill; or try a bay leaf or two, or some favorite whole pickling spice, or some onion rings, or even a few garlic cloves, peeled (but whole, so you can fish them out before serving).

Some rules advocate starting fermentation with a weak brine, but this procedure offers a loophole for too low a concentration of salt, and the likelihood of mushy food or even of spoilage instead of the desired acidity. Unless you're an old hand with sauerkraut and its relatives, you'll do well to stick to dry salting here.

As with vegetables preserved with salt earlier, you should never mix a fresh batch with one already fermenting.

Produce to be soured is not blanched: you want to encourage the micro-organisms that cause fermentation.

For fermenting you use 1/10 the amount of salt you needed for the preserving just described. This means 2½ percent of pure pickling salt by weight of the prepared food: 10 ounces (15 tablespoons or a scant 1 cup) of salt to 25 pounds of vegetables; 4 ounces (6 level tablespoons) of salt for 10 pounds of vegetables; 2 ounces (3 level tablespoons) of salt for 5 pounds of prepared food. This ratio of salt turns the sugar in the vegetables to lactic acid, and the desired souring occurs.

The vegetables should be kept between 68–72 F. during the fermenting period, which takes from 10 days to 4 weeks, depending on the vegetable being processed. Temperatures below 68 F. will slow down fermentation; above 72 F., and you court spoilage.

As a rough estimate, allow 5 pounds of prepared vegetables for each 1 gallon of container capacity, with the crock/jar holding a slightly greater weight of dense food that's cut fine. The instructions below use 10-pound batches, but you may want to deal with 25 or 30 pounds of cabbage or turnips at a time, using a 5-gallon crock.

Keep all souring vegetables covered with a clean cloth and weighted below the brine during fermentation. A top-quality vegetable should release enough juice to form a covering brine in around 24 hours; if it hasn't, bring the level above the food by adding a weak brine in the proportions of 1½ teaspoons of pickling salt for each 1 cup of cold water.

By the second day a scum will form on the top of the brine. Remove it by skimming carefully; then replace the scummy cloth with a sterile one, and wash and scald the plate/board before putting it back and weighting it.

Take care of this scum every day, and provide a sterile cloth and plate every day; otherwise the scum will weaken the acid you want, and the food will turn mushy and dark. If the brine gets slimy from too much warmth it's best not to tinker with it: do the simplest thing and decant the batch on the compost pile—and wait until cooler weather to start over again.

Fermentation will be continuing as long as bubbles rise to the top of the brine. When they stop, remove the cloth and weighted plate, wipe around the inside of the head-room; cover the vegetable with a freshly scalded plate/board, and put a close-fitting lid on the container. Then store the whole thing in a cool place at *c*. 38 F.

Dip out with a glass or china cup what you need for a meal, making sure that enough brine remains to cover the vegetable so it won't discolor or dry when exposed to air. Always keep the container well closed.

Fermented vegetables may be canned. See individual instructions below.

Sauerkraut (Fermented Cabbage)

Quarter each cabbage, cut out the core; shred fine and weigh. Using 2½ percent of pickling salt by weight—6 tablespoons to each 10 pounds of shredded cabbage (or see other amounts in the general method above)—pack the container with alternate layers of salt and cabbage, tamping every two layers of cabbage to get rid of trapped air and to start the juice flow—you don't need to get tough with it: just tap it gently with a clean wooden potato-masher or the bottom of a small jar. Top with a layer of salt. Cover with a sterile cloth and weight it down with a plate, etc.; hold at 68–72 F. while it ferments.

Follow the daily skimming procedure given above. When fermenting has stopped in about 2 weeks or so, the sauerkraut will be a clear, pale gold in color and pleasantly tart in flavor. It's a good idea to lay a clean plate on it to keep it below the brine's surface; at any rate cover the container with a close-fitting lid. Store in a cool place and use as needed.

If your storage isn't around 38 F., you'd better can it (q.v.).

Chinese Cabbage Sauerkraut

Follow the method for Sauerkraut. The result usually has more flavor than regular fermented cabbage does, thanks to more, and sweeter, natural juice.

Sour White Turnip (Sauer Rüben)

Peel and quarter young rutabagas or white turnips (rutabagas are usually firmer and juicier than turnips). Shred fine or chop with medium knife of a food grinder, catching stray juice in a bowl placed underneath. Pack with layers of salt as for Sauerkraut, but do not tamp down—there should be juice aplenty without tamping, and it's enough to press down on the topmost layer to settle the pack.

Proceed in every way as for Sauerkraut.

Sour Rutabagas, handle like White Turnips

Souring Other Lower-acid Vegetables

Even though properly created fermentation raises the *pH* acidity rating of lower-acid raw vegetables, *unless they are heat-processed for storage* they cannot be regarded as safe from spoilage or from growth of certain dangerous heat-sensitive bacteria.

So, because you should can them anyway for safe storage, it doesn't make much sense to go through the business of fermenting them as a preamble to putting them by for serving later as accompaniments to meat or whatever.

Especially when there are recipes for relishes and pickles made from corn, pumpkin, zucchini and beets in Chapter 4.

Especially when, if you're strapped for canning jars, the chances are that you can rig up some kind of heat to dry them with (see Drying).

BRINING TO PRESERVE VEGETABLES

This is not the same thing as the preparation for making pickles, which is designed either to crisp the ingredients or to season them as a base for adding vinegar, sugar, spices, etc.

What you're doing here is to reproduce the 25 percent, 1-to-4, rule for Dry Salting to Preserve, above—but you're making up for the scarcity of natural juice by using water with a heavy concentration of salt.

It's easier to gauge the amounts of brine and salt needed if you think in terms of 10 pounds of whole vegetables (or large pieces of cut ones) held in a 5-gallon container: there'll be extra room, but that doesn't matter —you'll weight the vegetables to keep them under the surface of the cure. If you'd rather work in 5-pound lots, use a 2-gallon container and halve the given amounts of brine and salt.

Weigh out 10 pounds of fresh, perfect vegetables, and to them add 2 gallons of brine of the strength given for the individual vegetable; this much brine should cover them, but make a bit more if it doesn't. Cover the brined vegetables with clean muslin or doubled cheesecloth, and on it lay a weighted plate to keep them submerged.

The next day you start gradually to increase the salt in the solution, thus compensating for the natural juice that's drawn out to weaken the brine.

First, for each 10 pounds of vegetable, pour 1 pound (1½ cups) of pickling salt on the wet cloth where it will dissolve slowly into the brine; replace the weight on the mound of salt.

One week later, put 4 ounces (6 tablespoons) more salt on the cloth to absorb. Repeat once a week for 3 or 4 weeks more—making a total of 1 to 1¼ pounds of salt added in the weekly doses.

The vegetable is cured in 4 to 5 weeks. Take away the cloth, cover the container with a close-fitting lid, and store at an ideal 38 F.

To cook, freshen the vegetable in several cold waters, and simmer in a very little water until tender, seasoning to taste.

Brined Green Peppers

Halve firm, crisp peppers, remove their seeds; weigh. For 10 pounds of peppers make 2 gallons of strong brine —around a 15 percent solution—by dissolving 4⅓ cups of pure pickling salt in 8 quarts of water. Pour it over the peppers in a 5-gallon crock (it should cover them, but make a bit more if it doesn't). Cover with cloth and weight down. About 24 hours later, put your first addition of salt on the cloth, carrying on with the general method described above.

Brined Cauliflower

Wash heads of fresh young cauliflower, remove leaves and core, and break apart the flowerets in c. 1-inch pieces. Continue as for Brined Green Peppers, above.

Brined Onions

Peel whole, fresh onions, not large ones—and not grown from sets (these can have an inner core wrapped in brown "paper"). Weigh. For every 10 pounds of onions prepare 2 gallons of very strong brine—around a 20 percent solution—by dissolving 4 pounds (6 cups) of pure pickling salt in 8 quarts of water.

From here on, follow the method for Brined Green Peppers, above.

Salting Meat

The four keys to successful salt-curing of any meat are (1) strictly fresh meat to start with, properly handled and chilled; (2) sanitation; (3) temperature control; and (4) salt content. The same quality, cleanliness and care required for canning or freezing meat (q.v.) obtain in the procedures described below, and we give in the specific instructions the exact proportions of salt required to do each job.

However, temperature control demands special emphasis here. The meat must be kept chilled—held as constantly as possible at 38 degrees Fahrenheit—before curing; this is why country-dwellers wait for winter weather to slaughter hogs and beeves for their own tables. Once in the cure, meat should be held between 36 and 38 F.; for the largest pieces this means a thermometer inserted to the center of the meatiest part.

Below 36 F., salt penetrates the tissues too slowly. If the temperature of the storage area drops below freezing and stays there for several days, *increase the days of salting time by the number of freezing days.*

Above 38 F., the chances of spoilage increase geometrically with each degree of rise in temperature, and the cure changes from a clear, fresh liquid to a stringy-textured goo. It is the rare modern home that has natural storage constantly cool enough for curing meat right. Indeed, failure to ensure good temperature control is the main cause of unsuccessful curing in town and country alike.

In general, home-frozen meats do not cure well: even when defrosted completely, their texture has been changed too much to allow the cure to penetrate the tissues uniformly.

The term "pickle" is used in some manuals to designate a sweetened brine that contains some sugar as well as salt; it is *not* the solution with added vinegar that is described for pickles in Chapter 4. "Sugar cure" usually means adding ¼ as much sweetener as there is salt in the mixture; this amount of sugar is important as food for benign flavor-producing bacteria during long cures.

Salt-curing of meats is almost always followed by exposure to smoke in order to dry the surface of the meat, to add flavor, and to discourage attacks by insects. Smoking procedures are described in detail in the section following this one.

Some valuable references and pictures about cutting meat can be found in Chapter 2 under "Cuts of Meat from Large Animals."

THE NITRATE/NITRITE WORRY

A number of the procedures below for curing meat include the words ". . . ounces of saltpeter—*optional.*"

We include the saltpeter because it's in the recipes in concentrations which, at this writing, have been passed by the Food and Drug Administration as being safe for human begins who eat these foods in normal amounts.

BUT we add *optional* because there's increasing protest over the use of this substance in commercially prepared meats like cold cuts, hot dogs, ham and bacon, etc.: and if you wouldn't dream of buying foods that contain it, then there's no earthly sense in adding it to food you put by at home.

The nitrates are currently under intense scrutiny by the FDA and the USDA, which have set up a government/industry task force to evaluate the problem they raise. Which is: most of the nitrate converts—in the tissues of the food and in our bodies—to a nitrite, which is known to combine with the red blood cells, and, if in sufficient quantity, could limit the oxygen-carrying power of the hemoglobin—thus hindering vital bodily functions.

What saltpeter is and why it's used

"Saltpeter" is either potassium nitrate or sodium nitrate (the latter often called "Chilean nitrate"). If you ask for saltpeter at the drugstore, you're likely to get potassium nitrate (labeled as a diuretic, actually), which is preferred over sodium nitrate because it doesn't absorb moisture during storage the way the sodium compound does. Only once in both the USDA meat-processing bulletins mentioned above is "saltpeter" defined—and then as potassium nitrate.

But if you buy a commercial ready-to-use-at-home mixture for curing your meat, the chances are that it will contain sodium nitrate (along with sodium nitrite and other preservatives, principally salt, NaCl).

For centuries saltpeter has been added to curing mixtures for meat as a means of intensifying and holding the red color considered so appetizing in ham and allied pork products, and in corned beef, etc. Its anti-microbial action is so incidental that we can declare: *the most important substance limiting bacterial growth in cured meat is SALT*.

So when saltpeter is included in a rule for home-cured meat it is being added for cosmetic reasons.

And whether you add it is up to you.

So what do we use instead?

Still, for many people, color is psychologically important in food.

Therefore we suggest that you *use pure crystalline ascorbic acid* (Vitamin C) instead of saltpeter, because pure ascorbic acid can't do you an iota of harm. Use it *at the rate of 0.05 percent of the weight of meat*, which works out to 1 gram of crystalline ascorbic acid for every 4.5 pounds; or rounding it off, to ¼ teaspon for 5 pounds, ½ teaspoon for 10 pounds, 1 teaspoon for 20 pounds, and so on. Add the crystalline ascorbic acid to the dry cure or the brine.

Ascorbic acid will improve the red color, and retard the loss of color as time goes by. But, unlike the nitrates, it will not protect the color indefinitely; and this is a backhanded comfort, too: it is highly unlikely that color fading would be delayed by ascorbic acid to such an extent that the meat would be spoiled by bacteria and *still look appetizing.*

Some commercially prepared foods contain lecithin as a means of retarding rancidity. *We do not recommend lecithin,* for the simple reason that we know nothing about using it in home-processed foods, and neither does anyone else that we can discover.

A final parenthesis: It's not fair to lay blame for using a possibly dangerous substance entirely at the door of what has been called "the establishment." We were beguiled by an advertisement in a leading publication in the natural/organic food movement. When we got the material we'd sent (and paid) for, we found that it included a recipe for home-made luncheon meat done in what the author called "the natural way"—*and it included, as a seasoning, eight times the amount of saltpeter as is used in a comparable USDA recipe.* P.S. In our version of this Belogna-style Sausage—it's in Chapter 10—we don't have any saltpeter. The sausage may not be a pretty pink, but it's mighty good eating. (This would be a good place to try ascorbic acid.)

STORING CURED MEAT

Their heavy concentration of salt protects Corned Beef and Salt Pork for several months if the brine in which they're held is kept below 38 F.

Freezer storage of sausage and cured meats is relatively limited: after more than 2 to 4 months at Zero F., the salt in the fat causes it to become rancid.

Note of warning. A home-cured ham is not the same as a commercially processed one that has been "tenderized," etc. *Home-cured pork is still RAW.*

SALTING BEEF

Because they lack what producers and butchers call "finish," veal or calf meat shouldn't be used to make corned or dried beef. The product is disappointing.

Corned Beef

Use the tougher cuts and those with considerable fat. Bone, and cut them to uniform thickness and size.

To cure 25 pounds of beef, pack it first in pickling salt, allowing 2 to 3 pounds of salt (3 to 4½ cups) for the 25 pounds of meat. Spread a generous layer of coarse pickling salt in the bottom of a clean, sterilized crock or barrel. Pack in it a layer of meat that you've rubbed well with the salt; sprinkle more salt over the meat. Repeat the layers of meat and salt until all the meat is used or the crock is filled to within a couple of inches below the top.

Let the packed meat stand in the salt for 24 hours, then cover it with a solution of 1 gallon of water in which you've dissolved 1 pound (2 cups) of sugar, ½ ounce (*c.* 1 tablespoon) of baking soda, and 1 ounce (*c.* 2 tablespoons) of *optional* saltpeter—or 1¼ teaspoons of pure crystalline ascorbic acid (also optional).

Put a weighted plate on the meat to hold every speck of it below the surface of the brine; cover the crock/barrel; and in a cool place—not more than 38 degrees F. —let the meat cure in the brine from 4 to 6 weeks.

The brine can become stringy and gummy ("ropy," in some descriptions) if the temperature rises above 38 F. and the sugar ferments. The baking soda helps retard the fermentation. But watch it: if the brine starts to get ropy, take out the meat and wash it well in warm water. Clean and sterilize the container. Repack the meat with a fresh sugar-water-etc. solution (above), to which you now add 1½ pounds (2¼ cups) of pickling salt; this salt replaces the original 2 pounds of dry salt used to pack the meat.

To store it, keep it refrigerated in the brine; or remove it from the brine, wash away the salt from the surface, and can or freeze it (q.v.).

Dried (Chipped) Beef

Dried beef—which has about 48 percent water when produced commercially—is made from whole muscles or muscles cut lengthwise. Select boneless, heavy, lean-muscled cuts—rounds are best—and cure as for Corned Beef (above) *except* that you add an extra ¼ pound of sugar (½ cup) for each 25 pounds of meat.

The curing is completed in 4 to 6 weeks, depending on size of the pieces and the flavor desired. After it has cured satisfactorily, remove the meat, wash it, and hang it in a cool place to air-dry for 24 hours.

Then it is smoked at 100 to 120 F. for 70 to 80 hours (see Smoking)—or until it is quite dry.

To store, wrap large pieces in paper and stockinette (tubular, small-mesh material, which holds the wrap close to the meat) and hang them in a cool (below 50 F.), dry, dark, insect-free room; certainly refrigerate small pieces. Plan to use all the dried beef before spring.

SALTING PORK

All parts of the pig may be cured by salting. Some—such as the fat salt pork for baked beans, chowders, etc.—are used as they come from the salting process. The choice hams, bacon and, perhaps, loins are carried one step further and are smoked (q.v.) after being cured.

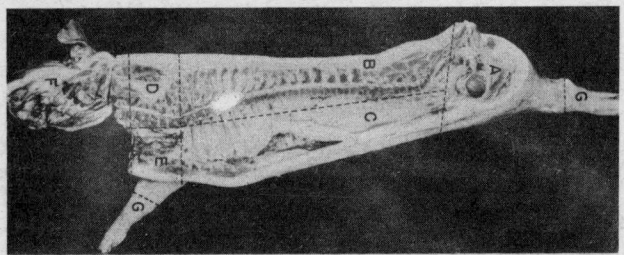

Parts of a pig: A—ham; B—loin and backfat; C—bacon strip with leaf fat; D—shoulder butt and plate; E—picnic shoulder; F—head; G—feet.

Have the meat thoroughly chilled, and hold it as closely as possible to 38 F. during the process of curing: salt penetrates less well in tissues below 36 F., and spoilage occurs with increasing speed in meat at temperatures above 38 F.

"Pumping"—i.e., forcibly injecting a strong curing solution into certain parts of a large piece of meat—is not included in the instructions below because we're leary of it: much safer to allow safe curing time than to try to speed the process by localized "spot" applications of the cure.

Allow 25 days as the minimum curing time for Dry Salted pork, with some of the larger pieces with bone taking longer. Allow at least 28 days for Sweet Pickled (Brined) pork, and more for the larger pieces. The days-per-pound are given for each cut cured by each method.

Before smoking or storing large pieces containing bone, run a skewer up through the meat along the bone, withdraw the skewer and sniff it. If the odor is sweet and wholesome, fine—proceed with the smoking or storing; but if there's any "off" taint, any whiff of spoilage, destroy the entire piece of meat, because it is unsafe to eat.

Dry Salting Large Pieces (Hams and Shoulders)

For each 25 pounds of hams and shoulders mix together thoroughly 2 pounds (*c.* 3 cups) of coarse-fine pickling salt, ½ pound (*c.* 1 cup) sugar—and an *optional* ½ ounce (*c.* 1 tablespoon) of saltpeter *or* 1¼ teaspoons crystalline ascorbic acid. Rub ½ the mixture in well on all surfaces of the meat. Poke ½ generously into the shank ends along the bone (you can even make a fairly long internal slit with a slender boning knife inserted at the shank, and push the mixture up into it: this is better than relying on "pumping" a strong solution to such areas where the salt must penetrate deeply). Plan to leave an ⅛-inch layer of the mixture on the ham face (the big cut end), with a thinner coating on the rest of the ham and on the shoulders.

Fit the salt-coated meat into a clean sterilized barrel or crock, taking care lest the coating fall off. Cover with a loose-fitting lid or cheesecloth and let cure in a cold place, 36 to 38 F.

One week later, remove the meat, re-coat it with the remaining half of the curing mixture, and pack it again in the barrel/crock.

Curing time: At least 25 days. Allow 2 or 3 days for each 1 pound of ham or shoulder, being sure to leave them in the curing container even after all surface salt is absorbed.

Then smoke them (q.v.).

Heavy salting for heavy cuts, left; then into the barrel/crock to cure.

Dry Salting Thin Cuts (Bacon, "Fat Back," Loin, etc.)

For each 25 pounds of thin cuts of pork, mix together thoroughly 1 pound (*c.* 1½ cups) of pickling salt, ¼ pound (½ cup) of brown or white sugar, and 1½ teaspoons of saltpeter—*optional: see "Nitrate Worry," above*—or 1¼ teaspoons of ascorbic acid.

Coat the cuts, using all the mixture. Pack the meat carefully in a sterilized crock or barrel, and cover it with a loose-fitting lid or layer of cheesecloth, and let it stand at 36 to 38 F. for the *minimum* total curing time of 25 days; allow 1½ days per pound. Thin cuts do not require an interim salting—that's why you used all the mixture in the first place. And leave them in the crock even after the surface salt has been absorbed.

All but the "fat back" (Salt Pork) is then smoked (q.v.). Wrap the Salt Pork in moisture-vapor-proof material; refrigerate what is intended for immediate use, and freeze the rest.

"Sweet Pickle" Salting Large Pieces (Brining Hams, etc.)

Curing hams and shoulders in brine is slower than the Dry Salting treatment just described, and therefore is well suited to colder regions of the country.

Pack the well-chilled (38 F.) hams and shoulders in a sterilized crock or barrel. For every 25 pounds of meat, prepare a solution of 2 pounds (*c.* 3 cups) of pickling salt, ½ pound (1 cup) of sugar, ½ ounce (*c.* 1 tablespoon) of saltpeter—*optional: see "Nitrate Worry" above*—and 4½ quarts of water. Dissolve all thoroughly, and pour over the meat, covering every bit of it: even a small piece that rises above the solution can carry spoilage down into meat submerged. Put a weighted plate or board over the meat to keep it below the brine, and cover the barrel/crock. Hold the storage temperature to 38 F.

After 1 week, remove the meat, stir the curing mixture, and return the meat to the crock/barrel, making sure that every bit of it is weighted down below the surface of the brine.

Remove, repack, and cover with the stirred brine at the end of the second and fourth weeks.

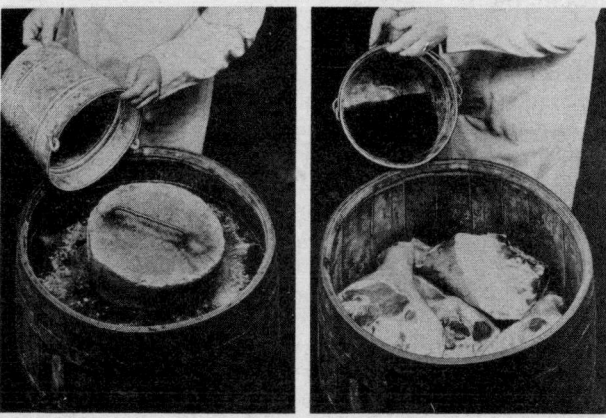

Cover with cold brine, then weight it all down below the surface.

If at any time during the cure you find that the brine has soured or become ropy or sirupy, remove the meat, scrub it well, and clean and scald the barrel/crock. Chill the container thoroughly, and return the meat, covering it with a fresh, cold curing solution made like your original brine, except that you *increase the water to 5½ quarts.*

Curing time: A minimum of 28 days; allow 3½ to 4 days for each 1 pound of ham or shoulder.

"Sweet Pickle" Salting Small Pieces
(Bacon, Loin, "Fat Back")

Pack the pieces in a sterilized crock or barrel, and cover with a brine like that for large pieces, except in a milder form: use 6 quarts of water, rather than 4½ quarts.

Proceed as for hams and shoulders, keeping the pieces well submerged, and overhauling the contents as above at the end of the first, second, and third weeks.

Curing time: A minimum of 15 days for a 10-pound piece of bacon, allowing 1½ days per pound; but 21 days for heavier pieces of bacon, or for the thicker loins.

Pork that is not to be smoked may be left in the brine until it is to be used—but it will be quite salty.

Smoking

Without intending either to pun or to discuss the pro/con of this traditional finishing process for many cured meats and a few cured fish, we feel duty-bound to note that smoking any food is under fire nowadays from some critics.

However, as we indicated in the individual salting instructions above, meats may be left in brine or dry salt until they're ready to be used. Or remove them from the cure, scrub them well to remove surface salt, and hang them in a cool, dry, well-ventilated place for from several days to a week to let them dry out a bit before storage.

We do not recommend using so-called "liquid smoke" or "smoke salt" in place of bona fide smoking. Either smoke your meat or call it a day at the end of the salt cure.

WHAT "COLD SMOKING" DOES

We're not concerned here with what is known as "hot smoking"—which in effect is cooking in a slow, smoky barbecue for several hours, thus making the food partially or wholly table-ready at the end of the smoking period.

What we'll do is hold the food in a mild smoke at never more than 120 F., and usually from around 70 to 90 F., for several days to color and flavor the tissues, help retard rancidity and, in many cases, increase dryness— the actual length of time depending on the type of food.

The food is then stored in a cool, dry place, or is frozen, to await future preparation for the table.

MAKING THE SMOKE

Use only hardwood chips for the fire—never one of the evergreen conifers, whose resinous smoke can give a creosote-y taste, or other softwoods. Among the most popular woods are maple, apple and hickory.

Or use corncobs. These should be the thoroughly dried cobs from popcorn or flint corn that has dried on the ear: cobs saved from a feast of sweet corn-on-the-cob aren't the same thing at all. If you don't have your own cobs or can't get them from a neighboring farmer, look in the Yellow Pages for a handler of hardwood sawdust or shavings; such a dealer often has chopped cobs to use as a tumbling medium for polishing. Merrill Lawrence of Newfane, Vermont—whose family has been curing and smoking meat for generations—says 2 bushels of cut corncobs can produce 72 hours of smoke, or enough to do a whole ham in a small smoke-box.

Avoid chemical kindlers, either fluids or small bricks impregnated with flammable mixtures, because their fumes can take a long time to dissipate (you don't want your meat to taste of them). Small, dry hardwood pieces laid te-pee-fashion over crumpled pieces of milk cartons catch well, and form a good base for the fire. Get your fire well established and burning clean, but do not have it hot; keep it slow, just puttering along evenly so the meat is in no danger of cooking. Hang a thermometer beside the food closest to the fire: fish, which is so highly perishable (even when lightly salted for smoking), should be smoked at 40 to 60 F., and then for a relatively short time compared to the temperature for meats.

The fire can be made and held in any sort of iron or tin brazier suitable for the size of the smokehouse or box. If you use sawdust or fine chips or chopped corncobs, the smoke might also be maintained well enough by using an electric hot plate to fire a tin pie pan that's filled with the smoke-making material. Set the hot plate on High to start the pan of stuff smoldering, then reduce the heat to Medium or Low. Experiment.

No matter which smoke-making fuel you use, it's important to know how it burns, and how much air intake you need to keep it going or to quiet it down for the type of smoker you have, before you commit a batch of cured meat to the smoking process.

Smoking Meat

Because bacteria in meat grow fastest between 70 and 100 F., you should smoke meat in fairly cold weather, in late fall or early spring, when temperatures are between 30 and 50 F. during the day. However, really cold weather, down to Zero, is not for the beginner who's using the highly simplified smoke-boxes described below.

Smoking should be as sustained as is reasonably possible, simply because you want to get it over with and get the meat cooled and wrapped in moisture-vapor-proof material and stored in a cool, dry place (or frozen). But it won't suffer from the hiatus if you can't smoke at night: the weather will probably keep it cool enough without freezing so you can leave it in the smokehouse and just start your smoke-maker again in the morning.

If you have a sudden sharp drop in temperature, though, you had better bring inside to cool storage any meat that shows danger of freezing without the warmth of the smoke. Resume counting the total smoking time when the smokehouse is operating again.

PREPARING THE MEAT
FOR THE SMOKEHOUSE

Remove the meat from the salting crock, scrub off surface salt, using a brush and fresh lukewarm water. Then hang the meat in a cool, airy place for long enough to get the outside of it truly dry—up to 24 hours.

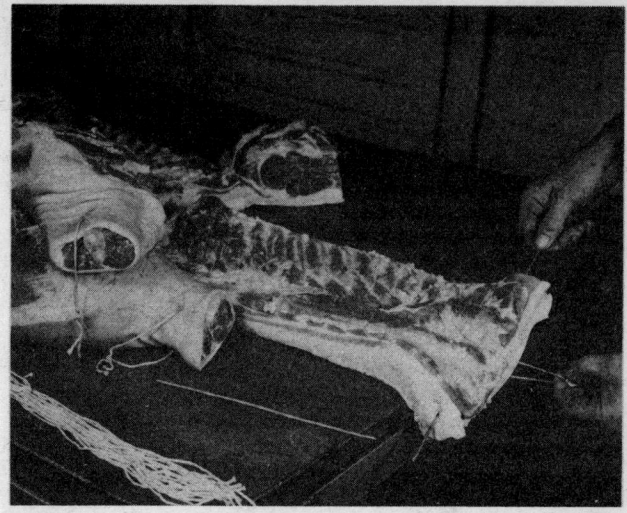

Run several thicknesses of bailing twine or a strong stainless-steel wire through each piece of meat several inches below one end; tie the string or double-twist the wire to form a loop that will hold the weight of the meat. Hams are hung from the shank (small) end. It's a nuisance to have the meat fall to the bottom of the smoke-box because either the loop or the meat has given way.

SMALL HOMEMADE SMOKE-BOXES

The USDA bulletin on processing pork, cited earlier, contains plans for a full-fledged smokehouse, and also a description of a barrel smoke-box. Here are our variations of the barrel, and a description for making use of a discarded refrigerator.

A barrel

You can get the smoking parts of half a 200-pound pig a 55-gallon steel barrel that you make into a "smokehouse. This means that one ham, one shoulder, and one side of bacon cut in pieces can be smoked at the same time (assuming that you've put a jowl and two feet into Head Cheese, q.v.).

Wooden barrels large enough to do the job are (1) hard to come by these days, and (2) their staves shrink when dried out (as they'd be after several days' worth of warm smoke) and open. So use a metal barrel with one head removed. If it's had oil in it, set the residue of oil on fire and let it burn out; then scour the drum thoroughly inside and out with plenty of detergent and water; rinse; scald the inside, and let it dry in the air.

OUTSIDE ON THE GROUND

In the bottom of the barrel cut a hole large enough to take the end of an elbow for whatever size of stovepipe you want to use (see the sketch). Set the barrel on a mound of earth—with earth banked high enough around it to hold it firm and steady—and dig a trench from it down to a fire-pit at least 10 feet away, and inclining at an angle of something like 30 degrees. Via the trench either connect the barrel to the pit with stovepipe, or build a box-like conduit (stovepiping is easier to remove and clean). You should have the

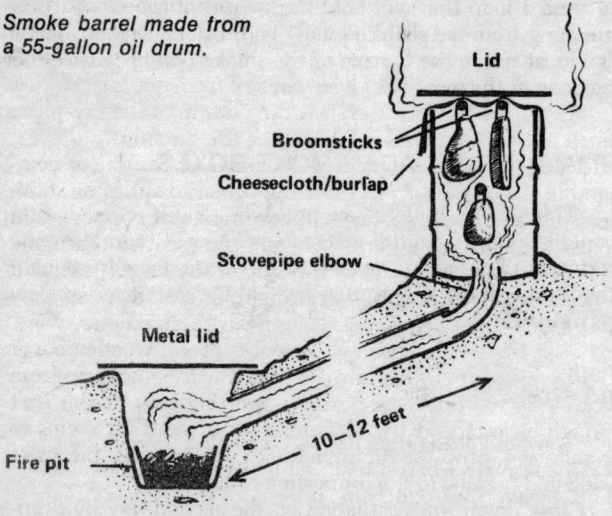

Smoke barrel made from a 55-gallon oil drum.

Lid

Broomsticks

Cheesecloth/burlap

Stovepipe elbow

Metal lid

10–12 feet

Fire pit

length in order to cool the smoke on its way to the meat, and the pitch to encourage the draft.

Put a cover of close-fitted boards over the fire-pit, arranged so it can be tilted to increase the draft when necessary.

ON THE PORCH, OR IN A GARAGE

The electric-plate/pan-of-sawdust arrangement should be used only in a dry place with fire-retardant material underneath it. This can be sheet metal, or a concrete floor.

Set your barrel on supports—cinder-blocks or trestles of some sort—to hold the elbow well away from the floor. Connect the stovepipe, and lead it from the barrel to the electric smoke-making unit. Make a wooden box, lined with fire-retardant material, to house the hot plate and the pan of sawdust, cut adequate slits for regulating air intake; and merely lift off the box when you want to add more fuel for making smoke.

SMOKING IN THE BARREL

Get the fire or smoke-maker well established and producing evenly before hanging the meat to smoke. There should be good ventilation from the top in order to carry off moisture the first day (to keep the fire from getting too hot, though, reduce the air intake at the bottom of the fire-pit as much as you can without letting the fire go out).

Hang the ham, shoulder and chunks of bacon (or comparable sizes of beef pieces) from broom-handles or stainless-steel rods—not galvanized, not brass, not copper—laid across the top of the barrel. Stagger the meat so that none of it touches other pieces or the side of the barrel; suspend smaller pieces on longer loops of strong steel wire so they drop below the large pieces. Hang your thermometer.

Over the whole business lay a flat, round wooden cover slightly bigger than the barrel's top. It will be held up from the rim by the thickness of the supporting rods. If this isn't enough clearance at the beginning, or if the draft seems to be faltering, prop the lid higher with several cross-pieces of wood laid parallel to the supporting rods.

Close down the ventilation on the second day by draping a piece of clean burlap or several thicknesses of cheesecloth over the supporting rods *under* the lid. The cloth will also protect the meat inside from debris, or from insects attracted to it if the smoke stops. Weight the lid down over the cloth with a good-sized rock to keep it in place.

HOW LONG?

Smokiness—color and flavor of the meat—is a matter for individual taste. If it's oversmoked, the meat is likely to be too pungent, especially on the outside. And you can always put the meat back for more smoking if the flavor isn't enough for you.

So try out your system in a small way. Give a shoulder of cured pork, say, 45 to 55 hours of smoking; take it out and slice into it—you may want to give it a few hours more: 60 hours is about average for a smoked shoulder.

The average ham takes about 72 hours of total smoking time.

Bacon, being a thinner piece of meat, is usually smoked enough in a total of 48 hours.

BUT ALL THESE TIMES ARE APPROXIMATE— they're mentioned merely as guides.

An old refrigerator as a smoke-box

This rig has the disadvantage of being likely to smoke hotter than the ideal 90 F. maximum. Still, the USDA bulletin says the smokehouse can be 100 to 120 F.—just warm enough to melt the surface grease on pork—so if you can keep the inside of the refrigerator box within this maximum you'll probably be all right. You can always extend the distance from the smoker into the box by using a short length of stovepipe to get cooler smoke.

Ask a friendly dealer in household appliances to give you a one-door electric refrigerator of the simplest type, and that's beyond repair. Or ask the people in charge of your community's sanitary landfill to save you one (and make sure the hinges were not damaged irreparably when the door was taken off before the box was dumped).

Remove the compressor unit at the bottom, and the ice-cube compartment; block off any places where ducts may have entered the interior of the box. Take out all galvanized inside parts. While you're at it, get rid of all plastic fittings too: who knows how they'll act if the smoke gets too hot? And you can always use the room.

Cut a hole at least 6 inches in diameter low on one side of the refrigerator, just above the floor of the box (see the sketch). Cover any ventilating grids snugly on front, back and sides of the space that once held the motor.

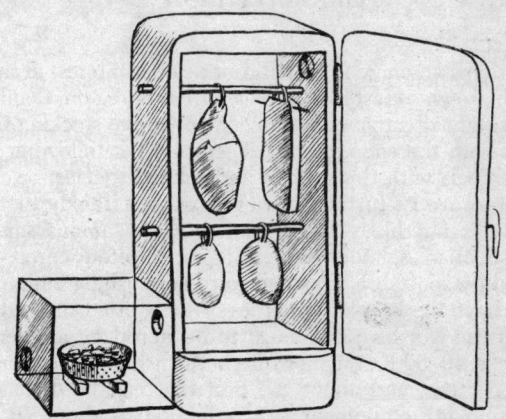

On the top, close to one corner, set in a butterfly vent that you can open in the early stage of smoking. This vent can also be used later to encourage a slight draft, and so keep the smoke from backing up and smothering the fire.

If you can rest cross-rods on the supports that formerly held the shelving, so much to the good. Otherwise you'll have to bore holes in opposite sides of the box to carry the rods from which you'll hang the food.

Set your smoke-maker (either an electric hot plate heating a pan of hardwood sawdust, or a brazier of smoldering chips or corncobs—see above) to one side of the refrigerator. House it in a box that sits over it and fits against the side of the refrigerator and over the hole you have cut without leaking any smoke. In a far corner of this housing box set a butterfly vent that can be used to regulate the air intake of the smoke-maker. Hang an oven thermometer on the side of the refrigerator near the smoke-hole, so you can move the box away for a reading—a rough way of gauging the smoking temperature, but the alternative is to install a thermometer through the refrigerator wall which lets you know the interior heat without opening the door.

Hang the meat inside the refrigerator, close the door and carry on as with the barrel smoke-box above.

Smoking Fish

Perhaps nowhere in North America has interest in smoking fish grown so fast as in the Great Lakes region, thanks to increasingly effective pollution control and to stocking these waters with the coho salmon. Therefore the following procedure deals with this salmon and related species.

Before we go further, however, we say frankly that the product is not likely to equal the world-famous smoked Nova Scotian salmon—which is most often cured and smoked by specialists attached to private fishing clubs. Nor will it have the keeping qualities of the storied smoked salmon of the Northwest Indians: to be stored for any length of time at 40–50 F., the moisture in the fish must be reduced to 20 percent, and about the best that can be done with home-smoking equipment is a reduction to about 40 percent. And meticulous handling and sanitation are vital, as was explained at length in "Canning Fish," Chapter 2.

Pre-smoking preparation

Dress, scrub and fillet your salmon, and hold the pieces as close to 32 F. as possible. In your largest kettle or tub— enameled, ceramic or wooden, *never* one that can corrode— prepare enough ice-cold brine to cover your fish, made in the proportion of 3 cups pure pickling salt dissolved in each 1 gallon of fresh drinking water. Depending on the thickness of the fillets, hold the fish in this 30–40-degree-F. brine for 1 to 2 hours, during which time diffused blood will be drawn out, the oil in the tissues will be sealed in to a large extent, and the flesh will be chilled so much that the following dry-salt cure will not penetrate too rapidly.

Remove the fillets, drain, and scrub away débris. Using pickling salt in the proportion of 3½ pounds (5¼ cups) for each 10 pounds of fish, dredge the pieces completely in salt and pack them in a large non-corrodible container with plenty of salt between the layers. Put the pieces skin-side down, except for the top layer; cover the top layer with salt. Keep the container as cool as possible, and hold the fish in it for 3 hours.

Remove fish, rinse well. Air-dry in single layers away from sun or heat for 1 to 3 hours until a thin shiny "skin," or pellicle, forms on the surface. The fish is now ready to smoke.

Cold-smoking the fish

Many beginners are confused by the term "hot-smoking," which is a sort of long-distance barbecue in which the flesh reaches an internal temperature of up to 180 F., after which it is eaten within a couple of days—as with any cooked food—or is frozen. We are not speaking here of this type of smoke-cooking.

Build your regular hardwood fire; after it is burning well, smother it with fine hardwood chips or sawdust to produce a very dense smoke with little heat—the temperature inside the smoke chamber ideally should never exceed 70 F. in order to inhibit growth of bacteria in this highly perishable food. Tend the fire night and day: smoking fish is a continuous process.

After the end of 4 full days of smoking, sample a piece of fish to see if its color, flavor and texture are what you want. If not, smoke it 24 hours more, and test again. When it is smoked to your satisfaction, air-dry the pieces in a cool place for several hours. Then package the fillets individually in plastic wrap and store between 32 and 40 F. for up to 3 months.

Freezing will cause salt in the tissues to deposit on the surface of the fillets. We do *not* recommend that smoked fish be canned at home.

Sprouting 8

One of the most nutritious of foods, and one recently "discovered" by the general public is the sprout: an infant plant still living off the nutrients formerly encased in its parent seed.

Part of the new interest in this ancient food is that sprouts are fun to grow, producing a crop of high-protein, vitamin-packed fresh vegetables with only a few minutes' attention a couple of times a day. They also require only simple equipment, of the sort available in every household; they usually take under a week from beginning to harvest: and, to top it all off, they are strikingly versatile in the hands of an imaginative cook.

Winston A. Way, Extension Agronomist at the University of Vermont, published a pioneering "brieflet" titled *Sprouting Seeds for Nutrition and Fun* in January 1973. Among the recent books, we have found the most helpful to be Esther Munroe's *Sprouts to Grow and Eat* (among other virtues it contains more than 150 recipes, ranging from appetizers to baby foods).

WHAT SPROUTING IS

Edible sprouts are the germinated seeds of an astonishingly wide variety of plants. You start with the dormant seed, give it the right combination of moisture, temperature, air and light to awaken it, and within a day the seed has swelled, and the coat has burst. Some species will produce cotyledons—the so-called "seed leaves"—and the finest possible hair-like roots by the third or fourth day. Some sprouts are eaten when they are about 1 inch long; others, from smaller seeds, will be ½ inch long when the crop is ready.

WHAT TO SPROUT

Some seeds are poisonous by nature and others have been made poisonous by people.

POISONOUS BY NATURE

Winston Way, who should know, includes in his list of poisonous species that "are definitely out" the castorbean, apple, stone fruits, sorghum, Sudan grass and potato (other writers add the tomato to this roster). He warns against sprouting seeds from any plant used for drugs.

MADE POISONOUS BY MAN

In order to prevent infestations and mold—even modern storage conditions can't always be guaranteed as ideal— virtually all seeds sold for planting have been treated with some chemical pesticide or fungicide that is highly toxic to human beings as well. Also, imported seeds are required by law to be stained with various colors, but their packets do not always contain the information that the seeds are dyed and therefore are taboo as human food. Therefore:

NEVER SPROUT SEEDS THAT HAVE BEEN TREATED CHEMICALLY, OR DYED. SPROUT ONLY THE SEEDS THAT HAVE BEEN CERTI-FIED EXPLICITLY AS EDIBLE.

And we add a corollary: Never sprout for eating any certified-as-edible *seeds that have even the smallest amount of mold.* Reason: Mycotoxins, the poisons produced by growing molds, and they can mean bad trouble.

AND THEN THERE ARE DUDS

Even if you know they're safe, "hard seed" plants like birdsfoot trefoil don't always sprout satisfactorily for our eating purposes, and therefore too many seeds would be wasted.

You can sprout beans, lentils, chick-peas (the list is amplified in a minute), and of course these legumes are food-pure when sold in the supermarket. However, they've usually been dried so much and bashed around so hard that the yield in sprouts is disappointing. Cook them up instead.

And of course seeds must contain living embryos (even though dormant, they'll soak up their first moisture and burst into life). So make sure that the package label tells you the guaranteed percent of germination.

WHERE TO GET SPROUTING SEEDS

Health- and natural-food stores usually carry seeds for sprouting, plus information and equipment for the whole procedure.

A number of the large garden-seed companies list sprouting seeds and equipment in their catalogs: make sure the seeds you order are the ones they certify as edible.

And you can always grow your own—next year. Plant the varieties you want this coming growing season: use chemically treated seed for sowing. Raise your crop, harvest the seeds when fully mature. They are ready when the pods (using legumes as the illustration, because they are most commonly grown for sprouting) are dry, but just *before* they split open.

Dry the seeds in a warm, dry, airy place, preferably in the shade. If you must dry them indoors, *don't use the oven:* the seeds will get too hot and the embryo will be killed. If you use a dryer with a mild heat and fan, keep the temperature well below a killing range of 100–180 F.; keep the seeds away from the heating element.

Check and store—*without pasteurizing*—in sterile, airtight containers. Here's a fine place to use the one-trip mayonnaise and pickle jars that you should not re-use for canning. Keep the containers in a cool, dry, dark place.

WHAT KIND TO GET

Sprouts have flavors as distinctive as the plants from which the seeds come. The following are popular, with the legumes leading the list. *Legumes:* alfalfa, clover, soybean, mung bean, adzuki bean, chick-pea (garbanzo). *Grains:* wheat, barley, millet, corn, buckwheat. *Vegetables:* radish, cress, celery, beet, pumpkin, squash. Herbs, wild greens and oil seeds are also sprouted.

Start small, though. First we'll try alfalfa. From there on it's up to you.

GENERAL PROCEDURE

There are many kinds of equipment and a great variety of seeds, but as a "starter set" we recommend this: certified-as-edible alfalfa seeds; clean, scalded 1-quart canning jar, preferably wide-mouth, with a screwband (no lid) or a stout rubber band; and a scrupulously clean nylon stocking. Time enough for deluxe fittings and dozens of different seeds after you become an expert in flavors and technique. Room temperature is usually best for all sprouting. Some seeds do well below the 65–70 F. level; almost all seeds seem to grow faster at warmer temperatures, but anything above 70 F. also helps to promote the growth of mold (see Chapter 1).

(1) Put no more than 2 tablespoons of alfalfa seeds in your quart jar; fill half full of lukewarm water, and let stand overnight.

(2) Make a strainer for the jar by stretching one layer of nylon stocking over the mouth, and fasten it snugly with the screwband or rubber.

(3) In the morning, pour the water off the seeds (don't worry about browning). Rinse twice in lukewarm water, drain completely.

(4) Hold the jar horizontally, tapping it to distribute the swollen seeds evenly; keep them away from the nylon: this is where the sprouts will get air to breathe. Lay the jar on its side in a dark cupboard that's not too warm (see above).

(5) At midday and again at night take the jar from the cupboard and rinse the seeds (through the nylon) with lukewarm water, drain well. Rinsing is vital: the infant sprouts are undergoing metabolic change as they grow, and they throw off gases and substances that make the sprouts slimy if they're not drained. (Some sprouts also exude a mucilage as they develop.)

(6) By the end of the second day in the cupboard—having been rinsed well and drained three times a day—little sprouts will appear, and in 12 hours more the sprouts will have roots $\frac{1}{2}$ inch long.

(7) After the morning rinse on the third or fourth day, place the jar in a sunny window. The cotyledons will turn green with amazing swiftness. By now most of the seed coats will come loose. Remove the netting and fill the jar with water to overflowing; shake it a little and many of the seed coats will float to the top and wash away. Let those in the bottom stay there. Seed coats, like popcorn hulls, are all right to eat but for appearance's sake you may want them removed.

(8) Drain well. (You no longer need the nylon—the mass of sprouts will not fall out if you hold your fingers over the jar opening.) Store drained sprouts in the refrigerator in tightly covered containers. Like lettuce and celery, sprouts will stay sweet and fresh for four to six days.

RANDOM TIPS

Using unglazed earthenware/flowerpots. You can make unglazed sprouters very like expensive store-bought units out of two brand-new bulb pots (with the hole in the bottom one plugged with a rubber stopper, and a wooden drawer-pull bolted with a washer through the hole in the upper pot, which thereby becomes the lid). Scrupulously clean of course. And soak the bottom pot in water to cover for an hour before its first use, and scrub off loose clay dust; before subsequent uses, soak the cleaned pot until it's saturated—without presoaking, the clay will absorb precious moisture from the seeds resting in it, to the detriment of the sprouting procedure.

The giant clay saucers that hold big flowerpots also work well, and have no drainage-holes to plug.

Such arrangements make handy sprouters because they don't need to be kept in a dark cupboard.

For extra-delicious salads. Let the sprouts of alfalfa (and clover, rye, wheat, etc.) reach up to 2 inches in length; expose the sprouts to indirect sunlight for about 3 hours for the chlorophyll to do its work—and enjoy their sweet "green" taste in salads.

To freeze. For an unexpected surplus of sprouts that is not to be canned (see page 167), wash, remove husks, blanch in one layer at a time over vigorous steam for 3 minutes; cool quickly in ice water, drain, pack.

Babies on solid food thrive on sprouts cooked (with chicken or liver or whatever their diets allow), all whirled in a blender to the palatable consistency. Extra servings can be frozen in ice-cube trays, removed and wrapped separately for easy heating later; the packets are then bagged and stored in the deep freeze for use in a few weeks.

The Roundup 9

Most of the procedures and recipes in this section involve materials that could, through not knowing or through affluence, be ignored. All the instructions are here because they didn't seem to fit easily anywhere else.

RENDERING LARD

Lard is in disrepute these days because of blanket indictments against animal fats as cholesterol-makers. This is a pity, because active, sensible eaters shouldn't automatically forgo the special flakiness and tenderness that good lard gives pastry and cakes—unless their doctor has advised them to do so.

Lard is fresh pork fat, rendered—i.e., melted—at no more than 255 F., and then congealed quickly and cold-stored for later use. The fine lard generally considered best by the purists comes from the "leaf" next to the bacon strip; backfat from along the loin and plate fat from the shoulder also are good (as in the photographs; and see also the labeled parts of a pig in "Salting Pork" in the Curing section). The caul fat (attached at the stomach) and the ruffle fat (around the intestines) usually make darker lard than the other parts do, and so are rendered separately.

A live 225-pound hog will yield slightly more than 25 pounds of all fats usable for lard; obviously only a small part of the total can come from the leaf (but it gives up a higher percentage of its weight in lard than the other sections do). So stop now to decide how you want to use the fat: to render the leaf separately; to combine leaf, loin, plate and trimmed fat; to combine caul, ruffle and trimmed fat, rendering them for soapmaking later.

Cut the fat in small pieces. Put a little of it in a large kettle over *low* heat and stir. When melting starts, add the

480

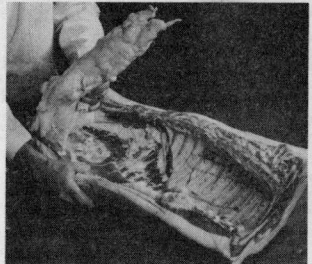
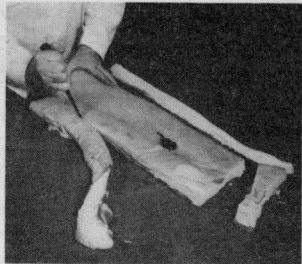

Main sources of lard: leaf fat, left above; trimming from bacon, right; plate fat from the shoulder, below left; back fat from the loin.

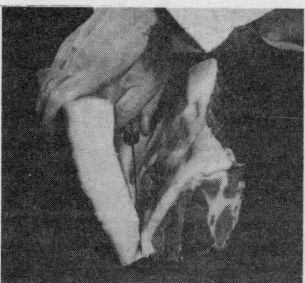
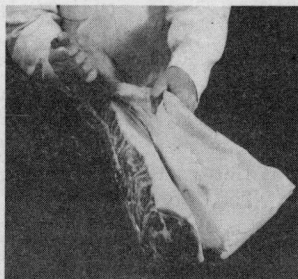

rest of the fat gradually—but don't fill the kettle more than to 3 inches from the top, lest the fat boil over. Stir often and keep the heat low: the cracklings—bits of tissue—can stick and scorch.

Cracklings that brown and float to the top will settle to the bottom later on when they lose fat and moisture. As the water in the tissues steams away, the temperature of the fat will rise slowly above 212 degrees Fahrenheit. Use a deep-frying thermometer to make sure the heat doesn't get above 255 F., which is the optimum temperature for thorough rendering.

Let the rendered lard cool a bit so all sediment and cracklings will settle to the bottom. Carefully dip the clear liquid from the top of the kettle into containers, filling them to the brim; cover tightly and seal. Label and store immediately at freezing or below to make a fine-grained product.

Strain the lard at the bottom of the kettle through several

layers of cheesecloth to remove the settlings; pack and store as above.

Air and light make stored lard rancid, and spoil it irrevocably for cooking purposes. Thorough rendering (to 255 F.) eliminates the moisture that causes souring during storage.

SOAPMAKING

The Fundamental Rule

Never, *never*, NEVER make soap with young children around. Or pets.

Even a cold lye solution $\frac{1}{6}$ as strong as the one below for soapmaking may taste sweetish (much like a dose of baking soda in water to settle one's stomach) for a split second—but long enough for a child to swallow it; and then it burns painfully, with damage.

The heavy concentration of lye used in soap can strip off skin as easily as the peel is rubbed from a blanched peach.

Antidotes. Slosh eyes with cold water, follow immediately with cold boric-acid solution—which you'll do well to have handy by.

For hands, etc., wash immediately with cold water, followed by generous bathing with vinegar.

If taken internally, the sufferer should be made to drink salad oil (but don't pour a bit of it down the windpipe) AND TAKEN TO A DOCTOR AS FAST AS POSSIBLE.

Labels on containers give directions for other first-aid measures.

Lye can ruin a garbage disposal unit in the sink drain. Lye can raise Cain with the working of a septic tank.

Lye reacts violently with aluminum and is used in commercial drain-openers for just this reaction.

Equipment for Soapmaking

First-aid supplies—cold water, boric-acid solution, vinegar, salad oil

An enameled or granite kettle for warming the fat (in a small batch)

(A big iron kettle and/or a wooden tub for larger batches)

A sturdy enameled kettle for dissolving the lye in water

Floating or dairy thermometer

Wooden paddles or long-handled wooden spoons for stirring

A shallow wooden box, large enough to hold about 2½–3 quarts of liquid; soak the box to make it tight, and line it with clean wet muslin that's big enough to drape generously over all the sides

A board to cover the box

A dripping pan big enough to hold the wooden box

An expendable rug or blanket to keep the new soap warm for a while

Tough string or fine wire (for cutting the soap in bars)

Glass measuring cups

What Lye Is (and Does for Soap)

NOT THIS

Liquid or crystalline products that designate themselves as drain-openers contain nitrates, activating aids (like aluminum filings) and stabilizers. Such products are not remotely suitable for making soap.

O.K. COMMERCIAL LYES

The several crystalline preparations sold simply as "lye" are sodium hydroxide; these also can double as drain-openers, but they don't have the additives that make bad soap. They come in 13-ounce cans. If you handle the lye immediately after the can is opened you may use ounces in weight and ounces in volume interchangeably when faced with instructions that call for lye either in pounds or in less than 13 ounces. The stuff readily absorbs moisture from the air, expanding in the process, so work quickly to get reasonably accurate measurements. Cap any remainder the minute you're through measuring.

This usable type of lye is mixed with cold water—whereupon it reacts and heats the water, giving off fumes that make ammonia seem bland in comparison. When the solution stops acting, it begins to cool; but it's still frighteningly caustic.

When a solution of the right strength and temperature is poured into the right amount of clean warm fat at the right temperature, the fat saponifies—changes character and hardens.

OLD-TIME "POTASH WATER"

Another kind of lye can be made by leaching cold water through pure hardwood ashes.

Remove the head from a water-tight wooden barrel and set it, open end up, on trestles high enough to allow a wooden pail to be set under it. Bore a small hole in the barrel's bottom near the front; make a tight-fitting plug for it and drive it in. Put several bricks in the bottom of the barrel, arranged to hold a protecting board over the hole; these plus a layer of straw board and straw will prevent the ashes from packing too tightly on the bottom or fouling the hole.

Fill the barrel with firmly packed hardwood ashes. Make an opening in the center of the ashes, and in it pour all the cold water the barrel will hold. Cover the barrel and raise it at the rear so the liquid will run toward the front. Add more water the next day if the barrel will hold it.

After 3 weeks, remove the plug and draw off the lye into the wooden pail. The liquid will be, in the traditional saying, "strong enough to float an egg"—strong enough to make soap with. (Even if it's not, there's no worry: excess water can be boiled off when you make the soap.)

Fats Good for Soapmaking

Best for soap is tallow—rendered from beef or mutton fat as you rendered lard (q.v.); next best is lard, and a combination of lard and tallow; then comes olive oil, followed by other vegetable oils. Poultry fat doesn't do well alone. Mineral oil won't saponify at all.

All fats must be fresh, clean and salt-free to avoid disappointing results.

Master Recipe for Soap

Having the right temperature for both the lye solution and the fat is one way to ensure fine soap (the others are accurate measurement and fresh, clean, salt-free fat). Here are the temperatures:

Fresh, sweet lard at 80–85 F., lye solution at 70–75 F.

Half lard and half tallow at 100–110 F., with lye solution at 80–85 F.

All tallow at 120–130 F., with lye solution at 90–95 F.

Melt 6 pounds (c. 13½ cups) of clean rendered fat in a large enameled kettle until it's a clear liquid; cool it to the right temperature until it thickens; stir it occasionally to keep the fat from graining as it gets near room temperature.

Meanwhile, in a large enameled pot mix 1 just-opened can (13 ounces) of suitable crystalline household lye with 5 cups of cold water. Stir with a wooden paddle until it has dissolved—but stand clear of the fumes and avoid spatters of the caustic liquid: it will boil up like mad of its own accord. Let the solution cool to the proper temperature for the fat you're using.

When both fat and lye are the right heat, very carefully pour the warm lye into the fat in a *thin* steady stream, stirring the mixture very slowly and evenly (another pair of hands would be a great help here). If the combining temperatures are too cold or too hot, or if the fat and lye are combined too fast or stirred too hard, the soap will separate —and then you must boil the separated mixture with much extra water to make it get together again.

In 10 to 20 minutes of careful stirring the fat and lye will have blended thoroughly and become thick. Pour the saponified mixture at once into the soaked, cloth-lined box; set the box in the dripping pan to catch leaked soap. Lay the protecting board over the box and cover the whole she-bang with a rug or blanket to hold the heat while the soap develops a good texture.

Don't disturb the soap for 24 hours. Then grasp the overhanging edges of the cloth liner and raise the soap block from the mold. Cut it by cinching strong twine or wire around the soap and pulling tight. Let the bars of young soap dry at room temperature in a draft-free place for 10 days or 2 weeks. Don't let the soap freeze during this time.

When the soap has conditioned, wrap each piece in paper and store it in a cool dry place. Aging improves soap.

Variations

Make floating soap by folding air into the thick saponified mixture before you pour it in the box-mold.

Make soap jelly by shaving 1 pound of soap into 1 gallon of fresh water; boil for 10 minutes; cool. Store in a closed container in a cool place. This is good for shampooing hair.

Make your own soap flakes by gently shaving 3-day-old soap against the medium holes of a vegetable grater. Stir the shavings gently as they air-dry. Store in closed containers.

For more information write to Household Products Information, the Pennwalt Corporation, 3 Parkway, Philadelphia, Pennsylvania 19102.

FARMERS' SAUSAGE

See earlier instructions for canning sausage or freezing ground meat.

20 pounds of fresh pork trimmings—⅔ lean, ⅓ fat	3 teaspoons ground ginger *or* ground nutmeg, if desired
½ cup pickling salt	2½ cups ice water (if you're freezing it in rolls like cookie dough)
6 tablespoons ground sage (omit if canning: it gets bitter)	
3 tablespoons ground pepper	

Put the meat through the food grinder twice; add seasonings and put through a third time. Mix it all thoroughly on a large bread-board.

If you plan to freeze it—or just to refrigerate it—rolled in bulk, knead the ice water in gradually until the liquid is absorbed and the sausage is doughy. The water keeps the sausage from crumbling when it's sliced for frying.

BOLOGNA-STYLE SAUSAGE

This rule makes about 25 pounds of sausage. If canned, *it must be Pressure-processed at 15 pounds* (see Chapter 2).

Prepare. Using the coarse knife of a food chopper, grind 20 pounds of lean beef and 5 pounds of lean pork; add seasonings, and re-grind with the fine knife. Usually ½ pound of salt and 2½ ounces (a standard small box) or less of pepper are enough. To help hold the meat's color, mix 1¼ teaspoons pure crystalline ascorbic acid (Vitamin C) with the salt and pepper. Garlic juice may be added, or finely minced garlic; or coriander or mace.

For this quantity of seasoned meat, add 3 to 4 pints of very cold water, and, using your hands, mix it all thoroughly until all the water is absorbed into the meat. Stuff tightly, without air pockets, into beef casings or rounds that have been soaked in cold water (such casings are available from butcher-supply houses). Use short casings; or tie long ones at intervals short enough to allow you to cut sections suitable for family use. Tie ends with stout string. Smoke the sausage for 2 to 3 hours at 60 to 70 degrees Fahrenheit (see Smoking).

Next, simmer the smoked bologna at about 200 F. until it floats. A full-size sausage will take a whale of a kettle; if you don't have a pot big enough, cut the thing in sections where you've tied it off, and cook it piecemeal, holding any extra pieces at 30 to 38 F. until they can go into the pot.

Storage. Store in the refrigerator any sausage you'll use quite soon. Canning is best for longer-term storage at home (see above). Because of its fat content and seasonings, it doesn't hold flavor well when frozen.

FARM-WIFE'S PORK HEAD CHEESE

1 hog's head,	salt
thoroughly trimmed	pepper
and then quartered	

Put the well-scrubbed quarters in a large kettle and cover with unsalted water. Simmer until meat falls from the bones—about 3 hours. Remove from heat and let cool until the meat can be handled, but don't let the broth get so cold that the fat congeals. Drain away the liquid, strain it well and return it to heat to reduce by ⅓ or ½.

Put the meat—which has been picked from the bones and gristle—in a wooden chopping bowl and chop it very fine. Add enough reduced broth to make a wet mixture and season all to taste with salt and pepper. Pack in several standard-size bread tins about ¾ full. Nest one pan on top of another with waxed paper in between and weight the top pan to press all the meat properly; if extra juice is pressed out it can be saved for stock if desired. Set the stack in the refrigerator overnight. In the morning slide each loaf from its pan and wrap it in moisture-vapor-resistant paper; freeze.

Of course you may add herb seasoning if you like; but do try this simple rule first before you jazz it up. This head cheese is nothing like the commercial product. It's grand in a sandwich; and served with slices of cold Liver Loaf and homemade Bologna Sausage (both q.v.) it makes the best platter of cold cuts ever.

PASTEURIZING MILK

When you count the tests and immunizations a milch cow must have before she's shown at even a small accredited fair or her milk is offered for public sale, you see why pasteurizing is a mandatory safeguard for our store-bought milk. And this doesn't guarantee that new spoilers can't attack it if it's handled carelessly once it's brought home.

For pasteurizing milk you need a dairy thermometer, a sterilized large spoon for stirring, and an out-size double boiler (rig one up by setting a large pot/kettle in an old-fashioned enameled dishpan).

Put milk in the inner pot, which you set in the outer one (the dishpan). Pour water in the outer container until it reaches the level of milk in the inside pot.

Put the whole thing on the stove and heat it until the hot water in the dishpan raises the temperature of the milk to 145 degrees F. Hold the milk there for 30 minutes, stirring to heat the milk evenly; check the thermometer in the milk pot to make sure you're maintaining a steady 145 F.

Then take the milk from its hot-water bath and cool it by setting the milk pot in cold water. Get it down to 50 F. as soon as possible, then refrigerate.

COUNTRY-STYLE COTTAGE CHEESE

Cottage cheese is something you make at home only if you have considerable surplus fresh milk and more time than money. If you buy milk at supermarket prices to make cottage cheese, you'll be spending around 25 percent more for it per pound than you pay for the stuff ready made. However.

This type of cottage cheese has a small curd; it's well drained but not washed, so it's rather more soured-tasting than a cheese whose whey has been rinsed out; and the original rule has been translated from raw to pasteurized milk. The proportions are for 1 gallon of milk.

Bring the milk to 72 degrees F.—average room temperature in the kitchen anyway—and add ¼ to ⅓ cup of fresh buttermilk. (Or add ⅛ to ¼ cup of lactic "starter" made from rennet according to directions on the packet; and there's probably no reason why yogurt culture wouldn't work well too—experiment with proportions.)

Cover the kettle and keep the milk at 72 F. overnight—at least 16 hours, or until the milk has curdled into a jellylike substance, rather firm, with some watery whey on top.

In the kettle, cut the curd in ¼-inch pieces with a long-bladed knife. Slice the curd straight up and down in cuts about ¼ inch apart; then give the kettle a 90-degree turn and cross-hatch another series of up-and-down slices. Now the trick is to cut these square little columns of curd into something like cubes, and you can do a fair job by cutting through the curd again at a 45-degree slant, considering the bottom of the kettle as the base of the angle, then giving the kettle a quarter-turn and making another series of angled slashes.

Set the pot of cut curd in a larger container of water, and *slowly*—so slowly that it takes 30 minutes to raise the milk from 72 to 100 F.—heat the milk over hot water, stirring gently every 5 minutes or so to distribute the heat. When the curd has reached 100 F. you can raise its temperature more quickly to 115 F., and hold it there for about 15 minutes more. Test a curd: if it still breaks easily, bring the heat up to 120 F.; if it's firm but not tough, remove it from the hot water so it starts to cool. The higher the temperature, the tougher the curd.

Dip off what whey you can, then pour the curd into a cloth-lined colander to drain for 5 minutes. The curd compacts if you let it drain a long time. People who want to wash the curd may do so now, by gathering the corners of the straining cloth to form a bag and dunking the bag of curd in fresh cold water, then letting it drain well.

Add salt if you want to counteract the sourish-acid taste of the unwashed curd: 1 teaspoon of salt for 1 pound of cheese (1-gallon-of-milk worth). Add 4 to 6 tablespoons of cream if you like.

Store in closed containers in the refrigerator.

With raw milk. If you're using safe raw milk you don't need buttermilk or a lactic starter to make the milk curdle: it will sour/clabber all by itself at room temperature. When it's satisfactorily curdled, cut the curd and proceed as for pasteurized curd above.

Cottage cheese made from raw milk is more likely to get a yeasty, sweetish, "off" taste than cheese from pasteurized milk does in cold storage, so don't let it hang around.

WATERGLASSING EGGS

Refrigeration is the best temporary storage for fresh eggs, with freezing (q.v.) recommended for longer holding.

Waterglass is sodium silicate, and you buy it from building-supply places nowadays, to seal concrete floors or to stick insulation around heating pipes. It's already dissolved, rather sirupy, and tastes like washing soda. It fills the pores of the eggshells, thus preventing moisture loss from the inside or air damage from the outside. It does well enough for a *maximum of 3 months* if the eggs are kept in cold storage.

Select nest-fresh eggs with perfect shells; discard dirty ones. Wipe eggs with a clean dry cloth (washing would remove natural protective film) and put them in a crock. Cover them 2 or 3 inches above the top-most layer with diluted waterglass, using 1 part waterglass to 11 parts of cooled, boiled water (⅓ cup waterglass to each 1 quart of water). Cover the crock and hold it under 40 F.

Remove eggs as needed for cooking and wash them before breaking. The whites will have lost a good deal of viscosity (they'll be worthless for meringue, for example), but the eggs are O.K. for long-cooked dishes calling for whole eggs (see Freezing). Examine each egg as it's broken before adding it to the food mixture.

Fertile eggs. Farmers, being experienced with hens and their proclivities, would isolate the rooster from the flock for at least 1 month before putting any new-laid eggs in waterglass.

DRYING PRODUCE IN A WOOD-STOVE OVEN

If you're familiar with the workings of an old wood-burning black iron cookstove, you know that the oven heats *from the top:* therefore you dry food below the center of the oven, rather than higher.

Dry with an old, banked fire—air intake at the bottom of the firebox closed, drafts up through the stack closed. Feed the fire only when absolutely necessary, and then with very large hardwood, one billet at a time. You may even have to tilt the stove lids to cool the top of the oven; open the cross-draft regulator at the top of the firebox to keep the fire as low as you can. Hang an oven thermometer on the *upper* rack of food to keep track of maximum temperature, and do your drying with the oven door halfway or fully open.

RE-CANNING MAPLE SIRUP

To deal with the haze of mold that can form on maple sirup in a large container that's been opened repeatedly, pour the sirup into a deep enameled kettle, bring it quickly to boiling over high heat. Remove from heat, immediately skim the surface clean, and pour the boiling-hot sirup into hot sterilized pint jars, leaving ½ inch of headroom; cap firmly with sterilized lids. Process in a Boiling-Water Bath (212 F.) for 10 minutes to ensure the seal. Remove jars, cool naturally. Refrigerate opened sirup.

Recipes 10

The dishes in this chapter have been chosen for two
reasons: they emphasize ways to use food that's put by, and
they are things not often found in today's save-time-in-the-
kitchen-at-any-cost cookbook. There are gaps in the recipes,
because in limited space they couldn't extend beyond the
limit of what was—or is—on hand as the result of old-
fashioned foresight, effort and thrift. There's only one salad,
for example, and it's just a tangy mixture of canned beans,
but there's a way to stretch chicken into a company dinner
that's good enough for a visiting preacher or a group of
experts on *bonne femme* cuisine. Most of the people who
sent them to us never made hollandaise sauce in their lives,
but they knew how to get kids to eat liver and what to do
with salt codfish and why a dark sleety day calls for rhubarb pie.

Bea Vaughan contributed many recipes from her heirloom collection of cookery books, many handwritten and
containing the elliptical memoranda that were all that an
expert home-maker needed for a new dish or a different
method. For the most part, Mrs. Vaughan translated these
"receipts" into modern usage—even to telling temperatures
in Fahrenheit. But she left in such things as baking soda in
tomato cream soup (to keep it from curdling, for who ever
had heard of Vitamin C?), and lard in pastry (no dawn-to-
dusk farmer knew about cholesterol, nor would he have
given a hoot if he did), and white sugar (a luxury to be
proud of, for the alternative was hunks of truly raw sugar,
chipped from the barrel with a special hatchet and then
crushed fine with a rolling pin).

Today's nutrient-minded cooks can always substitute in these and other cases. It will still, all of it, make good eating. J.G.

Never taste any suspect canned food: boil it hard for 15 minutes to destroy any hidden toxins (corn, greens, meat, poultry and seafood require 20 minutes), adding liquid if necessary. If it looks spoiled or foams or has an off-odor during boiling, destroy it completely so it can't be eaten by people or animals.

Soups

Washday Soup

2 cups dried mixed beans, any kind
½ cup dried split peas
2 tablespoons rice
2 tablespoons barley
2 quarts cold water
2 large stalks celery, sliced
2 medium onions, peeled and sliced
2 medium potatoes, peeled and cubed
1 cup diced turnip
2 cups undrained canned tomatoes
1 teaspoon salt
¼ teaspoon pepper
ham bone (with some meat left on it)

Cover beans, peas, rice and barley with the cold water and let stand overnight. Bring to a boil in the same water, adding vegetables, salt and pepper. Simmer about 1 hour. Add ham bone and simmer about 1 hour longer. If water cooks away, add a bit more from time to time. Remove ham bone, pick off bits of meat and return them to soup. Serves 6 to 8.

Aunt Chat's Tomato Cream Soup

4 cups canned tomatoes, undrained
1 medium onion, peeled and chopped
2 stalks celery, chopped fine
4 cups scalded milk
about 1 tablespoon flour for thickening
pinch of baking soda (⅛ teaspoon)
1 tablespoon butter or margarine
salt and pepper to taste

Simmer tomatoes, onion and celery for 30 minutes. Thicken scalded milk with the flour, which has been mixed with a little cold water to make a paste. Stir over low heat until milk thickens slightly. Just before serving, add baking soda to the tomato mixture (to prevent curdling); as frothing subsides, pour all through a strainer into the hot milk, stirring constantly. Add butter and season to taste with salt and pepper. Serve at once. Serves 5 to 6.

The recipe may be doubled, but *never increase the amount of soda.*

Fish Chowder

about 1½ pounds haddock or cod, fresh or frozen	3 medium potatoes, peeled and diced
3 cups water	2 medium onions, peeled and diced
1 teaspoon salt	4 cups milk
3 (¼-inch) slices fat salt pork, diced	1 tablespoon butter or margarine
	salt and pepper to taste

Put the fish in a kettle, add the water and salt. Poach slowly until the fish can be broken apart with a fork, but not long enough to get soft. Remove from the water—but *save the water to boil the potatoes in.* Boil the potatoes until tender. While potatoes are cooking, fry the salt pork in a small skillet until golden brown, add diced onions and cook slowly for 5 minutes longer. Remove skin and bones from fish (which is now cool enough to handle easily). To the kettle of undrained cooked potatoes, add the fish, then add pork and onions with the fat from the frying pan. Add milk. Let come just to boiling, add the butter, then pepper and salt to taste. Serve with pilot biscuit or common crackers. Makes 6 generous servings.

Vegetable Dishes

Wild Fiddleheads

True fiddlehead greens, the young stalks of the Ostrich Fern (*Struthiopteris germanica*), appear in moist, fertile North Temperate woods in April and May, when American cowslips bloom in the marshes. Unlike most other ferns, they are *not* fuzzy. Rather, their characteristic is a paper-dry, parchment-like sheath that is scaling off. Gather them when they're about 4 inches above ground—3 inches of stalk, 1 inch of tight-curled "fiddlehead" tip. Before they have a chance to wilt, rub the dry scales from tips and stalks and wash well. Sprinkle lightly with salt and cook covered in a small amount of water until tender. Drain and serve with melted butter, with seasoned hot cream as a dressing, or on toast.

They're almost as good up to 6 inches or so: strip off the uncurled leaflets when scaling them, and be prepared to cook them a little longer.

Stewed or Creamed Milkweed

Milkweed (*Asclepias syriaca*) comes along nearly a month after fiddleheads, and may be collected along any back-country road that's not been treated with chemical de-icers (and get far enough off the road so lead from car exhausts hasn't contaminated the vegetation). Gather the young stalks when they're about 5 inches high and the leaves are pressed closely together like praying hands. Rub the greens between your hands to remove the wool-like fuzz, then wash well. Because it is necessary to remove the bitter milk juice, several changes of cooking water are necessary. Cover milkweed with boiling water and boil five minutes; drain, boil for five minutes more in fresh water; repeat process again, this time adding salt to taste to the fresh water. Boil about 30 minutes in all—or until tender. Drain well and serve buttered and seasoned like fresh asparagus; or cover with hot seasoned cream or hot cream sauce and serve on toast as a main dish.

Plain Dandelion Greens

Early in the spring, even before the Ostrich Fern is producing its edible fiddleheads, come dandelion greens. Gather them when they are only a rosette of leaves without any blossom-stalk showing. Look for them in meadows, unkempt lawns, former barnyards that are growing up to grass, in pastures, and avoiding places frequented by dogs. Cut them off just below the surface of the soil—use a broad-bladed knife, or a piece of strap-iron with one end notched and the notch beveled to make a sharp cutting edge. Any tight little buds deep in the heart of the cluster are a dividend.

Wash well in deep water, and lift out, so grit and dirt sink to the bottom of the pan.

Handle like spinach for canning or freezing.

Only if you want to remove any bitter tang (delicious to many people) do you boil in two waters: cover with boiling water, cook 5 minutes, lift out greens and discard water; return greens to the pan with fresh water, salted, and cook until tender. Drain and serve hot, as is or garnished to taste.

Four-Bean Salad

2 cups drained canned whole green beans
2 cups drained canned yellow wax beans
2 cups drained canned red kidney beans
2 cups drained canned garbanzos (chick peas)
2 medium onions, peeled
1 medium green pepper, seeded
3 medium stalks celery, trimmed
1 cup vinegar
½ cup water
1 cup sugar
½ cup salad oil
1 teaspoon salt

Cover drained green and yellow beans with cold water. Simmer 5 minutes; drain. Put drained kidney beans in colander and rinse with fresh water; drain. Combine vinegar, water and sugar, and boil 2 minutes. Remove from heat and add oil and salt, mixing well. Combine all beans in a large bowl. Cut onions, pepper and celery in thin slices, then add to beans. Pour vinegar mixture over beans and mix in well.

Cover and let stand in refrigerator several hours or overnight. The longer they stand, the better the flavor. Stir occasionally. Makes 8 to 10 servings.

Dandelion Greens, Potatoes and Cornmeal Dumplings

3 pounds dandelion greens*	¼-pound chunk of salt pork
1 teaspoon salt	6 medium potatoes

Pick over and wash dandelions thoroughly in several changes of water. Place in a kettle, cover with unsalted boiling water and bring to boiling over medium heat. Cook 10 minutes, then drain to remove any lingering grains of sand (this also will take away a little of the bitter tang, if desired). Drain dandelions and half cover with fresh boiling water, adding the salt. Slice the salt pork down to the rind so that it will still be in one piece and add it to the dandelions. Cover the kettle and simmer 1 hour. Add potatoes and cook 30 minutes longer. Make Cornmeal Dumplings and drop by small spoonfuls on top of boiling greens and potatoes. Cover tightly and cook 15 minutes. Remove dandelions, potatoes and dumplings to a serving platter; finish cutting the slices of pork so they can be arranged around greens. Serves 6.

* Fresh greens are best for this good old one-dish meal. You can use salted greens (q.v.) after they've freshened overnight in cold water. If you use canned or frozen greens, adjust the cleaning/cooking accordingly.

Cornmeal Dumplings

1 cup flour	1 cup cornmeal
1¼ teaspoons salt	1 beaten egg
2 teaspoons baking powder	about ¾ cup milk

Sift flour with salt and baking powder. Add the cornmeal and then the beaten egg, which has been combined with the milk. Stir well but do not beat. Drop by teaspoonfuls on top of boiling greens. Cover tightly and cook 15 minutes. Serve hot. Serves 6.

Saturday Night Baked Beans

1 pound yellow-eye or pea beans	½ cup maple syrup*
½ teaspoon baking soda (optional)	½ teaspoon dry mustard
	½ pound salt pork
1 teaspoon salt	1 small onion (optional)

Cover beans with cold water and let stand overnight. In the morning, drain and cover with fresh water. Add the soda. Cover and simmer until beans are just tender: *don't overcook.* (A good test is to spoon out several beans and blow on them; if the skin cracks, the beans are ready.) Drain and put beans in a pot or casserole. Add salt, maple syrup and mustard. Some people like the flavor of onion in their baked beans; if you do, tuck the whole peeled onion well down into the center. Add boiling water just to the top of the beans. Score the salt pork and place on the surface of the beans. Cover the pot and bake slowly in a 275 oven for at least 4 hours—you can scarcely bake beans too long, so don't worry about overcooking. But good baked beans should never be dry; inspect them occasionally and add a bit of boiling water if they appear dry. Six servings.

* Substitute ½ cup sorghum; or ⅓ cup raw sugar and ⅓ cup water; or ⅓ cup of corn syrup and 3 tablespoons molasses—or whatever combination of sweeteners you like, in the amount you like.

Eggplant Party Casserole

This good main-dish vegetable casserole is a cousin to ratatouille, but the cheeses add an Italian touch and it is always served hot.

1 quart (4 cups) frozen eggplant slices*
about 2 tablespoons salt (to draw out eggplant juice)
about ¼ cup vegetable oil (olive is best here)
2 medium onions, thinly sliced
2 cloves garlic, chopped fine
2 small unpeeled zucchini squash (about 7 inches), cut in ½-inch slices
1 quart (4 cups) canned whole tomatoes,** drained, de-seeded and chopped

2 good-sized ribs of celery, chopped small
1 tablespoon fresh basil (or ½ teaspoon dried)
¼ cup chopped fresh parsley
scant ½ cup grated Parmesan cheese (1½ to 2 ounces)
1½ cups ½-inch fresh bread cubes (about 4 slices)
about 2 tablespoons more oil (for pan-toasting bread cubes
1 cup (¼ pound) coarsely grated Mozzarella cheese

* Or peel and slice 2 fresh medium-sized eggplants, sprinkle with salt, let stand; rinse and pat dry, and cube as above.
** Or 4 cups peeled, seeded and coarsely chopped fresh tomatoes—about 3 pounds.

Defrost eggplant slices (this is much easier if they've been frozen between double layers of plastic film), spread them in a large cookie pan and sprinkle them rather generously with salt. Let stand for 10 minutes to draw out excess moisture, then rinse away the salt and pat each slice very dry. Cut slices in ½-inch cubes and brown them lightly in hot oil in the bottom of a heavy pot over moderate heat. Add onions, garlic and zucchini, and cook them together for 2 to 3 minutes, stirring and adding a bit more oil as needed. Add celery, tomatoes and basil, cover and simmer until the squash is tender—about 7 to 10 minutes—stirring occasionally. Remove the pot from heat, stir in the Parmesan cheese and parsley; add salt and pepper to taste.

To bake the whole recipe (it makes about 10 cups), pour all the vegetable mixture into a 3- to 4-quart casserole. Top with bread cubes, which have been lightly browned in about 2 tablespoons of hot oil in a heavy skillet, and sprinkle with the grated Mozzarella cheese over all. Bake uncovered in a preheated 375 oven until it's bubbly and browned—about 30 minutes. Serves 12 to 15.

To freeze: If the whole recipe is frozen in one big chunk, it would take ages to defrost at room temperature, or would need a too-long first baking to thaw and get hot enough for addition of the bread-and-cheese topping—so divide the batch in half and freeze it in two parts. Put each half of the thoroughly cooled vegetable mixture in a 1½- to 2-quart baking dish, and follow directions as given for packaging in the introduction to "Freezing Convenience Foods."

To bake the whole recipe, defrost both halves at room temperature, pour all into a 4-quart casserole, cover, and preheat in a 375 oven until warm through before adding bread cubes and Mozzarella cheese.

To bake half the recipe without defrosting, drop the frozen chunk into its buttered casserole, and bake *covered* in a 375 oven until the mixture is lukewarm—30 minutes or so. Remove the cover and top the vegetables with ¾ cup of pan-toasted bread cubes and ½ cup of coarsely grated Mozzarella cheese. Return to the oven *uncovered* and continue baking until brown and bubbling—about 30 minutes more. Makes 6 good servings.

Mixed Vegetable Casserole

3 strips bacon (or salt pork), diced	1½ cups corn kernels, fresh, frozen or canned
1 cup thinly sliced onions	1 cup chopped toma- toes—canned and drained, or fresh
1 clove garlic, minced	
2 small zucchini (about 7 inches), thinly sliced or cubed	1 teaspoon salt
	¼ teaspoon ground black pepper
3 cups cut or frenched green beans (or a 1½-pint freezer container or 1 pound fresh)	½ teaspoon sugar
	1 teaspoon lemon juice

Fry the bacon (or salt pork) in a skillet until lightly brown. Drain off all except 2 tablespoons of the fat. Add the onions and garlic and sauté until soft—about 5 minutes. Combine the remaining ingredients in a large bowl and stir in the onion-bacon mixture. Spoon into a greased casserole, cover and bake in a 350 oven about 50 minutes. Stir occasionally and gently with a wooden spatula. If more convenient, cook in a heavy pan over low top-of-stove heat for about 45 minutes. In either case remove the cover for the last 5 minutes of cooking. Should the dish need more liquid during cooking, add a little water.

To freeze: The uncooked mixture may be frozen in its casserole for later cooking. Or a casserole may be lined with heavy foil, the mixture poured in and quick frozen; then removed from the dish. In either case tightly overwrap and seal with moisture-vapor-proof covering and return to the freezer.

To cook, defrost, wrapped, in the refrigerator. Bake as usual. Or put still frozen food in freezer-to-oven casserole and bake for about half again the usual time.

Quick Pickled Beets

1 quart canned sliced beets, drained	2 teaspoons mustard seed
1 cup vinegar	½ teaspoon celery seed
¾ cup sugar	1 large onion, peeled and thinly sliced
½ teaspoon salt	

Drain beets, reserving 1 cup of the liquid. Combine liquid with the vinegar, sugar, salt and spices. Bring to boiling over moderate heat; add beets and onion slices. Bring again just to the boiling point. Remove from heat and cool. Cover and refrigerate for 24 hours for flavors to blend. Serve cold. Serves 12.

Baked Cabbage, Tomato and Cheese

3 cups finely shredded cabbage	2 teaspoons light brown sugar (raw)
1½ cups undrained canned tomatoes	1 cup shredded sharp cheese (¼ pound)
¾ teaspoon salt	1 cup dry breadcrumbs
⅛ teaspoon pepper	2 tablespoons butter or margarine
⅛ teaspoon paprika	

Cook cabbage in boiling salted water for 5 minutes; drain. Combine tomatoes, salt, pepper, paprika and sugar. Bring just to boiling point. Arrange ½ of the tomato mixture in a buttered baking dish. Cover with ½ of the cabbage, then ½ the cheese and breadcrumbs. Repeat layers, and dot the top with the butter. Bake in a 325 oven for about 30 minutes—or until browned. Serves 6.

Basic Vegetable Soufflé

1 cup hot cream sauce	3 eggs, separated
1 cup minced, drained cooked vegetables (or canned)	salt and pepper to taste

Combine hot cream sauce with the vegetables and heat to just below boiling point. Beat egg yolks until fluffy, then beat in the hot mixture. Season to taste with salt and pepper. Fold in stiffly beaten egg whites. Bake in a buttered 1½-quart casserole in a 325 oven for about 45 minutes. Serve hot. Serves 4.

Carrots, peas, cauliflower, asparagus, celery, eggplant, celery and onions may all be used in this way.

Tomato-Corn Pudding

2 cups canned cream-style corn	pinch of pepper
1 cup canned tomatoes, undrained	1 tablespoon minced onion
1 beaten egg	1 tablespoon minced green pepper
¼ cup saltine cracker crumbs	1 tablespoon melted butter or margarine
½ teaspoon seasoned salt	

Combine all ingredients in order, then turn into a buttered 1½-quart baking dish. Bake uncovered in a 400 oven for about 30 minutes. Remove from oven and let stand 10 minutes before serving. Serves 6.

Salsify "Oyster" Patties

6 large salsify roots (2 cups mashed)	1 tablespoon butter, melted
1 egg, beaten	flour for dusting
½ teaspoon salt	fat for frying
pinch of pepper	

Wash and scrape salsify; slice and boil until tender—about 30 minutes. Drain; mash well and mix with seasonings, beaten egg and melted butter. Shape in small cakes (about 2 tablespoonfuls to each cake), dust on both sides with flour, and fry in a small amount of hot fat, turning once, until brown. Serve hot. Serves 4.

Rinktum Tiddy

1 tablespoon butter
1 tablespoon chopped
　onion
1 quart canned stewed
　tomatoes, drained of
　their juice

2 well-beaten eggs to bind
　everything together
1 pound grated cheddar
　cheese (4 cups)
salt, cayenne pepper
　and Worcestershire
　sauce to taste

In a 4-quart kettle simmer the onion in the butter until it is soft, add the drained tomatoes and continue to cook slowly for 10 minutes. Meanwhile beat the egg, stir a bit of the hot tomato mixture into it, then return it to the tomato mixture along with the grated cheese. Stir over medium heat until the cheese is melted. Add the salt, cayenne pepper and Worcestershire sauce in amounts to please your taste. To serve, ladle over toast squares, crackers or cooked noodles. Serves 6 easily.

Garden Medley

⅓ cup butter or
　margarine
1 small onion, peeled
　and grated
⅓ cup flour
2 cups milk
1 cup chicken broth
1 teaspoon salt
　pinch of pepper

1 (4-oz.) can sliced
　mushrooms, drained
2 cups cooked
　asparagus (or
　canned)
2 cups cooked green
　peas (or canned)
2 cups cooked sliced
　carrots (or canned)
1 beaten egg yolk

Melt butter, add onion and flour. Combine milk and chicken broth, then blend into butter mixture. Stir and cook over low heat until thickened and smooth. Season with the salt and pepper. Add the vegetables. Stir ¼ cup of the hot sauce into the beaten egg yolk, then add to mixture in skillet. Heat very hot *but don't boil*. Serve over hot toast, biscuits or rice. Serves 6.

Meat and Fish Dishes

Traditional Boiled Dinner

4 pounds corned beef
1 pound salt pork,
    unsliced
¼ cup brown (or raw)
    sugar*
1 large cabbage,
    quartered

10 large whole carrots,
    peeled
1 large turnip, peeled
    and cut in large
    chunks
8 large potatoes, peeled
8 parsnips, peeled
12 small beets, unpeeled

Simmer corned beef and salt pork for 1 hour in unsalted water to cover, to which you've added the ¼ cup of sugar. After an hour, add turnip, carrots and parsnips. Cook 1 more hour, then add potatoes and cabbage. Meanwhile cook beets separately in salted water until tender. When corned beef is very tender, remove to a platter with the salt pork. Drain vegetables and arrange around the meat. Drain beets, immerse briefly in cold water, and slip off skins; add to vegetables on platter. Serve at once. Serves 8.

* This boosts the flavor of corned beef, smoked shoulder or ham, but is not discernable in itself. Substitute ⅓ cup maple syrup or sorghum for the ¼ cup brown/raw sugar.

Red Flannel Hash

4 cups corned beef (left-
    overs of Boiled
    Dinner)
4 cups cooked potatoes

other leftover vegetables
    from Boiled Dinner
salt to taste
bacon fat for browning

The amounts given here are intended to show proportions rather than actual amounts, since leftovers seldom come out evenly. The rule-of-thumb is to have equal amounts of corned beef and potatoes, and enough beets to make the hash dark red. Put corned beef, potatoes and other vegetables through the meat grinder, using a coarse knife. Season to taste with salt and pepper, and brown in melted bacon fat in a large frying pan. Put on a low rack under the broiler for a few minutes to brown the top.

Hearty Vegetable-Beef Stew

about 2 pounds cubed stewing beef, dredged and browned	1 teaspoon salt
	¼ teaspoon pepper
	4 medium potatoes, peeled and cubed
1½ quarts water	2 cups undrained canned tomatoes
1 bay leaf	
½ cup diced celery	¼ cup tomato catsup
¼ cup diced green pepper	1 cup canned tomato juice
4 medium carrots, peeled and sliced	¼ medium head of cabbage
1 large onion, peeled and diced	1 teaspoon sugar

Combine first nine ingredients. Cover and simmer 2 hours. Add remaining ingredients and simmer 1 hour longer. Add additional salt and pepper, if needed. Serves 8.

Roast Venison

In her classic *The Venison Book—How to Dress, Cut up and Cook Your Deer,* Audrey Alley Gorton points out that deer meat that has been properly dressed, skinned, trimmed and chilled should be treated like a prime cut of beef except for one difference. Which is: because venison fat is strong-tasting and acts like mutton fat, it must be trimmed away, and therefore the roast must be cooked with added lard, salt pork or bacon to prevent dryness. Otherwise, season it and cook it by your favorite method for roast beef, including degree of doneness.

However, if you have doubts about the flavor of your venison, marinate and cook it as follows:

5-pound haunch roast	8 cups buttermilk
¼ pound salt pork, cut in strips	4 medium onions, peeled and chopped
1 teaspoon salt	3 bay leaves
⅛ teaspoon pepper	

Add bay leaves, onions and salt to the buttermilk to make a marinade. Marinate the venison refrigerated for 48 hours, turning it night and morning. Discard marinade and lard venison well with the strips of salt pork tucked into deep slits cut in the meat. Place in a roasting pan and roast uncovered in a 350 oven until tender, allowing 20 minutes to the pound.

Alpine Stewed Venison (or Beef)

This is quite a large recipe, so half—or all of it, in 2 easy-to-handle amounts—can be frozen. If you do plan to freeze it, note the several small variations in handling below.

3 pounds boned venison (or beef chuck), frozen or fresh	1 tablespoon paprika
	1 cup water
¼ cup flour*	1 tablespoon tomato paste (or ½ cup drained, canned, stewed tomatoes, sieved)
2 teaspoons salt	
¼ teaspoon ground black pepper	
3 to 4 tablespoons vegetable oil	1½ cups dry red wine
1 cup chopped onions	1 cup sour cream, more or less according to taste
2 cloves garlic, minced	

* If you plan to freeze this, use the same amount of waxy corn flour or waxy rice flour (since wheat-thickened gravies and sauces tend to separate from freezing).

Cut the meat in 1½-inch cubes and dredge it well in a mixture of the flour, salt and pepper. In a heavy skillet or a Dutch oven, brown the meat in the oil. Add onions and garlic and continue cooking over moderate heat for 10 minutes, stirring often. Add paprika, water, wine and tomato paste (or sieved tomatoes). Bring to the boil, cover, reduce heat and cook slowly, stirring occasionally, until the meat is tender—about 2½ hours. Take care lest it cook dry: add a little hot water if needed. At serving time remove the meat to a warmed serving dish, and stir the sour cream into the gravy; heat to just below a simmer, and pour it over the meat. Serves 10 to 12.

To freeze: In addition to substituting for the wheat flour, as noted, *omit adding the sour cream.* Pack the well-cooled cooked meat and its gravy in an adequate freezer box (or, if you're freezing the whole recipe, in 2 boxes to cut down on defrosting time), seal, label and freeze.

To serve, defrost the meat overnight in the refrigerator, and heat it slowly in a casserole or Dutch oven. Remove the meat to a warm serving dish, *now add the sour cream* to the gravy; heat, and pour over the meat, as above.

Hamburger and Onion Shortcake

1 tablespoon melted butter or margarine	2 eggs, beaten
4 medium onions, sliced thin	1 cup sour cream paprika
½ pound lean hamburger	½ batch Buttermilk Biscuits (1 cup flour)
1 teaspoon salt	
¼ teaspoon pepper	

Simmer butter and onions over low heat for about 10 minutes—until tender but not brown. Add hamburger, crumbling with a fork, and cook it for about 5 minutes until meat loses its red color. Beat eggs, sour cream and seasonings well, and combine with meat mixture. Make half the rule for Buttermilk Biscuits (q.v.), roll out on a floured board to about ¼ inch thick. Press the pastry into a lightly greased 9-inch pie tin. Pour the meat mixture over the biscuit pastry, sprinkle with paprika, and bake in a 375 oven for about 35 minutes. Serve hot, cut in wedges. Four hearty servings for hungry people.

Italian Liver

Among the many accomplishments of the Italians is their invention of what might be called fried liver and onions—but how different *fegato alla Veneziana* is from the usual "smothered" flap of sole-leather! The tomatoes and green pepper may be omitted: cooking method is the main thing here.

1 pound sliced frozen or fresh calf's liver°	1 teaspoon chopped fresh basil (or ¼ teaspoon dried)
1 medium onion, sliced very thin	½ teaspoon salt
2 firm, ripe, medium-sized tomatoes°°	⅛ teaspoon pepper
½ medium-sized sweet green pepper, sliced small	3½ to 4 tablespoons good olive oil

In a heavy skillet, start the sliced onion cooking over low heat in a scant 2 tablespoons of oil. After a few minutes add the ½ green pepper, seeded and cut small. While these are cooking, peel the tomatoes, squeeze out most of their seeds and juice, and cut small; add to the onion mixture with the salt, pepper and basil, and cook until the vegetables are soft. Meanwhile pat the sliced liver dry; trim it as needed, and cut it in pencil-thick strips, then cut the strips in pieces about 1½ inches long. Remove the skillet of vegetables from heat, but let them keep warm in their pan. In another heavy skillet *over high heat* sauté the liver pieces, stirring and turning constantly so they don't stick, until they are cooked through —about 3 minutes. Immediately dump the liver into the waiting skillet of warm onions, etc., return it to heat for a minute to let flavors combine, then serve. Serves 4 generously.

° Baby beef or pork—even regular beef—liver are good too; small pieces and very quick cooking prevent the older liver from becoming tough.

°° Or about ⅔ cup well-drained canned stewed tomatoes.

Baked Liver Loaf

1½ pounds beef liver, cut in strips	pinch each of marjoram, rosemary and basil
½ cup dry breadcrumbs	¼ teaspoon sage
1 large onion, ground	1 egg, slightly beaten
¾ pound sausage	⅓ cup light cream
1 teaspoon salt	4 large strips of bacon
pinch of pepper	

Dredge liver strips in the breadcrumbs, then put them through the medium knife of a food grinder, with the onion. Mix with sausage and seasonings. Add cream to the egg and mix thoroughly in with the meat mixture. Pack firmly into a standard-size bread tin. Cut each bacon strip in half and lay it across the top, forming a solid layer of bacon. Bake in a 375 oven for about 1 hour. Serves 6 to 8.

This tasty loaf is also good sliced cold, like a portion of pâté, or served cold with potato salad.

Baked Stuffed Heart

1 large beef heart	2 cups Sausage Stuffing
8 cups cold water	(about ½ batch)
1 cup vinegar	page 511
1 teaspoon salt	½ cup water

Soak the heart for 1 hour or longer in the cold water, to which the vinegar has been added. Drain and cut away fat and tubes. Parboil for 1 hour in fresh water to which the 1 teaspoon of salt has been added. Remove and drain well. Stuff the heart with Breadcrumb Sausage Stuffing (q.v.) and sew it together at the top. Place in a heavy baking pan that has a lid, add the ½ cup fresh water. Cover and bake in a 350 oven until tender—about 2 or 3 hours. Serves 6.

Two veal or pork hearts may be used for the same yield. If the smaller hearts are used, parboiling may be omitted and the salt added to the ½ cup of water in the pan for baking.

Breadcrumb Sausage Stuffing

3 cups coarse dry bread-crumbs	⅛ teaspoon pepper
2 teaspoons sage or poultry seasoning	½ pound sausage meat
	¼ cup melted sausage fat
1 teaspoon salt	½ cup hot water

Add seasonings to breadcrumbs. Fry sausage meat gently for about 10 minutes, but do *not* let it get crisp and brown. Remove meat and drain, reserving ¼ cup of the fat. Add fat to breadcrumbs and seasonings, then add sausage meat. Stir in the hot water. Mix well but do not beat. Let cool before stuffing a bird; do not pack tightly, or it will be soggy. About 4½ cups.

Fried Salt Pork and Gravy

½ pound salt pork, sliced very thin	2 tablespoons flour for thickening gravy
flour and cornmeal for dredging	2 cups milk (or sour cream)

Cover pork slices with boiling water and let stand 3 minutes: *do not allow to cook*. Remove slices and drain well. In a mixture of equal amounts of flour and cornmeal, dredge the slices. Fry in a heavy frying pan in their own fat until crisp and golden brown. Remove slices and drain on a paper towel: keep hot. Pour off enough fat from the pan to leave only 2 tablespoonfuls. Return the pan to heat and blend the flour into the fat. Add the milk. Stir and cook over low heat until smooth and thickened, making a creamy gravy speckled with tiny crumbs of pork. Serve pork slices with boiled potatoes, the gravy in a bowl on the side. Serves 4 to 5.

Olden cooks sometimes made the gravy with sour cream and omitted the flour, since the cream needs no additional thickening. The sour-cream gravy has an unusual and good flavor.

Ham and Dried Apples with Egg Dumplings

(This fine old Pennsylvania Dutch rule is based on using uncooked home-cured ham and dried apples: there's
nothing "instant" about it, and it's wonderful eating.)

end of ham with
    bone, about 3 pounds
2 cups dried apples
	2 tablespoons light brown
	    sugar (or raw)

Soak dried apples for several hours (or overnight if
they're quite leathery) until they've softened a bit and be-
gun to plump. Meanwhile cover ham with cold water and
bring to a boil, covered. Simmer about 2 hours, then add
drained apples; simmer about 1 hour longer. Remove ham
to a platter, lift apples from the pot with a slotted spoon,
and place them around the ham; keep all hot while the
dumplings are cooking. Serves 6.

Egg Dumplings

1½ cups sifted flour
3 teaspoons baking
    powder
½ teaspoon salt
	1 tablespoon butter or
	    margarine
	about ¼ cup milk
	1 beaten egg

Sift flour with baking powder and salt. Rub in the butter
with your fingertips and stir in enough milk to make a soft
dough. Stir in the beaten egg. Drop from a spoon into the
boiling ham broth. Cover tightly and simmer about 12 minutes. Arrange dumplings on the platter around the ham and
apples, and serve.

Sausage Yorkshire Pudding

1 pound sausage (bulk or link) 1 cup milk	2 eggs, very well beaten 1 cup sifted flour ½ teaspoon salt

Make small patties of sausage and arrange them in the
bottom of a 9-inch-square cake pan. (If link sausage is used,
prick each link several times with a fork, and arrange in the
pan.) Bake for 15 minutes in a 425 oven. Meanwhile beat

milk into eggs, slowly add flour sifted with salt, and continue beating until big bubbles rise from the bottom of the batter. Take sausage from the oven, and drain off enough fat to leave about ⅛ inch of fat covering the bottom of the pan. Quickly pour batter onto hot sausage and fat and return to the oven. Bake 30 minutes more at 425—or until puffed high with brown peaks but not too crisp. Serve at once, cut in squares. Serves 4 to 6.

Wonderful "Stretched" Chicken

This takes more time than some dishes, but it makes 10 delicious servings—and is really kind to the budget.

1 fowl (*c.* 5 pounds), cut up (or 2 small cut-up chickens from the freezer)	3 tablespoons soft butter or margarine
	salt and pepper to taste
1 small onion, peeled salt*	1 cup finely diced celery
	5 eggs
7 cups whole-wheat breadcrumbs	4 tablespoons flour
	1 chicken bouillon cube

*If you're using cut-up fryers, add 2 bouillon cubes to the water to make a more flavorful broth, and reduce the amount of salt in the water.

Cover chicken with lightly salted water, add the onion and simmer until tender; save the broth. Remove the bones and skin and cut the chicken in small pieces (about 1-inch cubes). Season the crumbs with salt and pepper and blend in the butter; reserve 1 cup for the topping. Boil the celery for 10 minutes in 1½ cups of the broth, then add it, with its cooking liquid plus enough more broth to make 3 cups, to 6 cups of the seasoned crumbs; stir to mix, and add 2 of the eggs, slightly beaten. Set the mixture aside. Dissolve the bouillon cube in 3 cups of the broth and carefully blend in the flour; stir and cook gently until the broth is slightly thickened. Remove from the heat and quickly whisk in the remaining 3 eggs, well beaten, return to the heat and simmer and stir until thick; season to taste with salt and pepper; cool. In a large, oblong baking pan (about 10 × 13 inches) make alternate layers of crumb dressing, cut-up chicken, and the custard-like sauce; top with the reserved 1 cup of crumbs. Bake in a 375 oven until slightly bubbly—about 30 minutes. Serves 10.

Cadillac Codfish

about ⅔ cup salt codfish pieces	2 tablespoons butter
	pepper to taste
1½ cups lukewarm water	1½ cups Tomato Sauce
4 boiled medium potatoes, cold, sliced	1 cup buttered cracker crumbs (or breadcrumbs)

Break the salt cod into small pieces; cover with the lukewarm water, simmer until soft, drain. In a buttered 1½-quart baking dish arrange alternate layers of potatoes and fish; sprinkle each with bits of butter and pepper. Pour 1½ cups Tomato Sauce over all and sprinkle with the buttered crumbs. Bake 20 minutes in a moderate oven (375 to 400 F.). Serves 4.

Tomato Sauce

1¾ cups fresh or canned stewed tomatoes	3 tablespoons butter
1 generous slice of onion, chopped	2½ tablespoons flour
	¼ teaspoon salt
	¼ teaspoon pepper

Simmer onion with the tomatoes for 15 minutes, then rub through a sieve. Blend together the butter, flour and seasonings in a small saucepan and cook gently for 3 minutes, combine with sieved tomato mixture and cook until it thickens—about 5 minutes.

Codfish Cakes

1 cup shredded salt codfish	1 tablespoon butter or margarine
6 medium potatoes, peeled	⅛ teaspoon pepper
	1 beaten egg
	fat for frying

Soak codfish in cold water for 1 hour. Drain and cover with boiling water to cover well. Add potatoes and cook until tender. Drain well. Mash with a fork and add butter and pepper. Stir in beaten egg. Mix lightly but well. Shape into balls and fry in deep fat, or brown in shallow fat in skillet until browned on both sides. Makes about 8 good-sized balls, or 4 hearty servings.

Codfish Gravy

Allow 1 cup hot unsalted cream sauce to each 1 cup of freshened salt codfish. Freshen fish in cold water overnight, then drain. Cover with fresh water and simmer 5 minutes. Drain well and add to cream sauce. Serve hot with boiled potatoes.

Fish Hash

2 cups flaked cooked fish	peeled and chopped
2 cups diced cooked	fine
potatoes, cooled	1 teaspoon salt
2 tablespoons melted	pinch of pepper
bacon fat	1 teaspoon Worcester-
1 small onion, peeled	shire sauce
and minced	(optional)
2 hard-cooked eggs,	⅔ cup milk

Sauté onion in the melted fat until light golden—about 5 minutes. Meanwhile stir gently together the fish, potatoes, chopped eggs, seasonings and milk. Mix lightly, then turn into the skillet with the onion and fat. Cover and cook over moderate heat until the bottom is crusty and brown (lift an edge to see). With a pancake turner, fold half the hash over the other half, as with a puffed omelet, and serve at once. Serves 5.

Breads, Cakes and Cookies

Quick Pumpkin Bread

1½ cups sugar
½ cup vegetable oil
2 eggs
1⅔ cups flour
¼ teaspoon baking powder
1 teaspoon baking soda
¾ teaspoon salt
¼ teaspoon cinnamon
½ teaspoon cloves
½ teaspoon nutmeg
1 cup strained canned pumpkin
½ cup chopped nuts
1 cup raisins

Thoroughly mix together the first three ingredients. Sift together the dry ingredients and add to the sugar mixture. Mix well. Add pumpkin, nuts and raisins. Pour into a standard bread pan and bake in a 350 oven for 1½ hours. Remove from the pan and cool on a rack.

Apple Bran Bread

2 cups bran
2 teaspoons salt
½ cup sugar
2 tablespoons melted butter or margarine
1½ cups sweetened applesauce
1½ cups boiling water
1 yeast cake (or envelope dry yeast)
⅓ cup lukewarm water
about 8 cups sifted flour

Combine the first five ingredients, then stir in the boiling water. Let stand until lukewarm. Soften yeast in the lukewarm water and add to the first mixture when it has cooled to lukewarm. Add flour to make a smooth, easily handled dough. Knead slightly on a floured board, place in a large bowl; cover and let rise in a warm place until doubled. Knead again until smooth, then shape in 2 loaves. Place in lightly greased bread tins and cover with a towel; let rise until doubled. Bake in a 350 oven for about 50 minutes or until bread tests done. Turn out on a wire rack and cool thoroughly before slicing.

Country Buttermilk Biscuits

2 cups flour
1 teaspoon salt
1 teaspoon any baking powder
½ teaspoon baking soda

4 tablespoons lard (or vegetable shortening)
about ¾ cup buttermilk

Sift flour with baking powder, salt and soda. Quickly rub in lard, then add enough buttermilk to make a soft dough that yet can be handled. Turn out on a floured board and knead lightly a few strokes, then gently roll out to about ¾ inch thick. Cut in rounds, bake on a greased sheet in a 425 oven for 15 minutes. Makes about 15 biscuits.

Substitute shortening: An equal amount of any solid vegetable shortening may be used instead of the lard. So may oil—but it does not give the same lightness and texture that solid shortening does.

Salt Pork Cake

Sometimes called "Poor Man's Fruitcake" because it is economical and calls for ingredients that deep-country housewives had ready on hand, this cake is dark, moist and mighty good. It is from Mildred Wallace's grandmother's rule, which has been expanded from the lady's abbreviated instructions for her own use, and updated for modern kitchens.

½ pound salt pork, finely ground or diced small
1 cup brown sugar, firmly packed
1 cup molasses
1 cup boiling water
2 beaten eggs (optional) about 5 cups flour
1 teaspoon baking soda
1 teaspoon cream of tartar
¼ teaspoon ground cloves

½ teaspoon ground cinnamon
⅛ teaspoon ground nutmeg
2 cups seeded raisins, chopped
½ cup finely chopped candied citron— about ¼ pound (optional)
½ cup dried currants (optional)
1 cup broken nutmeats (optional)

In a large bowl mix the salt pork well with the brown sugar, then add the molasses and the boiling water. Stir together and add the beaten eggs at this time (if you use them). Sift together about 4 cups of the flour, the baking soda, cream of tartar and spices, and mix these dry ingredients in well with the pork mixture, a little at a time. In a separate bowl, toss ½ cup of the remaining flour together with the raisins and optional fruits and nutmeats until they are well coated; add them to the batter, using a little more flour if needed—the batter should be stiff. Spoon the batter 2½ inches deep into a 10-inch greased tube pan lined with cut-to-shape greased waxed paper; put the surplus in a similarly prepared 4½-×-8-inch loaf pan (either size will yield a 2½-pound cake).

Bake the cakes in a preheated slow oven (275 F.) until a toothpick pushed into the center of each comes out clean— about 2½ to 3 hours. Place a shallow pan of water in the oven during the first 2 hours of baking to keep the cakes from drying. Cool the cakes in their pans on a wire rack for 20 to 30 minutes, then remove the cakes and peel off their paper liners to finish the cooling. Wrap thoroughly cooled cakes securely in aluminum foil and store them in a cool, dry, safe place for at least 2 weeks before cutting.

Variation: Wrap the cooled cakes in brandy-soaked cheesecloth and overwrap well with foil; then store in an earthenware crock in a cool place for at least 2 weeks before cutting.

Dried Apple Fruit Cake

3 cups dried apples	3 eggs, beaten
3 cups light molasses (right: there's no other sweetening)	1 teaspoon baking soda
	1 teaspoon salt
	1 teaspoon cinnamon
1 cup seeded raisins	½ teaspoon nutmeg
3 cups flour	¼ teaspoon cloves
1 cup softened shortening	

Soak apples overnight in just enough water to cover. In the morning cut apples quite fine, add molasses, and cook until apples are very tender. Add raisins and cook 5 minutes more. Remove from heat; cool. Add shortening and eggs. Sift dry ingredients together and add. Blend well, then pour into 2 standard-size bread tins lined with waxed paper. Bake in a 350 oven for 1 hour—or until a toothpick poked in the center comes out clean. Cool on a rack, remove from pans and peel off the paper. Wrap and store.

The bits of apple in this very old cake taste like citron.

Applesauce Cake with Praline Topping

2¾ cups sifted flour	½ teaspoon ground
1⅓ cups sugar	cloves
2 teaspoons baking	½ cup softened
powder	shortening
¼ teaspoon baking soda	2 cups canned smooth
½ teaspoon salt	applesauce
1½ teaspoons ground	2 eggs, to be beaten in
cinnamon	1½ cups seedless raisins

Sift flour with sugar, baking powder, baking soda, salt, cinnamon and cloves. Add shortening and applesauce. Beat with electric beater until blended, about 2 minutes. Add eggs and beat 1 minute longer. Stir in raisins. Bake in a greased 13 × 9 cake pan in a 350 oven for about 35 minutes —or until cake tests done. Spread cake with Praline Topping and place under broiler on the lowest rack for 5 minutes. Topping should be golden brown and bubbling when done. Cool cake in the pan set on wire rack. Serve slightly warm or cold. Makes 20 servings.

Praline Topping

½ cup softened butter or	¼ cup undiluted evapo-
margarine	rated milk
¾ cup firmly packed light	¾ cup chopped nuts
brown (raw) sugar	⅔ cup flaked coconut

Blend butter and sugar, then beat in evaporated milk. Blend thoroughly, stir in nuts and coconut. Spread on cake, broil as directed.

Blueberry Buckle

¾ cup sugar
¼ cup soft vegetable
    shortening (or
    butter)
1 egg
½ cup milk
2 cups all-purpose flour
½ teaspoon salt
2 teaspoons double-

acting baking
    powder
optional: ¾ teaspoon
    grated lemon rind
2 cups drained canned
    blueberries (or 1
    pint defrosted
    frozen berries or 2
    cups fresh)

Preheat oven to 350 or 375 F.° Sift together the dry ingredients and set aside. In a large mixer bowl cream shortening with the sugar; when smooth beat in the egg and lemon rind. At low speed blend in the milk, then the sifted dry ingredients; when smooth stir in, by hand, the blueberries. Pour into a greased and floured 7½-×-12-inch baking pan and sprinkle topping (below) over the batter.

°If the pan is metal, bake this in a 375 oven for 45 to 50 minutes; if the pan is glass, bake in a 350 *oven* for the same time. Cool, in the pan, on a wire rack. This is best served warm, but if time is a problem it does reheat quite well. Serves 10 to 12.

Crumb Topping

½ cup sugar (or brown
    sugar, firmly
    packed)
⅓ cup flour

½ teaspoon cinnamon
¼ cup soft butter or
    margarine

Measure all ingredients into a small bowl and blend them with a fork or a pastry blender until the mixture is almost the consistency of cornmeal. Sprinkle it over the Blueberry Buckle batter before baking it. This is also a fine topping for a simple white or yellow cake instead of the usual icing.

Pumpkin Cookies

⅓ cup shortening	2½ cups flour
1 cup sugar	4 teaspoons baking
2 eggs	powder
1 teaspoon vanilla	¼ teaspoon salt
1 teaspoon lemon	½ teaspoon ginger
extract	½ teaspoon nutmeg
1 cup mashed cooked	1 cup seeded raisins
pumpkin (or	½ cup chopped nuts
canned)	

Cream shortening and sugar, beat eggs in well. Stir in the vanilla and lemon extract. Put pumpkin through a sieve and add, mixing well. Sift dry ingredients and add with the raisins and nuts. Mix thoroughly. Drop by the teaspoonful onto a greased cookie pan and bake about 15 minutes in a 375 oven. Makes 4 dozen.

Carrot-Orange Cookies

1 cup mashed carrots	½ teaspoon vanilla
(2 cups canned	1 egg
sliced, drained)	2 cups sifted flour
1 cup sugar	2 teaspoons baking
1 cup softened shorten-	powder
ing	½ teaspoon salt
1 tablespoon grated	
orange rind	

Combine carrots, sugar, shortening, grated orange rind, vanilla and the egg and beat hard until thoroughly mixed. Sift flour with the baking powder and salt; stir into the batter. Mix well. Drop by heaping teaspoonfuls onto a greased cookie pan. Bake in a 375 oven for about 20 minutes. Cool on a wire rack. Makes 4 dozen.

Puddings and Pies

Old Settler Indian Pudding

5 tablespoons yellow cornmeal	2 eggs, beaten
4 cups milk, scalded	1 teaspoon cinnamon
2 tablespoons butter	¾ teaspoon ginger
1 cup maple sirup*	1 teaspoon salt
	1 cup cold milk

Add cornmeal to hot milk, and cook over low heat until thickened, stirring constantly. Remove from heat, add butter, maple sirup, beaten eggs, spices and salt. Mix well and pour into a buttered baking dish. Bake in a 300 oven for about 1 hour. Stir and add the 1 cup of cold milk and continue baking for 1 more hour. Serve warm with plain or whipped cream, even vanilla ice cream. Serves 8 to 10.

* ½ to ⅔ cup brown (raw) sugar and ¼ cup molasses may be substituted. Experiment.

Frances Bond's Rhubarb-Berry Fool with Yogurt

This light and delicious dessert really gains from being made with frozen fruit. Mrs. Bond's only sweetening is in the rhubarb sauce, but a bit more sugar—or a touch of mild-flavored honey—may be folded into the mixture before it is chilled in the refrigerator if you like it sweeter.

2 cups frozen pre-sweetened rhubarb sauce (1 pint container)	sweetened strawberries (or raspberries)
1 to 2 cups frozen un-	1 cup plain yogurt*

*Whipped cream can be substituted for the yogurt—½ cup heavy cream makes 1 cup whipped—but this dessert loses lightness and tang when made with cream.

Defrost the fruit completely at room temperature. Press fruits through a sieve, or smooth them in a blender. Combine them with the yogurt, mixing well until the result is creamy. Put it in a pretty serving dish (or in individual sherbet glasses) and chill in the refrigerator for at least 1 hour before serving to allow flavors to blend. Serves 6.

Basic Rule for Fruit Cobblers

around ¾ cup sugar
1 tablespoon cornstarch
1 cup boiling water
3 cups prepared fresh fruit

1 tablespoon butter or margarine
¼ teaspoon cinnamon (optional)
⅛ teaspoon nutmeg (optional)

Topping

1 cup sifted flour
1 tablespoon sugar
1½ teaspoons baking powder

½ teaspoon salt
¼ cup shortening
½ cup milk

Prepare fruit by washing, stemming, peeling and slicing as necessary. Combine sugar and cornstarch, blend with boiling water. Stir over moderate heat until it boils, then cook 1 minute more. Add prepared fruit, and pour all into a buttered 10 × 6-inch baking dish and dot with butter. Sprinkle with spices, if desired (cinnamon is especially good with fresh blueberries and blackberries). Put the dish in a 400 oven to keep hot while you make the topping. Combine sifted flour, sugar, baking powder and salt; sift together. Rub in shortening and stir in milk to make a soft dough. Drop by spoonfuls on to surface of the hot fruit. Bake at 400 for 30 minutes. Serve warm with plain cream. Serves 6.

Be sure the fruit mixture is bubbling hot when you add the dough, otherwise the topping will have raw spots underneath.

Adjust the amount of sugar according to tartness of the fruit used.

Canned Fruit Cobblers

Following Basic Cobbler recipe, use 1 quart of Open-Kettle Canned fruit *but omit* the boiling water, and *cut down* the sugar to ½ cup, or to taste.

Steamed Carrot Pudding

1 cup grated raw carrots (about 3 medium)	1½ cups flour
	1 teaspoon salt
	1 teaspoon ground cinnamon
1 cup grated raw potatoes (about 2 medium)	½ teaspoon ground nutmeg
1 teaspoon baking soda	½ teaspoon ground cloves
½ cup butter or margarine, melted	1 cup seeded raisins
1 cup brown (raw) sugar	1 cup heavy cream for whipping

Dissolve soda in combined grated potatoes and carrots. Stir in melted butter and sugar. Sift flour with salt and spices and add to first mixture. Stir in raisins. Steam 3 hours in a tightly covered, greased pudding mold. Serve hot with the whipped cream, sweetened to taste. Serves 6 to 8.

Steaming

Round tins (1-pound coffee cans are fine) with tight-fitting covers do well if you have no pudding mold. Heavy waxed paper or foil held firmly in place with a rubber band will serve as a cover in a pinch.

Fill the well-greased mold about ⅔ full of batter. Cover it, and put it on a trivet or rack in a pan of boiling water that comes halfway up the sides of the mold. If the mold tries to float, put a weight on it.

If water level drops, add more boiling water, and do *not* let the water stop boiling.

Betsy Pudding

½ cup firm jam	1 tablespoon melted butter or margarine
1 egg	
2 tablespoons sugar small pinch of salt	2 cups scalded milk
¼ teaspoon nutmeg	1 cup coarse dry bread-crumbs
	1 teaspoon vanilla

Spread jam over the bottom of a buttered small casserole. Beat the egg with sugar, salt, spice and melted butter, then beat in the hot milk. Stir in the breadcrumbs and vanilla. Mix well, then pour it over the jam. Bake in a 350 oven for about 40 minutes—or until puffed and brown. Serve warm or cold. Serves 4 to 6.

Basic Berry Pie

3 cups fresh berries	2 tablespoons margarine
about 1 cup sugar	or butter
¼ cup sifted flour	¼ cup cold water
small dash of salt	pastry for 2-crust pie

Turn the berries into a bowl. Sift sugar, flour and salt together and gently mix with the berries. Turn into a pastry-lined 9-inch pie pan. Dot with the margarine, then pour the cold water over all. Cover with the top crust, seal and crimp the edges and cut a vent in the top as usual. Bake in a 450 oven for 10 minutes; reduce heat to 375 and bake about 30 minutes longer. Cool before serving. Serves 8.

Unsweetened, whole frozen berries can be used in this good old rule; so may Cold-Water Canned blueberries, drained. Blackberries and blueberries take kindly to ¼ teaspoon of cinnamon and 1 tablespoon of lemon juice. The amount of sugar depends on the tartness of the berries.

Old Newfane Inn Squash Pie

1 cup canned, strained winter squash	⅛ teaspoon ground cinnamon
1½ teaspoons flour	⅛ teaspoon ground nutmeg
1 cup sugar	
2 medium eggs	⅛ teaspoon salt, if wanted
1 pint (2 cups) milk	
1 tablespoon butter	pastry for a 1-crust pie (large)

In the large bowl of a mixer or in a blender combine the strained squash, sugar, eggs and spices and beat until smooth. Scald the milk with the butter in it; add to the squash mixture and pour into a pastry-lined 9-inch pie plate. Bake in a 425 F. oven for 15 minutes, reduce heat to 350 F.

and bake for 30 minutes longer or until a knife blade inserted about 2 inches from the center of the pie comes out clean. Remove to a wire rack to cool. Serves 6 to 8 depending on your generosity; you'll be tempted to hide the extra cut for yourself.

Applesauce Custard Pie

1 cup sweetened apple-sauce, very smooth	pinch of salt
1 cup milk	¼ teaspoon ground nut-meg
2 beaten eggs	pie pastry for 1 crust

Beat eggs well, and combine with applesauce, milk and seasoning. Continue beating together until thoroughly blended, then pour into a pastry-lined, 9-inch pie pan and bake in a 425 oven for 10 minutes; reduce heat to 350 and continue baking for 25 minutes longer—or until the custard filling is set. (Test by inserting a table-knife blade near the center of the filling: if it comes out clean, the custard is done.) Cool. Serves 6 to 8.

Mince Pie: Use 3 cups Mincemeat or Green Tomato Mincemeat (in Preserves) and your favorite 2-crust rule for a generous 9-inch pie. Bake 10 minutes at 425, reduce to 350 and bake 25 minutes more. Serve with natural Cheddar. Serves 8.

Rhubarb Pie

3 cups sliced uncooked rhubarb	few grains salt
1 cup sugar	1 beaten egg
2 tablespoons flour	pastry for 2-crust, 9-inch pie

Combine first five ingredients. Put the mixture in a pastry-lined pie pan. Cover with the top crust, seal around the edge and cut or prick vents in the center. Bake in a 400 oven for about 50 minutes. Cool before cutting. Serves 8.

Oddments

Bea Vaughan's Spiced Blackberry Cordial

This pleasant old "company" refreshment makes a delightful Holiday present, especially for the older folks we love.

4 cups home-canned blackberry juice*	2 tablespoons whole cloves
2 cups sugar, or its sweetening equivalent in honey**	¼ ounce whole allspice about 1 pint unflavored brandy, or cognac
½ stick of cinnamon	

Tie the spices in a small double-layer cheesecloth bag, add to the juice and sugar, and bring all to boiling. Reduce heat, and simmer for about 20 minutes to get the benefit of the spices. Remove the spice bag and strain the hot juice into a hot sterilized pitcher, preferably enameled-ware. Stir in the brandy, and pour the cordial, still very warm, into sterilized bottles or small canning jars, leaving no headroom. Cap immediately with sterilized tops. Protect the join of cap and container with sealing wax or pressure-sensitive tape, unless you are using conventional 2-piece screwband lids for the jars.

Pretty slender-necked bottles that sirups or salad-dressings came in work well. If they have screw-tops, remove each lining disc before sterilizing the lids; put a small double fold of good-quality household plastic wrap over the rim of the bottle, and screw the cap on tight. Or buy appropriate corks from the hardware store, sterilize them, and push them down in the necks of the bottles (having left adequate headroom), and cover the seals with sealing wax.

*If the juice is presweetened, cut down on the added sweetening: the cordial doesn't want to be sirupy.

**How sweet you like the cordial is a matter of taste. Only mild-flavored honey should be used, and it should not be substituted measure-for-measure for sugar (see "Sweeteners" in Chapter 1).

Hay-time Switchel

1 cup light brown sugar (raw)	1 tablespoon ground ginger
1 cup apple cider vinegar	1 quart cold water
½ cup light molasses	

Combine and stir well. Makes about 6 seven-ounce glasses.

This can be refrigerated, but old-timers made it with cold, cold spring water, and said nothing quenched a thirst or cooled a dusty throat in haying time so well as this homey drink. (It's a good energy-restorer without promoting "cotton mouth" in athletes; cross-country skiers or snowshoers should drink it during a tour race with the chill off, though.)

Rhubarb Punch

8 cups diced uncooked rhubarb, not peeled	3 lemons
5 cups water	a few drops of red food coloring, if needed
about 2 cups honey or sugar	1 quart pale dry ginger ale, chilled
6 oranges	

Simmer the rhubarb in the water until it's quite mushy. Strain (use a muslin jelly bag if you want it really clear), and measure the liquid into an enameled kettle. Add ⅓ cup sugar or honey for each 1 cup of rhubarb juice, stirring over low heat until the sweetening has dissolved. Cool. Add strained juice of oranges and lemons plus food coloring; chill. Just before serving, add ginger ale carefully. Serve over ice cubes. Makes about 3½ quarts of punch before icing.

In olden days, variations of this very good cooler were made for hill-country weddings where oranges and lemons were rare and ginger ale was virtually unheard of.

Postscript

(Update on Some Canning Worries)

The comments on canning-jar lids and mushy canned green beans are based on information gathered by *PFB* during the 1975 canning season from May (when the Second Edition went to press) to October (in time for the third printing). We present them as a postscript, cross-referred to the relevant earlier pages, for three reasons: (1) they have a bearing on food safety; (2) they are too long to be added to the main text without throwing all the page references out of kilter; and (3) some parts of the problems may have resolved themselves by the spring of '76.

Newcomers in the Canning-lid Crunch

New entries in the home-canning-*jar* market started appearing in our corner of New England in early summer of 1975, and the half-dozen recent brands of modern masons (a type described on page 57) which we found, all appear to be made by manufacturers of glass containers for the food industry. The *glass* in all the jars—five made in the U.S.A. and one from Canada—seems to be of a standard weight and strength suitable for use both in Pressure and in Boiling-Water Bath canners.

However, the *closures* that came with the boxed jars did not do their job with uniform effectiveness for us; and because we believe that "by their seals you shall know them," we include the 2-piece screwband lids on these jars in the following general discussion of the new proprietary lids and caps that have been sold separately.

THE NEW CLOSURES SOLD SEPARATELY

While jars seemed to be in adequate supply, during the summer of 1975 even non-canners were aware of a critical scarcity of replacement lids for the familiar 2-piece screw-band closures for modern masons. Obviously the dearth of replacement lids would prompt a number of makers to enter the home-canners' market—and they did.

Consensus among food technicians in and out of government is that the supply of *safe* canning-jar closures is likely to be ample for 1976. For one thing, the familiar brands of lids should once more be in general distribution. For another, the good newcomers, bolstered by further independent testing for home use, are likely to stay on the market. And last, the opportunistic fly-by-night entries may drop out of the market altogether.

Meanwhile, though, as a help to people who have collected recently marketed closures for use next year, we offer the checklist below. For fairness, we give no brand names, and our remarks will be quite general.

CHECKLIST FOR SAFE CANNING-JAR CLOSURES

GOOD DIRECTIONS FOR USE?

The most experienced hands at safe home-canning pay attention to the instructions for using any new product, and they read with really critical attention if the product has an unusual feature.

First, *are* there any directions, either printed on the container or on a separate sheet/folder inside the package?
   (A *responsible* manufacturer—one who knows home-canning, and has food safety in mind—will want us to use his closure in the way that gets good, wholesome results which store well.)

Next, if there are directions, do they make sense? Although they may be terse, do they cite with helpful clarity the steps required for using that *particular closure successfully?*

Then, do the directions say how to handle the particular closure after the processed jar of food comes from the canner? Do they tell us explicitly *how this closure looks* when the cooled jar has a good seal? Or whether, if it's a conventional 2-piece screwband type, the band may/may not be removed before storing?

(One brand of jars newly marketed in our area never mentioned the distinctive appearance which the top of the lid should have after the seal was accomplished. This sort of lapse is no help to a first-time canner who might think a jar has sealed when it hasn't, and store it away—to spoil. Phooey to such directions.)

And finally, do the directions say clearly that the steps they give cover only the procedure for using *that closure?* Do they therefore tell the home-canner to use an established guide for complete instructions on packing and processing specific foods?

(Makers of canning-jar closures have an obligation to the people who use their products, and it takes only a few lines of print to give an address we can write to for a government bulletin—or for their own home-canning booklet, if they've been in the business long enough to have one. See pages 29–32 in Chapter 1 for sources of information and some good publications.)

ACCEPTABLE CHARACTERISTICS FOR ITS TYPE?

If the new entry seems to be of a design almost identical to the familiar stand-bys for modern mason jars described on pages 57–59, it has a separate metal lid with a ring of sealing compound bonded to its underside, and during processing (either in a Pressure or in a Boiling-Water Bath canner) a screw-on band holds it firmly in place on the sealing rim of the jar. The lid may not be re-used: and it was the replacements for once-used lids that were so scarce during the 1975 canning season.

Among the individual tops we found and used were a couple of newly marketed *one*-piece screw-on closures that incorporate the conventional metal lid and the band. Therefore the characteristics noted below for an acceptable, separate lid would also hold true for such one-piece closures.

Is it well made and of durable material, able to withstand high heat for a maximum processing period without distorting?

(A sleazy lid is unlikely either to behave well in a Pressure Canner at 240 F./116 C. or to retain a seal after months on the jar.)

Does it have a ring of sealing compound that's evenly applied? And sufficiently wide to cover *amply* the jar's sealing rim?

Does the top have a degree of flexibility—meaning springiness—in the center?

(On a well-sealed modern mason, all lids of this type will be slightly depressed in the center, which is pulled down by the strong vacuum created as the food cools after correct processing.)

Does it have a tiny down-turning flange that lets it seat securely, and properly centered, on the jar's rim *before* the lid is held down by the band?

(A piping-hot lid that slides back and forth on the jar requires some extra manipulation when we put on a piping-hot screwband. And the lid section of every two-piece closure *must* be set separately on the jar rim before the band is applied; even the most terse manufacturers' directions make a point of doing this. *Never* clap the lid inside the band and then screw the whole business on as if it were the top to a peanut-butter jar, because the ring of sealing compound on the separate lid may not be wide enough to compensate for such treatment— which could force the lid so badly off-center that it fails to hold a good seal.)

HOW THEY SEEMED TO WORK

To see how well our collection of new lids/caps worked in average home conditions, we ran up a whopping batch of chili con carne (fresh tomatoes + low-acid vegetables + ground beef), precooked it, and Hot-packed it in new 1-pint jars with appropriate headroom; capped each jar with a different closure; and Pressure-processed the whole thing for the requisite time. Where there were directions for applying and tightening the closure before the jar went into the canner, we followed them; if there were no instructions we handled the lid like the ones we knew well.

When the test jars had cooled overnight, all but one *appeared*—that is, just by looking at them—to be satisfactorily sealed. The exception was a non-metal separate lid with no original springiness to speak of, and which was held in place with a spare metal screwband: it had bulged upward in the center so much as to look almost rounded.

The next step was to remove each closure, noting these things as we did so: How easily did the closures unscrew? Was there seepage of food from under the sealing compound? How strongly did the sealing section of the closure seem to be seated on the jar's rim? And, as each lid was removed from the jar, how strong was the *whoosh!* of incoming air that rushed to fill the vacuum?

2-PIECE (LID + BAND) RESULTS

The conventional-looking *metal* lids among the new 2-piece screwband closures all had sealed, with the strength of the seals ranging from medium-good to very good indeed. These lids had been depressed to the extent of individual flexibility, with the springiest lids pulled downward the most, etc. There was no seepage of food from under any of them.

The *non-metal lid failed* to meet our test. A good deal of food had been forced out during processing—even though the headroom was the same as for the other jars; the sealing compound had disintegrated, with some of it having been forced out too; and the rounded bulge in the inflexible top turned out to be a true indication of no seal.

Not only would we not use this non-metal lid in a Pressure Canner, we wouldn't use it in a Boiling-Water Bath.

THE "REVOLUTIONARY" GOLDEN CAP

In our home-style test we also used two new caps that combined the sealing lid with the screw-on band. One—silver-colored, without much spring, and having its sole instruction imprinted on it (it was sold in bulk) to cap the filled jar in a "normal manner"—was a negligible performer for us.

But the other—a gold-colored cap with its broad band of sealing compound on the lid section, which was notably springy—made a very good seal. And it's worth particular mention here because it is offered as *re-usable*, a concept that is truly revolutionary (see page 59).

This golden one-piece cap came boxed, with directions printed on the container and a sheet inside describing a professional testing laboratory's results after using the new closure in both a Pressure Canner and a B-W Bath, with contents either Hot-packed or Raw-packed.

As for its characteristics, the lid section was very springy, clicking down audibly to become noticeably depressed as the jars cooled. We say "jars" because we capped two jars with this closure: on one, the cap was screwed down as firmly tight as is our custom with the conventional 2-piece screwband unit, and on the other we screwed down the cap a shade more gently, to follow the directions' "Screw cap on *gently, by hand. Do NOT* force or use a wrench." Our more gentle application provided the stronger seal.

In a letter in response to some of our questions, the manufacturer says, "There is more to fear from the overzealous pre-processing cap-tightener than there is from the 97-year-old arthritic grandma," because a cap wrenched tight could become deformed in the lid section. We saw one of these caps that had been distorted, and its top clicked down when pressed—and stayed down, attesting to a sealed vacuum inside the jar; removed, it looked deformed again. The directions should describe the looks of a cap that can't be re-used.

As for the cap's unique re-usability, only time will tell how many canning seasons it is good for. At this writing it has been on the market too short a time to have amassed significant reports on extended shelf-life or on the number of times it can be used again with satisfactory results. Meanwhile we recommend that the home-canner put jars with this closure on the front of the storage shelf, and check them often—as one would do when using any new product or method.

THE DANGEROUS "SEALER" IN SHEETS

This isn't something we found and tried, but we've had reports from several State Extension Service food scientists about it.

Toward the end of summer 1975 there appeared for sale on this side of the Atlantic some sheets of material containing *polyvinyl chloride* that were offered as a partial solution to the jar-lid shortage—we were supposed to cut it to fit and tie it on jars of preserves, etc.

But *DON'T!* Polyvinyl chloride is under fire from the Food and Drug Administration as questionable for use in contact with food that is prepared or packaged commercially.

Certainly no one would want to use such a material on food put by at home.

About Mushy Canned Green Beans

Every so often we hear from home-canners who complain that their green/snap/string/wax beans are "mushy," and they want to know what they can do about it.

First and always, never, *never, NEVER* cut down on the Pressure-processing time (see pages 136–141). This warning comes from all parts of the United States—from places where varieties, growing conditions, water hardness/softness are different.

But what can be done? Let's check some of the contributing factors.

VARIETY

On pages 119–120 we tell how hard it is to make sweeping recommendations about specific strains of tomatoes, and this also holds true for beans, thanks to variations in rainfall, soil chemistry and length of growing season from one locality to another.

If you plant your own, you might choose an old-fashioned pole bean. And you can always ask your County Agricultural Agent for varieties that do well in your area for canning, because tender new hybrids can get too soft from the necessary processing.

Variety is really the least important factor for the average householder, though.

A BETTER GUIDE: MATURITY

Several plant scientists offered a sensible idea when we called them: simply pick the beans a little later for canning than you'd pick them for immediate table use or for freezing.

Leave them on the plant until they have signs of bumpiness, which means that the bean seeds are starting to develop and fill the pods.

AND CONSIDER THE WATER

In addition to the discussion of water as an ingredient in processing food at home (pages 16–17 and 344), there are the following considerations.

Very soft water—rain water, or water that's been mechanically or chemically softened—will cause the beans to "slough," to soften excessively, when brought abruptly to a full rolling boil and held there when being cooked for the table. (Commercial canners sometimes give their beans a *low-temperature* blanch at around 165 F./74 C. for 5+ minutes to set the calcium pectate in the beans' outer tissues, before proceeding to Hot-pack the beans and Pressure-process them; but this extra handling might discourage some home-canners.)

On the other hand, water harder than average is likely to toughen the beans: certainly blanching in *hard* water as a step before freezing (pages 254–255 and 259) is often the cause of tough-ish green beans that were picked young and tender.

Therefore if you have a choice of water, use water on the hard side for canning beans. But don't tinker with too-soft water by adding chemicals; and even calcium chloride, a popular firming agent with commercial canners (pages 28–29), should be used sparingly if a bitter aftertaste is to be avoided.

And please remember that 10 pounds' pressure for the full time stipulated for processing is designed to destroy such dangers as the spores of *Clostridium botulinum,* it's not merely to cook the things. The best way to avoid overprocessing is to vent the Pressure Canner (pages 47–48) for 7 to 10 minutes, depending on its size (don't let it start venting strongly and then go vacuum the living room: it can build up unaccounted-for pressure of from 1 to 3 pounds if it vents too long). And don't be afraid of the "afterboil" that is still bubbling along in the jars when the canner has returned to Zero; remove the containers immediately and let them start cooling.

Index

ABOUT THE AUTHORS

RUTH HERTZBERG taught Home Economics in New England high schools and spent 15 years as a county agent in Home Economics with the Vermont Extension Service. She did postgraduate work in related home and family subjects at Cornell, and the Universities of Massachusetts, New Hampshire and Maine. As a county agent, Mrs. Hertzberg worked continually "out in the field" with young people starting their families and with older experienced homemakers, both in town and in rural areas. She has also been an instructor in the federal Expanded Food and Nutrition Education Program.

Many of the recipes in "The Preserving Kettle" section and in Chapter 10 are from the collection of heirloom recipes gathered by the late BEATRICE VAUGHAN, author of nine cookbooks.

JANET GREENE, co-author/editor of *Putting Food By*, has written and edited numerous books and articles on diversified Americana subjects (including how to cook on her antique wood range). "To 'put by,'" she says, "is an old-time country expression meaning to 'save what you can't use now against the time when you'll need it.'"